The Concepts of the Calculus

The Concepts of the Calculus

A CRITICAL AND HISTORICAL
DISCUSSION OF THE DERIVATIVE
AND THE INTEGRAL

———

By CARL B. BOYER

New York: Morningside Heights

COLUMBIA UNIVERSITY PRESS

1939

THE AMERICAN COUNCIL OF LEARNED SOCIETIES HAS
GENEROUSLY CONTRIBUTED FUNDS TO ASSIST IN THE
PUBLICATION OF THIS VOLUME

Preface

SOME ten years ago Professor Frederick Barry, of Columbia University, pointed out to me that the history of the calculus had not been satisfactorily written. Other duties and inadequate preparation at the time made it impossible to act upon this suggestion, but my studies of the past several years have confirmed this view. There is indeed no lack of material on the origin and subject matter of the calculus, as the titles in the bibliography appended to this work will attest. What is wanting is a satisfactory critical account of the filiation of the fundamental ideas of the subject from their incipiency in antiquity to the final formulation of these in the precise concepts familiar to every student of the elements of modern mathematical analysis. The present work is an attempt to supply, in some measure, this deficiency. An authoritative and comprehensive treatment of the whole history of the elementary calculus is greatly to be desired; but any such ambitious project is far beyond the scope and intention of the dissertation here presented. This is not a history of the calculus in all its aspects, but a suggestive outline of the development of the basic concepts, and as such should be of service both to students of mathematics and to scholars in the field of the history of thought. The aim throughout has therefore been to secure clarity of exposition, rather than to present a confusingly elaborate all-inclusiveness of detail or to display a meticulously precise erudition. This has necessitated a judicious selection and presentation of such material as would preserve the continuity of thought, but it is to be hoped that historical accuracy and perspective have not thereby been sacrificed.

The inclusion at the end of this volume of an extensive bibliography of works to which reference has been made has caused it to appear unnecessary to give full citations in the footnotes. In these notes author and title—in some cases abbreviated—alone have been given; titles of books have been italicized, those of articles in the periodical literature appear in Roman type enclosed within quotation marks. It is felt that the inclusion of such a list of sources on the subject may serve to encourage further investigations into the history of the calculus.

Preface

The inspiration toward the projection and completion of the present study has been due to Professor Barry, who has generously assisted in its prosecution through advice based on his wide familiarity with the field of the history of science. Professor Lynn Thorndike, of Columbia University, very kindly read and offered his competent criticism of the chapter on "Medieval Contributions." Professors L. P. Siceloff, of Columbia University, L. C. Karpinski, of the University of Michigan, and H. F. MacNeish, of Brooklyn College, have also read the manuscript and have furnished valuable aid and suggestions. Mrs. Boyer has been unstinting in her encouragement and promotion of the work, and has painstakingly done all of the typing. The composition of the Index has been undertaken by the Columbia University Press. Finally, from the American Council of Learned Societies came the subvention, in the form of a grant in aid of publication, which has made possible the appearance of this book in its present form. To all who have thus contributed toward the preparation and publication of this volume I wish to express my sincere appreciation.

CARL B. BOYER

BROOKLYN COLLEGE
January 3, 1939

Contents

Contents

I. Introduction

MATHEMATICS has been an integral part of man's intellectual training and heritage for at least twenty-five hundred years. During this long period of time, however, no general agreement has been reached as to the nature of the subject, nor has any universally acceptable definition been given for it.[1]

From the observation of nature, the ancient Babylonians and Egyptians built up a body of mathematical knowledge which they used in making further observations. Thales perhaps introduced deductive methods; certainly the mathematics of the early Pythagoreans was deductive in character. The Pythagoreans and Plato[2] noted that the conclusions they reached deductively agreed to a remarkable extent with the results of observation and inductive inference. Unable to account otherwise for this agreement, they were led to regard mathematics as the study of ultimate, eternal reality, immanent in nature and the universe, rather than as a branch of logic or a tool of science and technology. An understanding of mathematical principles, they decided, must precede any valid interpretation of experience. This view is reflected in the Pythagorean dictum that all is number,[3] and in the assertion attributed to Plato that God always plays the geometer.[4]

Later Greek skeptics, it is true, questioned the possibility of attaining any knowledge of such absolute character either by reason or by experience. But Aristotelian science had meanwhile shown that through observation and logic one can at least reach a consistent representation of phenomena, and mathematics consequently became, with Euclid, an idealized pattern of deductive relationships. Derived from postulates consistent with the results of induction from observation, it was found serviceable in the interpretation of nature.

[1] Bell, *The Queen of the Sciences*, p. 15. For full citations of works referred to in the footnotes see the Bibliography.
[2] See, for example, *Republic* VII, 527, in *Dialogues*, Jowett trans., Vol. II, pp. 362–63.
[3] See Aristotle, *Metaphysics* 987a–989b, in *Works*, ed. by Ross and Smith, Vol. VIII; Cf. also 1090a.
[4] Plutarch, *Miscellanies and Essays*, III, 402.

The Scholastic view, which prevailed during the Middle Ages, was that the universe is "tidy" and simply intelligible. In the fourteenth century came a fairly clear realization that Peripatetic qualitative views of motion and variation could better be replaced by quantitative study. These two concepts, with a revival of interest in Platonic views, brought about in the fifteenth and sixteenth centuries a renewal of the conviction that mathematics is in some way independent of, and prior to, experiential and intuitive knowledge. Such conviction is marked in the thinking of Nicholas of Cusa, Kepler, and Galileo, and to a certain extent appears in that of Leonardo da Vinci.

This conception of mathematics as the basis of the architecture of the universe was in turn modified in the sixteenth and seventeenth centuries. In mathematics, the cause of the change was the less critical and more practical use of the algebra which had been adopted from the Arabs, early in the thirteenth century, and then further developed in Italy. In natural science, the change was due to the rise of experimental method. The certitude in mathematics of which Descartes, Boyle, and others spoke was thus interpreted to mean a consistency to be found rather in the character of its reasoning than in any ontological necessity which it presented a priori.

The centering of attention on the procedures rather than on the bases of mathematics was emphasized in the eighteenth century by an extraordinary success in applying the calculus to scientific and mathematical problems. A more critical attitude was inaugurated in the nineteenth century by persistent efforts to find a satisfactory foundation for the conceptions involved in this new analysis of the infinite. Mathematical rigor was revived, and it was discovered that Euclid's postulates are not categorical synthetic judgments, as Kant maintained,[5] but simply assumptions. Such premises, it was found, may be so freely and arbitrarily chosen that—subject to the condition that they be mutually compatible—they may be allowed to contradict the apparent evidence of the senses. Toward the close of the century, as the result of the arithmetizing tendency in mathematical analysis, it was further discovered that the concept of infinity, transcending all intuition and analysis, could be introduced into mathematics without impairing the logical consistency of the subject.

[5] See *Sämmtliche Werke*, II, *passim*.

If the assumptions of mathematics are quite independent of the world of the senses, and if its elements transcend all experience,[6] the subject is at best reduced to bare formal logic and at worst to symbolical tautologies. The formal symbolic and arithmetizing tendency in mathematics has met with remarkable success in the study of the continuous. It has also led to stubborn paradoxes, a fact which has aroused increased interest in the nature of mathematics: its scope and place in intellectual life, the psychological source of its elements and postulates, the logical force of its propositions and their validity as interpretations of the world of sense perception.

The old idea that mathematics is the science of quantity, or of space and number, has largely disappeared. The untutored intuition of space, it is realized, leads to contradictions, a fact which upsets the Kantian view of the postulates. Nevertheless the mathematician is guided, although he is not controlled, by the external world of sense perception.[7] The mathematical theory of continuity originated in direct experience, but the definition of the continuum adopted in the end by the mathematician transcends sensory imagination. From this, mathematical formalists conclude that since we make no use of intuition in mathematical definitions and premises, it is not necessary that we should interpret the axioms or have any idea as to the nature of the objects and relations involved. The intuitionists, however, insist that the mathematical symbols involved should significantly express thoughts.[8] Although there are two (or more) views of the grounds for believing in the unassailable exactness of mathematical laws, the recognition that mathematical concepts are suggested, although not defined, by intuition thus easily accounts for the fact that the results of mathematical deductive reasoning are in apparent agreement with those of inductive experience. The derivative and the integral had their sources in two of the most obvious aspects of nature—multiplicity and variability—but were in the end defined as mathematical abstractions based on the fundamental concept of the limit of an

---

[6] Bertrand Russell has taken advantage of this disconcerting situation to define mathematics facetiously as "the subject in which we never know what we are talking about, nor whether what we are saying is true." "Recent Work on the Principles of Mathematics," p. 84.

[7] Bôcher, "The Fundamental Concepts and Methods of Mathematics."

[8] Brouwer, "Intuitionism and Formalism."

infinite sequence of elements. Once we have traced this development, the power and fecundity of these ideas when applied to the interpretation of nature will be easily understood.

The calculus had its origin in the logical difficulties encountered by the ancient Greek mathematicians in their attempt to express their intuitive ideas on the ratios or proportionalities of lines, which they vaguely recognized as continuous, in terms of numbers, which they regarded as discrete. It became involved almost immediately with the logically unsatisfactory (but intuitively attractive) concept of the infinitesimal. Greek rigor of thought, however, excluded the infinitely small from geometrical demonstrations and substituted the circumventive but cumbersome method of exhaustion. Problems of variation were not attacked quantitatively by Greek scientists. No method could be developed which would do for kinematics what the method of exhaustion had done for geometry—indicate an escape from the difficulties illustrated by the paradoxes of Zeno. The quantitative study of variability, however, was undertaken in the fourteenth century by the Scholastic philosophers. Their approach was largely dialectical, but they had resort as well to graphical demonstration. This method of study made possible in the seventeenth century the introduction of analytic geometry and the systematic representation of variable quantities.

The application of this new type of analysis, together with the free use of the suggestive infinitesimal and the more extensive application of numerical concepts, led within a short time to the algorithms of Newton and Leibniz, which constitute the calculus. Even at this stage, however, there was no clear conception of the logical basis of the subject. The eighteenth century strove to find such a basis, and although it met with little success in this respect, it did in the effort largely free the calculus from intuitions of continuous motion and geometrical magnitude. Early in the following century the concept of the derivative was made fundamental, and with the rigorous definitions of number and of the continuum laid down in the latter half of the century, a sound foundation was completed. Some twenty-five hundred years of effort to explain a vague instinctive feeling for continuity culminated thus in precise concepts which are logically defined but which represent extrapolations beyond the world of sensory experience. Intuition, or the putative immediate cognition of an element of experience which ostensibly fails of adequate expression, in the end gave way, as the result of reflective investigation, to those well-defined abstract mental constructs which science and mathematics have found so valuable as aids to the economy of thought.

The fundamental definitions of the calculus, those of the derivative and the integral, are now so clearly stated in textbooks on the subject, and the operations involving them are so readily mastered, that it is easy to forget the difficulty with which these basic concepts have been developed. Frequently a clear and adequate understanding of the fundamental notions underlying a branch of knowledge has been achieved comparatively late in its development. This has never been more aptly demonstrated than in the rise of the calculus. The precision of statement and the facility of application which the rules of the calculus early afforded were in a measure responsible for the fact that mathematicians were insensible to the delicate subtleties required in the logical development of the discipline. They sought to establish the calculus in terms of the conceptions found in the traditional geometry and algebra which had been developed from spatial intuition. During the eighteenth century, however, the inherent difficulty of formulating the underlying concepts became increasingly evident, and it then became customary to speak of the "metaphysics of the calculus," thus implying the inadequacy of mathematics to give a satisfactory exposition of the bases. With the clarification of the basic notions—which, in the nineteenth century, was given in terms of precise mathematical terminology—a safe course was steered between the intuition of the concrete in nature (which may lurk in geometry and algebra) and the mysticism of imaginative speculation (which may thrive on transcendental metaphysics). The derivative has throughout its development been thus precariously situated between the scientific phenomenon of velocity and the philosophical noumenon of motion.

The history of the integral is similar. On the one hand, it has offered ample opportunity for interpretations by positivistic thought in terms either of approximations or of the compensation of errors—views based on the admitted approximative nature of scientific measurements and on the accepted doctrine of superimposed effects. On the other hand, it has at the same time been regarded by idealistic metaphysics as a manifestation that beyond the finitism of sensory percipiency there is a transcendent infinite which can be but asymptotically approached by human experience and reason. Only the precision of their mathematical definition—the work of the past century—enables the derivative and the integral to maintain their autonomous position as abstract concepts, perhaps derived from, but nevertheless independent of, both physical description and metaphysical explanation.

At this point it may not be undesirable to discuss these ideas, with reference both to the intuitions and speculations from which they were derived and to their final rigorous formulation. This may serve to bring vividly to mind the precise character of the contemporary conceptions of the derivative and the integral, and thus to make unambiguously clear the *terminus ad quem* of the whole development.

The derivative is the mathematical device used to represent point properties of a curve or function. It thus has as its analogues in science instantaneous properties of a body in motion, such as the velocity of the object at any given time. When science is concerned with a time interval, the average velocity over this interval is suitably defined as the ratio of the change in the distance covered during the interval to the time interval itself. This ratio is conveniently represented by the notation $\dfrac{\Delta s}{\Delta t}$. Inasmuch as the laws of science are formulated by induction on the basis of the evidence of the senses, on the face of it there can be no such thing in science as an instantaneous velocity, that is, one in which the distance and time intervals are zero. The senses are unable to perceive, and science is consequently unable to measure, any but actual changes in position and time. The power of every sense organ is limited by a minimum of possible perception.[9] We cannot, therefore, speak of motion or velocity, in the sense of a scientific observation, when either the distance or the corresponding time interval becomes so small that the minimum of sensation involved in its measurement is not excited—much less when the interval is assumed to be zero.

If, on the other hand, the distance covered is regarded as a function of the time elapsed, and if this relationship is represented mathe-

[9] Cf. Mach, *Die Principien der Wärmelehre*, pp. 71–77.

matically by the equation $s = f(t)$, the minimum of sensation no longer operates against the consideration of the abstract difference quotient $\dfrac{\Delta s}{\Delta t}$. This has a mathematical meaning no matter how small the time and distance intervals may be, provided, of course, that the time interval is not zero. Mathematics knows no minimum interval of continuous magnitudes—and distance and time may be considered as such, inasmuch as there is no evidence which would lead one to regard them otherwise. Attempts to supply a logical definition of such an infinitesimal minimum which shall be consistent with the body of mathematics as a whole have failed. Nevertheless, the term "instantaneous velocity" appears to imply that the time interval is to be regarded not only as arbitrarily small but as actually zero. Thus the term predicates the very case which mathematics is compelled to exclude because of the impossibility of division by zero.

This difficulty has been resolved by the introduction of the derivative, a concept based on the idea of the limit. In considering the successive values of the difference quotient $\dfrac{\Delta s}{\Delta t}$, mathematics may continue indefinitely to make the intervals as small as it pleases. In this way an infinite sequence of values, $r_1, r_2, r_3, \ldots , r_n, \ldots$ (the successive values of the ratio $\dfrac{\Delta s}{\Delta t}$) is obtained. This sequence may be such that the smaller the intervals, the nearer the ratio r_n will approach to some fixed value L, and such that by taking the value of n to be sufficiently large, the difference $|L - r_n|$ can be made arbitrarily small. If this be the case, this value L is said to be the limit of the infinite sequence, or the derivative $f'(t)$ of the distance function $f(t)$, or the instantaneous velocity of the body. It is to be borne in mind, however, that this is not a velocity in the ordinary sense and has no counterpart in the world of nature, in which there can be no motion without a change of position. The instantaneous velocity as thus defined is not the division of a time interval into a distance interval, howsoever much the conventional notation $\dfrac{ds}{dt} = f'(t)$ may suggest a ratio. This symbolism, although remarkably serviceable in the carrying out of the operations of the calculus, will be found to have resulted from misapprehension on the part of Leibniz as to the logical basis of the calculus.

The derivative is thus defined not in terms of the ordinary processes of algebra, but by an extension of these to include the concept of the limit of an infinite sequence. Although science may not extrapolate beyond experience in thus making the intervals indefinitely small, and although such a process may be "inadequately adapted to nature,"[10] mathematics is at liberty to introduce the new limit concept, on the basis of the logical definition given above. One can, of course, make this notion still more precise by eliminating the words "approach," "sufficiently large," and "arbitrarily small," as follows: L is said to be the limit of the above sequence if, given any positive number ε (howsoever small), a positive integer N can be found such that for $n > N$ the inequality $|L - r_n| < \varepsilon$ is satisfied.

In this definition no attempt is made to determine any so-called "end" of the infinite sequence, or to deal with the possibility that the variable r_n may "reach" its limit L. The number L, thus abstractly defined as the derivative, is not to be regarded as an "ultimate ratio," nor may it be invoked as a means of "visualizing" an instantaneous velocity or of explaining in a scientific or a metaphysical sense either motion or the generation of continuous magnitudes. It is such unclear considerations and unwarranted interpretations which, as we shall see, have embroiled mathematicians, since the time of Zeno and his paradoxes, in controversies which often misdirected their energy. On the other hand, however, it is precisely such suggestive notions which stimulated the investigations resulting in the formal elaboration of the calculus, even though this very elaboration was in the end to exclude them as logically irrelevant.

Just as the problem of defining instantaneous velocities in terms of the approximation of average velocities was to lead to the definition of the derivative, so that of defining lengths, areas, and volumes of curvilinear configurations was to eventuate in the formulation of the definite integral. This concept, however, was likewise ultimately to be so defined that the geometrical intuition which gave it birth was excluded. As part of the Greek pursuit of unity in multiplicity, we

[10] Schrödinger, *Science and the Human Temperament*, pp. 61–62.

shall see that an attempt was made to inscribe successively within a circle polygons of a greater and greater number of sides in the hope of finally "exhausting" the area of the circle, that is, of securing a polygon with so great a number of sides that its area would be equal to that of the circle. This naïve attempt was of course doomed to failure. The same process, however, was adopted by mathematicians as basic in the definition of the area of the circle as the limit A of the infinite sequence formed by the areas $A_1, A_2, A_3, \ldots, A_n, \ldots$ of the approximating polygons. This affords another example of extrapolating beyond sensory intuition, inasmuch as there is no process by which the transition from the sequence of polygonal areas to the limiting area of the circle can be "visualized." An infinite subdivision is of course excluded from the realm of sensory experience by the fact that there exist thresholds of sensation. It must be banished also from the sphere of thought, in the physiological sense, inasmuch as psychology has shown that for an act of thought a measurable minimum of duration of time is required.[11] Logical definition alone remains a sufficient criterion for the validity of this limiting value A.

In order to free the limiting process just described from the geometrical intuition inherent in the notion of area, mathematics was constrained to give formal definition to a concept which should not refer to the sense experience from which it had arisen. (This followed a long period of indecision, the course of which we are shortly to trace.) After the introduction of analytical geometry it became customary, in order to find the area of a curvilinear figure, to substitute for the series of approximating polygons a sequence of sums of approximating rectangles, as illustrated in the diagram (fig. 1). The area of each of the rectangles could be represented by the notation $f(x_i)\triangle x_i$ and the sum of these by the symbolism $S_n = \sum\limits_{i=1}^{n} f(x_i)\triangle x_i$. The area of the figure could then be defined as the limit of the infinite sequence of sums S_n, as the number of subdivisions n increased indefinitely and as the intervals $\triangle x_i$ approached zero. Having set up the area in this manner with the help of the analytical representation of the curve, it then became a simple matter to discard the geometrical intuition leading to the formation of these sums and to define the definite

[11] Enriques, *Problems of Science*, p. 15.

integral of $f(x)$ over the interval from $x = a$ to $x = b$ arithmetically as the limit of the infinite sequence of sums $S_n = \sum_{i=1}^{n} f(x_i)\Delta x_i$ (where the divisions Δx_i are taken to cover the interval from a to b) as the intervals Δx_i become indefinitely small. This definition then invokes, apart from the ordinary operations of arithmetic, only the concept of the limit of an infinite sequence of terms, precisely as does that of the derivative. The realization of this fact, however, followed only after many centuries of investigation by mathematicians. The very notation $\int_a^b f(x)dx$, which is now customarily employed to represent this definite integral, is again the result of the historical development of the concept rather than of an effort to represent the final logical

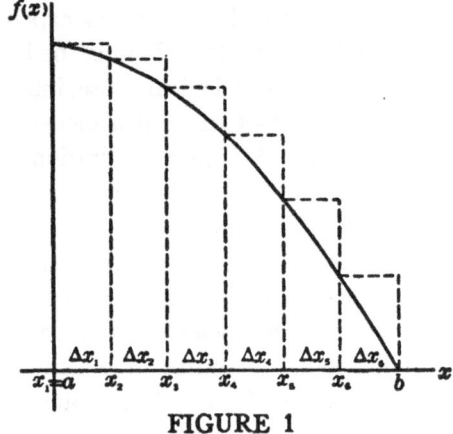

FIGURE 1

formulation. It is suggestive of a sum, rather than of the limit of an infinite sequence; and in this respect it is in better accord with the views of Leibniz than with the intention of the modern definition.

The definite integral is thus defined independently of the derivative, but Newton and Leibniz discovered the remarkable property constituting what is commonly known as the fundamental theorem of the calculus, viz., that the definite integral $F(x) = \int_a^x f(x)dx$ of the continuous function $f(x)$ has a derivative which is this very same function, $F'(x) = f(x)$. That is, the value of the definite integral of $f(x)$ from a to b may in general be found from the values, for $x = a$ and $x = b$, of the function $F(x)$ of which $f(x)$ is the derivative. This relationship between the derivative and the definite integral has been called "the root idea of the whole of the differential and integral calculus."[12] The function $F(x)$, when so defined, is often called the indefinite integral of $f(x)$, but it is to be recognized that it is in this case not a numerical limit given by an infinite sequence, as is the definite integral, but is a function of which $f(x)$ is the derivative.

The function $F(x)$ is sometimes also called the primitive of $f(x)$, and the value of $F(b) - F(a)$ is occasionally taken as the definition of $\int_a^b f(x)dx$. In this case the relationship $\int_a^b f(x)dx = \lim_{n \to \infty} \sum_{i=1}^{n} f(x_i) \triangle x_i$ is then the fundamental theorem of the calculus, rather than the definition of the definite integral.

Although the recognition of this striking inverse relationship, together with the formulation of rules of procedure, may be taken as constituting the invention of the subject, it is not to be supposed that the inventors of the calculus were in possession of the above sophisticated concepts of the derivative and the integral, so necessary in the logical development of the new analysis. More than a hundred years of investigation was to be required before the achievement of their final definition in the nineteenth century.

It is the purpose of this essay to trace the development of these two concepts from their incipiency in sense experience to their final elaboration as mathematical abstractions, defined in terms of formal logic by means of the idea of the limit of an infinite sequence. We shall find that the history of the calculus affords an unusually striking example of the slow formation of mathematical concepts by the emancipation from all sense data of ideas born of our primary intuitions. The derivative and the integral are, in the last analysis, synthetically defined in terms of ordinal considerations and not of those of continuous quantity and variability. They are, nevertheless, the results of attempts to schematize our sense impressions of these last notions. This explains why the calculus, in the early stages of its development, was bound up with concepts of geometry or motion, and with explanations of indivisibles and the infinitely small; for these ideas are suggested by naïve intuition and experience of continuity.

There is in the calculus a further concept which merits brief consideration at this point, not so much on account of any logical exigencies in the present structure of the calculus as to make clearer its historical development. The infinite sequences considered above in

[12] Courant, *Differential and Integral Calculus*, II, 111.

the definitions of the derivative and of the integral were obtained by continuing, in thought, to diminish ad infinitum the intervals between the values of the independent variable. Considerations which in physical science led to the atomic theory were at various periods in the development of the calculus adduced in mathematics. These made it appear probable that just as in the actual subdivision of matter (which has the appearance of being continuous) we arrive at ultimate particles or atoms, so also in continuous mathematical magnitudes we may expect (by means of successive subdivisions carried on in thought) to obtain the smallest possible intervals or differentials. The derivative would in this case be defined as the quotient of two such differentials, and the integral would then be the sum of a number (perhaps finite, perhaps infinite) of such differentials.

There is, to be sure, nothing intuitively unreasonable in such a view; but the criterion of mathematical acceptability is logical self-consistency, rather than reasonableness of conception. Such a view of the nature of the differential, although possessing heuristic value in the application of the calculus to problems in science, has been judged inacceptable in mathematics because no satisfactory definition has as yet been framed which is consistent with the principles of the calculus as formulated above, or which may be made the basis of a logically satisfactory alternative exposition. In order to retain the *operational* facility which the differential point of view affords, the concept of the differential has been logically defined, not in terms of mathematical atomism, but as a notion derived from that of the derivative. The differential dx of an independent variable x is to be thought of as nothing but another independent variable; but the differential dy of a function $y = f(x)$ is defined as that variable the values of which are so determined that for any given value of the variable dx the ratio $\dfrac{dy}{dx}$ shall be equal to the value of the derivative at the point in question, i. e., $dy = f'(x)dx$. The differentials as thus defined are only new *variables*, and not fixed infinitesimals, or indivisibles, or "ultimate differences," or "quantities smaller than any given quantity," or "qualitative zeros,"[13] or "ghosts of departed quantities," as they

[13] "Abominable little zeroes," they have been called by Osgood. See Osgood, "The Calculus in Our Colleges and Technical Schools"; and Huntington, "Modern Interpretation of Differentials."

have been variously considered in the development of the calculus.

Poincaré has said that had mathematicians been left the prey of abstract logic, they would never have gotten beyond the theory of numbers and the postulates of geometry.[14] It was nature which thrust upon mathematicians the problems of the continuum and of the calculus. It is therefore quite understandable that the persistent atomistic speculations of physical thought should have had a counterpart in attempts to picture, by means of indivisible elements, the space described by geometry. The further development of mathematics, however, has shown that such notions must be abandoned, in order to preserve the logical consistency of the subject. The basis of the concepts leading to the derivative and the integral was first found in geometry, for despite the apodictic character of its proofs, this subject was considered an abstract idealization of the world of the senses.

Recently, however, it has been more clearly perceived that mathematics is the study of relationships in general and must not be hampered by any preconceived notions, derived from sensory perception, of what these relationships should be. The calculus has therefore been gradually emancipated from geometry and has been made dependent, through the definitions of the derivative and the integral, on the notion of the natural numbers, an idea from which all traditional pure mathematics, including geometry, can be derived.[15] Mathematicians now feel that the theory of aggregates has provided the requisite foundations for the calculus, for which men had sought since the time of Newton and Leibniz.[16] It is impossible to predict with any confidence, however, that this is the final step in the process of abstracting from the primitive ideas of change and multiplicity all those irrelevant incumbrances with which intuition binds these concepts. It is a natural tendency of man to hypostatize those ideas which have great value for him,[17] but a just appreciation of the origin of the derivative and the integral will make clear how unwarrantedly sanguine is any view which would regard the establishment of these notions as bringing to its ultimate close the development of the concepts of the calculus.

[14] See Osgood, *op. cit.*, p. 457; Poincaré, *Foundations of Science*, p. 46.
[15] Russell, *Introduction to Mathematical Philosophy*, p. 4. See also Poincaré, *Foundations of Science*, pp. 441, 462.
[16] Russell, *The Principles of Mathematics*, pp. 325–26.
[17] Mach, *The Science of Mechanics*, p. 541.

II. Conceptions in Antiquity

THE PRE-HELLENIC peoples are usually regarded as pre-scientific in their attitude toward nature,[1] inasmuch as they palpably lacked the Greek confidence in its essential reasonableness, as well as the associated feeling that beneath the perplexing heterogeneity and ceaseless flux of events would be found elements of uniformity and permanence.

The search for universals, which the Greeks maintained so persistently, apparently held no attraction for the Egyptians and the Babylonians. So also the mathematical thought of these peoples—about whom, among those of the ancient world, we are best informed—bore no significant resemblance to ours, in that it lacked the tendency, essential to both mathematical and scientific method, toward the isolation and abstraction of certain samenesses from their confusingly varied concomitants in nature and in thought. Lacking these elements of invariance to serve as premises of inference, they were accordingly without appreciation of the characteristics which distinguish mathematics from science, namely, its logical nature and the necessity of deductive proofs.[2]

A large body of knowledge of spatial and numerical relations they did, however, acquire; and the more familiar their work becomes, the more it inspires our admiration.[3] It was, however, largely the result of empirical investigations, or at best of generalizations which were the result of incomplete induction from simple to more complicated cases. The Egyptian rule for computing the volume of a square pyramid—from which was obtained the most remarkable of all

[1] On this point, however, there are significant differences of opinion. Barry (*The Scientific Habit of Thought*, p. 104) places "the childhood of science" among the early Greeks; Burnet (*Greek Philosophy*, Part I, "Thales to Plato," pp. 4–5) says "natural science is the creation of the Greeks," and finds "not the slightest trace of that science in Egypt or even in Babylon." On the other hand, Karpinski ("Is There Progress in Mathematical Discovery," pp. 51–52) would regard the achievements of the Babylonians, Egyptians, and Hindus as "scientific in the highest sense."

[2] Milhaud, *Nouvelles études sur l'histoire de la pensée scientifique*, pp. 41–133.

[3] See Neugebauer, *Vorlesungen über Geschichte der antiken mathematischen Wissenschaften*, Vol. I, *Vorgriechische Mathematik*, for the best account of this work.

Egyptian results, the rule for determining the volume of a frustum of a square pyramid—was probably the result of this method of procedure.[4] That the demonstration could not be correct, in our understanding of the mathematical implications of this term, is clear from the fact that this result for the general case requires the use of infinitesimal or limit considerations[5] which, while constituting the point of departure for the story of the derivative and integral, are not found in any record before the Greek period.[6]

More fundamental than this lack of deductive proofs of inferred results is the fact that in all this Egyptian work the rules were applied to concrete cases with definite numbers only.[7] There was no conception in their geometry of a triangle as representative of all triangles,[8] an abstract generalization necessary for the elaboration of a deductive system. This lack of freedom and imagination is apparent also in Egyptian arithmetic, into which the abstract number concept, as such, did not enter,[9] and in which, with the exception of ⅔, all rational fractions were expressed as sums of unit fractions.[10]

Babylonian mathematics resembles the Egyptian more than the Greek, but with a stronger emphasis on the numerical side and thus a more highly developed algebra than that of Egypt. Here again we must not look for logical structure or proof, more complicated cases being reduced to simpler, and so "proved," or rather treated as analogous without proof;[11] and we must remember that this work, like the Egyptian, deals with concrete cases only. In connection with Babylonian astronomy, we find that problems involving continuous variation were studied, but only to the extent of tabulating the values of a function (such as the brightness of the moon, for example) for

---

[4] *Ibid.*, p. 128.

[5] See Dehn, "Ueber raumgleiche Polyeder," for proof that infinitesimal considerations cannot here be avoided.

[6] Neugebauer, *op. cit.*, pp. 126–28. There seems to be no basis for the implication made by Bell (*The Search for Truth*, p. 191) that the Egyptians used infinite or infinitesimal considerations in deductive reasoning.

[7] Neugebauer, *op. cit.*, p. 127.

[8] Luckey, "Was ist ägyptische Geometrie?" p. 49.

[9] Neugebauer, *op. cit.*, p. 203. See also Miller, "Mathematical Weakness of the Early Civilizations."

[10] For example, the Egyptians did not regard as a single number that number which we now represent by the symbol ⅔, but thought of it as the sum of the two fractions, ½ and ⅙.

[11] Neugebauer, *op. cit.*, pp. 203–4.

values of the argument (time) measured at equal intervals, and from this calculating the maximum (intensity) of the function.[12] The Greeks were the first, however, systematically to analyze[13] the idea of continuous magnitude and to develop concepts leading to the integral and the derivative.

Our information on the history of mathematics in the interval between that of the best Egyptian and Babylonian mathematics and the early work in Greece is unfortunately fragmentary. That these oriental civilizations influenced Greek culture is clear; but the nature and extent of their contribution is undetermined. However that may be, there is an obvious change in spirit in both science and mathematics, as these developed in Greece. The human mind was "discovered" as something different from the surrounding body of nature and capable of discerning similarities in a multiplicity of events, of abstracting these from their settings, generalizing them, and deducing therefrom other relationships consistent with further experience. It is for this reason that we consider mathematical and scientific method as originating with the Hellenic race;[14] but to say that Greek mathematics and science were autochthonous would be to forget the debt of subject-matter owed to Egypt and Babylon.[15] It is likely[16] that the new outlook of the Hellenes was the result of the flux of civilizations occurring at this time, this impressing upon the rising Greek fortunes the stamp of numerous cultures.

Thales is the first Greek mentioned in connection with this "intellectual revolution," which produced elementary mathematics and which was to reveal those difficulties in conception, the study and resolution of which were to produce within the next twenty-five hundred years the subject which we now call the calculus. He is said to have been a great traveler, to have learned geometry from the Egyptians and astronomy from the Babylonians, and upon his return to Greece to have instructed his successors in the principles of these

---

[12] Hoppe, "Zur Geschichte der Infinitesimalrechnung bis Leibniz und Newton."

[13] Neugebauer, *op. cit.*, p. 205.

[14] T. L. Heath, *A History of Greek Mathematics*, I, v, says, "Mathematics in short is a Greek science, whatever new developments modern analysis has brought or may bring."

[15] See Karpinski, "Is There Progress in Mathematical Discovery?" pp. 46–47. Cf. also Gandz, "The Origin and Development of the Quadratic Equations in Babylonian, Greek and Early Arabic Algebra," pp. 542–43.

[16] As Neugebauer suggests (*op. cit.*, p. 203).

subjects. Proclus says of his method of attack that it was "in some cases more general, in other cases more empirical." Thales' demonstrations may therefore have appealed to some extent to the evidence of the senses, and in fact his theorems were those the truth of which one would recognize by the execution of some practical construction.[17]

To Thales, nevertheless, is ascribed the establishment of mathematics as a deductive discipline.[18] He did not, however, construct a body of mathematical knowledge, nor did he apply his method to the analysis of the problem of the continuum. These tasks seem to have been performed by Pythagoras, the second Greek mathematician of whom we have substantial information. According to Proclus, he "transformed the study of geometry into a liberal education, examining the principles of the science from the beginning and probing the theorems in an immaterial and intellectual manner";[19] but, beyond admitting this, it is impossible to ascribe with any degree of certainty other mathematical or scientific accomplishments to Pythagoras the individual, since never in antiquity could he be distinguished from his school and it is hardly possible to do so now. The knowledge acquired by the school established by Pythagoras was held to be strictly esoteric, with the result that when the general nature of Pythagorean thought became apparent, after the death of the founder about 500 b. c., it was already impossible to attribute to a single member any one contribution. Nevertheless, the process of abstraction begun by Thales was evidently completed by this school.

A new difficulty, however, then entered into Greek thought, for the Pythagorean mathematical concepts, abstracted from sense impressions of nature, were now in turn projected into nature and considered to be the structural elements of the universe.[20] Thus the Pythagoreans attempted to construct the whole heaven out of numbers, the stars being units which were material points. Later they identified the regular geometrical solids, with which they were familiar, with the different sorts of substances in nature.[21] Geometry was regarded

---

[17] Paul Tannery, *La Géométrie grecque*, pp. 89 ff.

[18] T. L. Heath, *History of Greek Mathematics*, I, 128.

[19] *Ibid.*, I, 140. Cf. also Moritz Cantor, *Vorlesungen über Geschichte der Mathematik*, I, 137.

[20] See Brunschvicg, *Les Étapes de la philosophie mathématique*, pp. 34 ff.

[21] T. L. Heath, *History of Greek Mathematics*, I, 165.

by them as immanent in nature, and the idealized concepts of geometry appeared to them to be realized in the material world. This confusion of the abstract and the concrete, of rational conception and empirical description, which was characteristic of the whole Pythagorean school and of much later thought, will be found to bear significantly on the development of the concepts of the calculus. It has often been inexactly described as mysticism,[22] but such stigmatization appears to be somewhat unfair. Pythagorean deduction a priori having met with remarkable success in its field, an attempt (unwarranted, it is now recognized) was made to apply it to the description of the world of events, in which Ionian hylozoistic interpretations a posteriori had made very little headway. This attack on the problem was highly rational and not entirely unsuccessful, even though it was an inversion of the scientific procedure, in that it made induction secondary to deduction.

One very important result of the Pythagorean search for unity in nature and geometry was the theory of application of areas. This originated with the Pythagoreans, if not with Pythagoras himself,[23] and became fundamental in Greek geometry, in which it later led to the method of exhaustion, the Greek equivalent of our integration.[24] The method of the application of areas enabled them to say of a figure bounded by straight lines that it was greater than, equivalent to, or less than[25] a second figure. Such a superposition of one area upon another constitutes the first step in the attempt to make exactly definable the notion of area, in which a unit of area is said to be contained in a second area a given number of times. Modern mathematics has made fundamental the concept of number rather than that of congruence, with the result that the word "area" no longer calls vividly to mind that comparison of two surfaces which is essential in this connection and which was always uppermost in Greek thought. Greek mathematicians did not speak of the *area of one figure*, but of the *ratio of two surfaces*, a definition which could not, because of the problem of incommensurability, be made precise before a satisfactory concept of number had been developed. Such a concept the Pythag-

[22] See Russell, *Our Knowledge of the External World*, p. 19.

[23] T. L. Heath, *History of Greek Mathematics*, I, 150.

[24] It was basic also in the Greek solution of quadratic equations by geometrical algebra.

[25] Our names for the conic sections (ellipse, hyperbola, and parabola) were, incidentally, derived from the designations the Pythagoreans used in this connection.

oreans did not possess. This contribution was not made until the last
half of the nineteenth century, and it was to furnish, in the last
analysis, the basis of the whole of the calculus. However, to the Pythagoreans, in all probability, we owe the recognition of the need for some
such concept—a discovery which may be regarded as the first step, a
terminus a quo, in the development of the concepts of the calculus.

The inadequacy of the Pythagorean view of the ratio of magnitudes
was first made evident to the followers of this school on the application
of the doctrine, not to areas, but rather to the analogous comparison
of lines which is presupposed by our notion of length. Such investigations led the Pythagoreans to an intensely disconcerting discovery.
If the side of a square were to be applied to the diagonal, no common
measure could be discovered which would express one in terms of the
other. In other words, these lines were shown to be incommensurable.
Just when this discovery took place and whether it was made by
Pythagoras himself, by the early Pythagoreans, or by later members
of the school are moot points in the history of mathematics.[26] It has
also been maintained that Pythagoras owed his knowledge of the
irrational and of the five regular solids, as well as much of his philosophy, to the Hindus.[27]

The question as to how the incommensurability was discovered or
proved is also difficult to answer with any assurance. The method of
application would suggest as a form of proof the geometrical equivalent
of the process of finding the greatest common divisor, but there is
another aspect of Pythagorean thought which points to a different sort
of reasoning. A prevailing belief in the unity and harmony of nature
and knowledge had led the Pythagoreans not only to explain different
aspects of nature by various mathematical abstractions, as already
suggested, but also to attempt to identify the realms of number and
magnitude.[28] By the term number, however, the Pythagoreans did

[26] On this question see the following two papers by Heinrich Vogt: "Die Entdeckungsgeschichte des Irrationalen nach Plato und anderen Quellen des 4. Jahrhunderts": "Zur
Entdeckungsgeschichte des Irrationalen." Vogt concludes that the discovery was made by
the later Pythagoreans at some time before 410 B. C. T. L. Heath (*History of Greek Mathematics*, I, 157) would place it "at a date appreciably earlier than that of Democritus."

[27] See Schroeder, *Pythagoras und die Inder;* and Vogt, "Haben die alten Inder den
Pythagoreischen Lehrsatz und das Irrationale gekannt?"

[28] For a keenly critical account of the significance, from the scientific point of view, of
the Pythagorean problem of associating the fields of number and magnitude, see Barry,
The Scientific Habit of Thought, pp. 207 ff.

not understand the abstraction to which we give this name, but used it to designate "a progression of multitude beginning from a unit and a regression ending in it."[29] The integers were thus fundamental, numbers being collections of units, and, as was the case with their geometrical forms, they were immanent in nature, each having a position and occupying a place in space. If geometrical abstractions were the elements of actual things, number was the ultimate element of these abstractions and thus of physical bodies and of all nature.[30] This hypostatization of number had led the Pythagoreans to regard a line as made up of an integral number of units. This doctrine could not be applied to the diagonal of a square, however, for no matter how small a unit was chosen as a measure of the sides, the diagonal could not be a "progression of multitude" beginning with this unit. The proof of this fact, as given by Aristotle (and which possibly is that of the Pythagoreans),[31] is based on the distinction between the odd and the even, which the Pythagoreans themselves had emphasized.

The incommensurability of lines remained ever a stumbling block for Greek geometry. That it made a strong impression on Greek thought is indicated by the story, repeated by Proclus, that the Pythagorean who disclosed the fact of incommensurability suffered death by shipwreck as a result. It is demonstrated also, and more reliably, by the prominence given to the doctrine of irrationals by Plato and Euclid. It never occurred to the Greeks to invent an irrational *number*[32] to circumvent the difficulty, although they did develop as a part of geometry (found, for example, in the tenth book of Euclid's *Elements*) the theory of irrational *magnitudes*. Failing to generalize their number system along the lines suggested later by the development of mathematical analysis, the only escape for Greek mathematicians in the end was to abandon the Pythagorean attempt to identify the realm of number with that of geometry or of continuous magnitude.

The effort to unite the two fields was not given up, however, before intuition had sought another way out of the difficulty. If there is no

[29] T. L. Heath, *History of Greek Mathematics*, I, 69–70.

[30] Milhaud, *Les Philosophes géomètres de la Grèce*, p. 109.

[31] Zeuthen, "Sur l'origine historique de la connaissance des quantités irrationelles."

[32] Stolz, *Vorlesungen über Allgemeine Arithmetik*, I, 94; cf. also Vogt, *Der Grenzbegriff in der Elementar-mathematik*, p. 48.

finite line segment so small that the diagonal and the side may both be expressed in terms of it, may there not be a monad or unit of such a nature that an indefinite number of them will be required for the diagonal and for the side of the square?

We do not know definitely whether or not the Pythagoreans themselves invoked the infinitely small. We do know, however, that the concept of the infinitesimal had entered into mathematical thought, through a doctrine elaborated in the fifth century B. C., as the result of Greek speculation concerning the nature of the physical world. After the failure of the early Ionian attempts to find a fundamental element out of which to construct all things, there arose at Abdera the materialistic doctrine of physical atomism, according to which there is no one *physis*, not even a small group of substances of which everything is composed. The Abderitic school held that everything, even mind and soul, is made up of atoms moving about in the void, these atoms being hard indivisible particles, qualitatively alike but of countless shapes and sizes, all too small to be perceived by sense impressions.

There is nothing either logically or physically inconsistent in this doctrine, which is a crude anticipation of our own chemical thought; but the greatest of the Greek atomists, Democritus, did not stop here: he was also a mathematician and carried the idea over into geometry. As we now know from the *Method* of Archimedes, which was discovered as a palimpsest in 1906, Democritus was the first Greek mathematician to determine the volumes of the pyramid and the cone. How he derived these results we do not know. The formula for the volume of a square pyramid was probably known to the Egyptians,[33] and Democritus in his travels may have learned of it and generalized the result to include all polygonal pyramids. The result for the cone would then be a natural inference from the result of increasing indefinitely the number of sides in a regular polygon forming the base of a pyramid. This explanation would correspond to others involving similar infinitesimal conceptions, which we know Democritus entertained and which later influenced Plato.[34]

[33] Neugebauer, *Vorlesungen über Geschichte der Antiken mathematischen Wissenschaften*, p. 128.
[34] See Luria, "Die Infinitesimaltheorie der antiken Atomisten."

Aristotle and Euclid ascribe to him a mathematical atomism, and we know from Plutarch[35] that he was puzzled as to whether the infinitesimal parallel circular sections of which the cone may be considered to be composed are equal or unequal: if they are equal, the cone would be equal to the circumscribed cylinder; but if unequal, they would be idented like steps.[36] We do not know how he resolved this aporia, but it has been suggested that he made use of the idea of infinitely thin circular laminae, or indivisibles, to find the volumes of cones and cylinders, anticipating and using Cavalieri's theorem for these special cases.[37] Democritus seems to have discriminated clearly between physical and mathematical atoms—as did his later follower Epicurus, although Aristotle made no such distinction[38] and according to the much later account of Simplicius Democritus is said to have held that all lines are divisible to infinity.[39] However, since most of Democritus' work is lost, we cannot now reconstruct his thought. That he was interested in other mathematical problems bearing on the infinitesimal we know from the titles of works now lost, but which are referred to by Diogenes Laertius. One of these seems to have been on horn angles (the angles formed by curves which have a common tangent at a point), and another on irrational (incommensurable) lines and solids.[40] It may be inferred, therefore, that the Pythagorean difficulty with the incommensurable was probably familiar to him, and it may be that he tried to solve it by some theory of mathematical atomism. It has been maintained[41] that Democritus was too good a mathematician to have had anything to do with such a theory as that of indivisible lines; but it is difficult to imagine how a mathematical atom is to be conceived if not as an indivisible. At all events, whatever his conception of the nature of infinitesimals may have been, the influence of Democritus has persisted. The idea of the fixed infinitesimal magnitude has clung tenaciously to mathematics, frequently to be invoked by intuition when logic apparently failed to offer a solution, and finally to be displaced in the last century by the rigorous concepts of the derivative and the integral.

[35] Plutarch, *Miscellanies and Essays*, IV, 414–16.

[36] See also T. L. Heath, *History of Greek Mathematics*, I, 180; and Luria, *op. cit.*, pp. 138–40, for statements of this "paradox."

[37] Simon, *Geschichte der Mathematik im Altertum*, p. 181.   [38] Luria, *op. cit.*, pp. 179–80.

[39] Cf. *Simplicii commentarii in octo Aristotelis physicae auscultationis libros*, p. 7.

[40] T. L. Heath, *History of Greek Mathematics*, I, 179–81.   [41] *Ibid.*, p. 181.

That the infinitesimal was not eagerly welcomed into Greek geometry after the time of the Pythagoreans and Democritus may have been due largely to a school of philosophy that had risen at Elea, in Magna Graecia. The Eleatic school, although not essentially mathematical, was apparently familiar with, and probably influenced by, Pythagorean mathematical philosophy; but it became an opponent of the chief tenet of this thought. Instead of proclaiming the constitution of objects as an aggregate of units, it pointed out the apparent contradictions inherent in such a doctrine, maintaining against the atomic view the essential oneness and changelessness of the world. This stultifying monism was upheld by Parmenides, the leader of the school, with perhaps a touch of skepticism derived from his iconoclastic predecessor, the poet-philosopher Xenophanes.

In an indirect defense of this doctrine, the Eleatics proceeded to demolish, with skillful dialectic, the basis of opposing schools of thought. The most damaging arguments were offered by Zeno, the student of Parmenides. After presenting the obvious objection to the Pythagorean indefinitely small monad—that if it has any length, an infinite number will constitute a line of infinite length; and if it has no length, then an infinite number will likewise have no length—he added the following general dictum against infinitesimals: "That which, being added to another does not make it greater, and being taken away from another does not make it less, is nothing."[42] More critical and subtle than these, however, are his four famous paradoxes on motion.[43] There has been much speculation as to the purpose of Zeno's arguments,[44] lack of evidence making it impossible to decide conclusively against whom they were directed: whether against the Pythagoreans, or the atomists, or Heraclitus, or whether they were mere sophisms. That they were intended merely as dialectical puzzles may perhaps be indicated by the passage in Plutarch's life of Pericles:

> Also the two-edged tongue of mighty Zeno, who,
> Say what one would, could argue it untrue.[45]

[42] Zeller, *Die Philosophie der Griechen in ihrer Geschichtlichen Entwicklung*, I, 540.

[43] For an unusually extensive account of the history of Zeno's paradoxes, with bibliography, see Cajori's article on "History of Zeno's Arguments on Motion."

[44] On this subject see Cajori, "The Purpose of Zeno's Arguments on Motion." This article includes an account of the varying interpretations, as well as extensive bibliographical notes.

[45] Plutarch, *The Lives of the Noble Grecians and Romans*, p. 185.

On the other hand there is some reason to suppose that these arguments were presented in connection with a more significant purpose. It is not improbable that Zeno, although he was neither a mathematician nor a physicist, propounded the paradoxes to point out the weakness in the Pythagorean definition of a point as unity having position, and in the resulting Pythagorean multiplicity which did not distinguish clearly between the geometrical and the physical.[46] Pythagorean science and mathematics had been concerned with form and structure, and not with flux and variability; but had the Pythagoreans applied their philosophy to the aspects of *change* in nature rather than to those of *permanence*, the resulting explanation of motion would have been in terms of concepts attacked by Zeno in his third and fourth paradoxes (those of the *arrow* and the *stade*),[47] in which space and time are assumed to be composed of indivisible elements. The arguments would hold equally well, of course, against mathematical atomism. The first two paradoxes (the *dichotomy* and the *Achilles*[48]) are directed against the opposite conception, that of the infinite divisibility of space and time, and are based upon the impossibility of conceiving intuitively the limit of the sum of an infinite series. The four paradoxes are, of course, easily answered in terms of

[46] See Paul Tannery, *La Géométrie grecque*, p. 124; cf. also the same author, "Le Concept scientifique du continu. Zénon d'Elée et Georg Cantor." Milhaud (*Nouvelles études*, pp. 153–54) and Cajori concur in the view here presented. See further, Cajori, "Purpose of Zeno's Arguments."

[47] The argument in the *arrow* is as follows: Anything occupying space equal to itself (or in one and the same place) is at rest; but this is true of the arrow at every moment of its flight. Therefore the arrow does not move. (See *The Works of Aristotle*, Vol. II, *Physica* VI. 239b, for the statement of the paradoxes.) The argument in the *stade*, as given by Aristotle, is obscure (because of brevity), but is equivalent to the following: Space and time being assumed to be made up of points and instants, let there be given three parallel rows of points, A, B, and C. Let C move to the right and A to the left at the rate of one point per instant, both relative to B; but then each point of A will move past two points of C in an instant, so that we can subdivide this, the smallest interval of time; and this process can be continued ad infinitum, so that time can not be made up of instants.

[48] The argument in the *dichotomy* is as follows: before an object can traverse a given distance, it must first traverse half of this distance; before it can cover half, however, it must cover one quarter; and so ad infinitum. Therefore, since the regression is infinite, motion is impossible, inasmuch as the body would have to traverse an infinite number of divisions in a finite time. The argument in the *Achilles* is similar. Assume a tortoise to have started a given distance ahead of Achilles in a race. Then by the time Achilles has reached the starting point of the tortoise, the latter will have covered a certain distance; in the time required by Achilles to cover this additional distance, the tortoise will have gone a little farther; and so ad infinitum. Since this series of distances is infinite, Achilles can never overtake the tortoise, for the same reason as that adduced in the *dichotomy*.

the concepts of the differential calculus. There is no logical difficulty in the *dichotomy* or the *Achilles*, the uneasiness being due merely to failure of the imagination to realize, in terms of sense impressions, the nature of infinite convergent series which are fundamental in the precise explanation of, but not involved in our obscure notion of, continuity. The paradox of the flying arrow involves directly the conception of the derivative and is answered immediately in terms of this. The argument in this paradox, as also that in the *stade*, is met by the assumption that the distance and time intervals contain an infinite number of subdivisions. Mathematical analysis has shown that the conception of an infinite class is not self-contradictory, and that the difficulties here, as also in the case of the first two paradoxes, are those of conceiving intuitively the nature of the continuum and of infinite aggregates.[49]

In a broad sense there are no insoluble problems, but only those which, arising from a vague feeling, are not yet suitably expressed.[50] This was the position of Zeno's paradoxes in Greek thought; for the notions involved were not given the precision of expression necessary for the resolution of the putative difficulties. It is clear that the answers to Zeno's paradoxes involve the notions of continuity, limits, and infinite aggregates—abstractions (all related to that of number) to which the Greeks had not risen and to which they were in fact destined never to rise, although we shall see Plato and Archimedes occasionally straining toward such views. That they did not do so may have been the result of their failure, indicated above in the case of the Pythagoreans, clearly to separate the worlds of sense and reason, of intuition and logic. Thus mathematics, instead of being the science of possible relations, was to them the study of situations thought to subsist in nature.

The inability of Greek mathematicians to answer in a clear manner the paradoxes of Zeno made it necessary for them to forego the attempt to give to the phenomena of motion and variability a quantitative explanation. These experiences were consequently confined to the field of metaphysical speculation, as in the work of Heraclitus, or

[49] Accounts of the mathematical resolutions of Zeno's paradoxes are given in the works of Bertrand Russell, *The Principles of Mathematics* and *Our Knowledge of the External World*.

[50] Enriques, *Problems of Science*, p. 5.

to that of qualitative description, as the physics of Aristotle. Only the static aspects of optics, mechanics, and astronomy found a place in Greek mathematics, and it remained for the Scholastics and early modern scientists to establish a quantitative dynamics. Zeno's arguments and the difficulty of incommensurability had also a more general effect on mathematics: in order to retain logical precision, it was necessary to give up the abortive Pythagorean effort to identify the domains of number and geometry, and to abandon also the premature Democritean attempt to explain the continuous in terms of the discrete. It is, however, impossible satisfactorily to interpret the world of nature and the realm of geometry (spheres which for the Greeks were not essentially distinct) without superimposing upon them a framework of discrete multiplicity; without ordering, by means of number, the heterogeneity of impressions received by the senses; and without at every point comparing nonidentical elements. Thought itself is possible only in terms of a plurality of elements. As a consequence, the concept of discreteness cannot be excluded completely from the study of geometry. The continuous is to be interpreted in terms of successive subdivision, that is to say, in terms of the discrete, although from the Greek point of view the former could not be logically identified with the latter. The clever manner in which the method of successive subdivision was applied in Greek geometry, without the loss of logical rigor, will be seen later in the method of exhaustion—a procedure which was developed, not in Italy, but in and around the Greek mainland, whither many Pythagoreans wandered, on the breaking up of the school, toward the beginning of the fifth century B. C. Zeno likewise lived for a time in Athens, the rising center of Greek culture and mathematics. Here Pericles, the political leader of that city in its Golden Age, is said to have been one of his listeners.[51]

At Athens the great philosopher Plato, although himself not primarily a mathematician, was conversant with, and displayed a lively interest in, the problems of the geometers. He may not have contributed much original work in mathematics, but he advanced the subject, nevertheless, through his great enthusiasm for it. He is said to have paid particular attention to the principles of geometry—to the

[51] Plutarch, *Lives*, p. 185.

hypotheses, definitions, methods.[52] For this reason he was particularly concerned with the difficulties which led eventually to the calculus. In his dialogues he considered the Pythagorean problem of the nature of number and its relationship to geometry,[53] the difficulty of incommensurability,[54] the paradoxes of Zeno,[55] and the Democritean question of indivisibles and the nature of the continuum.[56]

Plato seems to have realized the gulf between arithmetic and geometry, and it has been conjectured[57] that he may have tried to bridge it by his concept of number and by the establishment of arithmetic upon a firm axiomatic basis similar to that which was built up in the nineteenth century independently of geometry; but we cannot be sure, because these thoughts do not occur in his exoteric writings and were not advanced by his successors. If Plato made an attempt to arithmetize mathematics in this sense, he was the last of the ancients to do so, and the problem remained for modern mathematical analysis to solve. The thought of Aristotle we shall find diametrically opposed to any such conceptions. It has been suggested that Plato's thought was so opposed by the polemic of Aristotle that it was not even mentioned by Euclid. Certain it is that in Euclid there is no indication of such a view of the relation of arithmetic to geometry; but the evidence is insufficient to warrant the assertion[58] that, in this connection, it was the authority of Aristotle which held back for two thousand years a transformation which the Academy sought to complete. A sound basis for either mechanics or arithmetic must be built upon the limit concept—a notion which is not found in the extant works of Plato nor in those of his successors. The Platonists, on the contrary, attempted to develop the misleading idea of indivisibles or fixed infinitesimals, a notion which the modern arithmetization of analysis has had cause to reject.

[52] To Plato are ascribed, among other things, the formulation of the analytic method and the restriction of Euclidean geometry to constructions possible with ruler and compass only. Hankel (*Zur Geschichte der Mathematik in Alterthum und Mittelalter*, p. 156) ascribes this limitation to Plato, but T. L. Heath (*History of Greek Mathematics*, I, 288) would place it earlier.

[53] *Republic* VII. 525–27.

[54] See, in particular, *Theaetetus* 147–48; *Laws* 819d–820c.

[55] *Parmenides* 128 ff.

[56] *Philebas* 17 ff.

[57] Toeplitz, "Das Verhältnis von Mathematik und Ideenlehre bei Plato."

[58] *Ibid.*, pp. 10–11.

Plato apparently did not give direct answers to the difficulties involved in incommensurability or in Zeno's paradoxes, although he expressed his opposition to the Pythagorean concepts of infinity, and of the monad as being unity having position,[59] and also to Democritean atomism. He was strongly influenced by both of these schools, but apparently felt that their views were too much the result of sense experience. Plato's criterion of reality was not consistency in experience but reasonableness in thought. For him, as for the Pythagoreans, there was no necessary distinction between mathematics and science; both were the result of deduction from clearly perceived first principles. The Pythagorean monad and the Democritean mathematical atomism, which gave every line a thickness, perhaps appealed too strongly to materialistic sense experience to suit Plato, so that he had recourse to the highly abstract *apeiron* or unbounded indeterminate. This was the eternally moving infinite of the Ionian philosopher, Anaximander,[60] who had suggested it in opposition to Thales' less subtle assertion that the concrete material element, water, was the basis of all things. According to Plato, the continuum, could better be regarded as generated by the flowing of the *apeiron* than thought of as consisting of an aggregation (however large) of indivisibles. This view represents a fusion of the continuous and the discrete not unlike the modern intuitionism of Brouwer.[61] The infinitely small was apparently not to be reached through a continued subdivision,[62] but was to be regarded, perhaps, as analogous to the generative infinitesimal of Leibniz, or the "intensive" infinitely small magnitude which appeared in idealistic philosophy in the nineteenth century. Mathematics has found it necessary to discard both views in making the infinitesimal subordinate to the derivative in the logical foundation of the calculus. However, the notion of the infinitesimal proved very suggestive in the early establishment of the calculus, and, as Newton remarked more than two thousand years later, the application to it of our intuitions of motion removes from the doctrine much of the harshness felt in the mathematical atomism of Democritus and later of Cavalieri. This, however, necessarily led to a loss both of precise logical defi-

---

[59] See T. L. Heath, *History of Greek Mathematics*, I, 293.

[60] Hoppe, "Zur Geschichte der Infinitesimalrechnung," p. 154; see also Milhaud, *Les Philosophes géomètres*, p. 68.

[61] Helmholtz, *Counting and Measuring*, pp. xxii–xxiv.   [62] Hoppe, *op. cit.*, p. 152.

nition and of clear sensory interpretation, neither of which Plato supplied.[63]

The belief that mathematics becomes "sterilized by losing contact with the world's work"[64] is widely held but is not easily justified. The conjunction of mathematics and philosophy, as found, for example, in Plato, Descartes, and Leibniz, has been perhaps as valuable in suggesting new advances as has the blending of mathematical and scientific thought illustrated by Archimedes, Galileo, and Newton. The disregard in Platonic thought of any basis in the evidences of sense experience has not unjustly been regarded, from the scientific point of view, as an "unmitigated misfortune." On the other hand, the successful development of his views would have given to mathematics—which is interested solely in relationships which are logically thinkable rather than in those believed to be realized in nature—a flexibility and an independence of the world of sense impressions which were to be essential for the ultimate formulation of the concepts of the calculus. One may therefore say, in a very general sense, that "we know from Plato's own writings that he was thinking out the solution of problems that lead directly to the discovery of the calculus."[65] It is, however, altogether too much to assert that "Indeed there are probably only four or five names of mathematical discoverers that stand between Plato on the one hand and Newton and Leibniz, the discoverers of the calculus, on the other hand."[66] We shall see that the calculus was the result of a long train of mathematical thought, developed slowly and with great difficulty by very many thinkers.

That the doctrines of the continuous and the infinitesimal did not develop along the abstract lines vaguely indicated by Plato was probably the result of the fact that Greek mathematics included no general concept of number,[67] and, consequently, no notion of a continuous algebraic variable upon which such theories could logically have been based. Disregard of the abstract idealizations which Plato

[63] Hoppe (*op. cit.*, p. 152) asserts that in Plato one finds the first clear conception of the infinitesimal. It is difficult, however, to perceive on what grounds such a thesis is to be defended.

[64] Hogben, *Science for the Citizen*, p. 64.

[65] Marvin, *The History of European Philosophy*, p. 142.

[66] *Ibid.*

[67] Miller, "Mathematical Weakness of Early Civilizations."

suggested, but never clearly defined, may also have been due in some measure to the opposition afforded by the inductive scientific views of Aristotle and the Peripatetic School. Aristotelian thought, while not destroying the rigorously deductive character of Greek geometry, may have preserved in Greek mathematics that strong reasonable and matter-of-fact cast which one finds in Euclid and which operated against the early development of the calculus, as well as against the Platonic tendency toward speculative metaphysics.

Although Plato did not solve the difficulties which the Pythagoreans and Democritus encountered, he urged their study upon his associates, inveighing against the ignorance concerning such problems which prevailed among the Greeks.[68] To Eudoxus he is said to have proposed a number of problems in stereometry which proved to be remarkably suggestive in leading toward the calculus. In this connection the demonstrations which Eudoxus gave of the propositions (previously stated without proof by Democritus) on the volumes of pyramids and cones led to his famous general method of exhaustion and to his definition of proportion. The achievements of Eudoxus are those of a mathematician who was at the same time a scientist, with none of the occult or mystic in him.[69] As a consequence, they are based at every point on finite, intuitively clear, and logically precise considerations. In method and spirit the later work of Euclid will be found to owe much more to Eudoxus than to Plato.

We have seen that the Pythagorean theory of proportion could not be applied to all lines, many of which are incommensurable, and that the Democritean view of infinitesimals was logically untenable. Eudoxus proposed means by which these difficulties could be avoided. The paths he indicated, in his theory of proportion and in the method of exhaustion, were not the equivalents of our modern conceptions of number and limit, but rather detours which obviated the necessity of using the latter. They were significant, however, in that they made it possible for the Greek mind confidently to pursue its attack upon problems which were to eventuate much later in the calculus.

The Pythagorean conception of proportion had been the result of the identification of geometrical magnitudes and integral numbers.

[68] See *Laws* 819d–820c.
[69] T. L. Heath, *History of Greek Mathematics*, I, 323–25; Becker, "Eudoxos-Studien," 1936, p. 410.

Two lines, for example, were to each other as the ratio of the (integral) numbers of units in each. With the discovery of the incommensurability of some lines with others, however, this definition could no longer be universally applied. Eudoxus substituted for it another which was more general, in that it did not require two of the terms in the proportion to be (integral) numbers, but allowed all four to be geometrical entities, and required no extension of the Pythagorean idea of number. Euclid[70] states Eudoxus' definition as follows: "Magnitudes are said to be in the same ratio, the first to the second and the third to the fourth, when, if any equimultiples whatever be taken of the first and third, and any equimultiples whatever of the second and fourth, the former equimultiples alike exceed, are alike equal to, or alike fall short of, the latter equimultiples respectively taken in corresponding order."[71] The theory of proportion thus stated involves only geometrical quantities and integral multiples of them, so that no general definition of number, rational or irrational, is necessary. It is interesting to see that after the development of mathematical analysis, the concept of proportion resembles the arithmetical form of the Pythagoreans rather than the geometrical one of Eudoxus. Even when the ratio is not expressible as the quotient of two integers, we now substitute for it a single number and symbol such as π or e. Although Eudoxus did not, as we do, regard the ratio of two incommensurable quantities as a number,[72] nevertheless his definition of proportion expresses the ordinal idea involved in the present conception of real number. The assertion that it is "word for word the same as the general definition of number given by Weierstrass"[73] will be found, however, to be incorrect, both literally and in its implications. The formulation of Eudoxus was, on the contrary, a means of avoiding the need of such an arithmetic definition as that of Weierstrass.

The method of exhaustion of Eudoxus shows the same abandonment of numerical conceptions which we have seen in his theory of proportion. Length, area, and volume are now carefully defined numerical entities in mathematics. After the time of the Pythagoreans, classic Greek mathematics did not attempt to identify number with

[70] Book V, Definition 5.
[71] *The Thirteen Books of Euclid's Elements*, trans. by T. L. Heath, II, 114.
[72] Stolz, *Vorlesungen über allgemeine Arithmetik*, I, 94.
[73] Simon, "Historische Bemerkungen über das Continuum," p. 387.

geometrical quantities. As a result no rigorous general definitions of
length, area, and volume could then be given, the meaning of these
quantities being tacitly understood as known from intuition. The
question, "What is the area of a circle?" would have had no meaning
to the Greek geometers. But the query, "What is the ratio of the
areas of two circles?" would have been a legitimate one, and the
answer would have been expressed geometrically: "the same as that
of squares constructed on the diameters of the circles."[74] The fact that
squares and circles are incommensurable with each other does not
cause any incongruity in the idea of their entering into the same proportion under the general definition of Eudoxus; but the proof of the
correctness of the proportion requires in this case the comparison of
squares with squares and of circles with circles.

Obviously the old Pythagorean method of the application of areas
cannot be employed in the case of circles, so Eudoxus had recourse
to an idea which had been advanced sometime before by Antiphon
the Sophist and again a generation later by Bryson. These men had
inscribed within a circle a regular polygon, and by successively
doubling the number of sides they seem to have hoped to reach a
polygon which would coincide with the circle and so "exhaust" its
area. It should, however, be borne in mind that we do not know just
what Antiphon (and later Bryson) said. The method of Antiphon has
been described[75] as equivalent at one and the same time to the method
of Eudoxus (as given in Euclid XII, 2), and to our conception of the
circle as the limit of such an inscribed polygon, but merely expressed
in different terminology. This cannot be strictly correct. If Antiphon
had considered the process of bisection as carried out to an infinite
number of steps, he would not have been thinking in the terms of
Eudoxus and Euclid, as we shall see. If, on the other hand, he did not
regard the process as continued indefinitely but only as carried out to
any desired degree of approximation, he could not have had our idea
of a limit. Furthermore, our conception of the limit is numerical,
whereas the notions of Antiphon and Eudoxus are purely geometrical.

The suggestive idea of Antiphon, however, was adopted by Bryson,
who is reputed not only to have inscribed a polygon within the circle

[74] Cf. Vogt, *Der Grenzbegriff in der Elementar-mathematik*, p. 42.
[75] T. L. Heath, *History of Greek Mathematics*, I, 222.

but also to have circumscribed one about it as well, saying that the circle would ultimately, as the result of continued bisection, be the mean of the inscribed and circumscribed polygons. Again we do not know exactly what he said, and cannot tell clearly what he meant.[76] Interpretations have been advanced[77] which would go so far as to see in the work of Bryson the concept of a "Dedekind Cut" or of the continuum of Georg Cantor, but the evidence would hardly appear to warrant such imputations. However, the idea which he suggested was developed by Eudoxus into a rigorous type of argument for dealing with problems involving two dissimilar, heterogeneous, or incommensurable quantities, in which intuition fails to represent clearly the transition from one to the other which is necessary to make a comparison possible.

The procedure which Eudoxus proposed has since become known as the method of exhaustion. The principle upon which this method is based is commonly called the lemma, or postulate, of Archimedes, although the great Syracusan mathematician himself ascribed it[78] to Eudoxus and it is not improbable that it had been formulated still earlier by Hippocrates of Chios.[79] This axiom (as given in Euclid X, 1) states that, given two unequal magnitudes (neither equal to zero, of course, since for the Greeks this was neither a number nor a magnitude), "if from the greater there be subtracted a magnitude greater than its half, and from that which is left a magnitude greater its half, and if this process be repeated continually, there will be left some magnitude which will be less than the lesser magnitude set out."[80] This definition (in which, of course, any ratio may be substituted in place of one-half) excluded the infinitesimal from all demonstrations in the geometry of the Greeks, although we shall find this banished notion entering occasionally into their thought as an explorative aid. From the fact that, on continuing the process indicated in the axiom of Archimedes, the magnitude remaining can be made as small as we please, the procedure introduced by Eudoxus came much later to be

[76] *Ibid.*, I, 224.

[77] See Becker, "Eudoxos-Studien," in particular, 1933, pp. 373–74; Toeplitz, "Das Verhältnis von Mathematik und Ideenlehre bei Plato," pp. 31–33.

[78] In his *Quadrature of the Parabola*. See T. L. Heath, *History of Greek Mathematics*, I, 327–28.

[79] Hankel, *Zur Geschichte der Mathematik in Alterthum und Mittelalter*, p. 122.

[80] Euclid, *Elements*, Heath trans., vol. III, p. 14.

called the method of exhaustion. It is to be remarked, however, that the word *exhaustion* was not applied in this connection until the seventeenth century,[81] when mathematicians somewhat ambiguously and uncritically employed the term indifferently to designate both the ancient Greek procedure and their own newer methods which led immediately to the calculus and which truly "exhausted" the magnitudes.

The Greek mathematicians, however, never considered the process as being literally carried out to an infinite number of steps, as we do in passing to the limit—a concept which allows us to interpret the area or volume as truly exhausted, or at least as defined as the limit of the infinite numerical sequence obtained in this manner. There was always, in the Greek mind, a quantity left over (although this could be made as small as desired), so that the process never passed beyond clear intuitional comprehension. A simple illustration will perhaps serve to make the nature of the method clear. The proposition, in Euclid XII, 2, that the areas of circles are to each other as the squares on their diameters will suffice for this purpose. The substance of this proof is as follows: Let the areas of the circles be A and a, and let their diameters be D and d respectively. If the proportion $a : A = d^2 : D^2$ is not true, then let $a' : A = d^2 : D^2$, where a' is the area of another circle either greater or smaller than a. If a' is smaller than a, then in the circle of area a we can inscribe a polygon of area p such that p is greater than a' and smaller than a. This follows from the principle of exhaustion (Euclid X, 1)—that if from a magnitude (such as the difference in area between a' and a) we take more than its half, and from the difference more than its half, and so on, the difference can be made less than any assignable magnitude. If P is the area of a similar polygon inscribed in the circle of area A, then we know that $p : P = d^2 : D^2 = a' : A$. But since $p > a'$, then $P > A$, which is absurd, since the polygon is inscribed within the circle. In a similar manner it can be shown that the supposition $a' > a$ likewise leads to a contradiction, and the truth of the proposition is therefore established.[82]

The method of exhaustion, although equivalent in many respects

[81] In particular in Gregory of St. Vincent, *Opus geometricum*, pp. 739–40.
[82] Cf. Euclid, *Elements*, Heath trans., vol. III, pp. 371–78.

to the *type of argument* now employed in proving the existence of a limit in the differential and the integral calculus, does not represent the *point of view* involved in the passage to the limit. The Greek method of exhaustion, dealing as it did with continuous magnitude, was wholly geometrical, for there was at the time no knowledge of an arithmetical continuum. This being the case, it was of necessity based on notions of the continuity of space—intuitions which denied any ultimate indivisible portion of space, or any limit to the divisibility in thought of any line segment. The inscribed polygon could be made to approach the circle as nearly as desired, but it could never become the circle, for this would imply an end in the process of subdividing the sides. However, under the method of exhaustion it was not necessary that the two should ever coincide. By an argument based upon the reductio ad absurdum, it could be shown that a ratio greater or less than that of equality was inconsistent with the principle that the difference could be made as small as desired.

The argument of Eudoxus appealed at every stage to intuitions of space, and the process of subdivision made no use of such unclear conceptions as that of a polygon with an infinite number of sides—that is, of a polygon which should ultimately coincide with the circle. No new concepts were involved, and the gap between the curvilinear and the rectilinear still remained unspanned by intuition. Eudoxus, however, had most ingeniously contrived to demonstrate—without resort to the logically self-contradictory infinitesimal previously invoked by vague imagination—the truth of certain geometrical propositions requiring a comparison of the curvilinear with the rectilinear and of the irrational with the rational.

There is no logical difficulty to be found in the argument used in the method of exhaustion, but the cumbersomeness of its application led later mathematicians to seek a more direct approach to problems in which the application of some such procedure would have been indicated. The method of exhaustion has, most misleadingly, been characterized as "a well-established algorithm of the differential calculus."[83] It is indeed true that the problems to which the method was applied were those which led toward the calculus. Nevertheless, it is not incorrect to say that the procedure involved actually directed

[83] Simon, "Zur Geschichte und Philosophie der Differentialrechnung," p. 116.

attention away from the discovery of an equivalent algorithm, in that it directed attention toward the synthetic form of exposition rather than toward an analytic instrument of discovery.[84] It did represent a conventional type of demonstration, but the Greek mathematicians never developed this into a concise and well-recognized operation with a characteristic notation. In fact the ancients never made the first step in this direction: they did not formulate the principle of the method as a general proposition, reference to which might serve in lieu of the argument by the ubiquitous double reductio ad absurdum.[85]

It was largely in connection with the search for some means of simplifying the arguments in the tedious methods of the ancients that the differential and integral calculus was developed in the seventeenth century. To trace this development is the purpose of this essay; but at this point it may not be amiss to anticipate the final formulation to the extent of comparing the nature of the basic concept of the calculus—that of the limit of an infinite sequence—with the view indicated in the method of exhaustion.

The limit of the infinite sequence $P_1, P_2, \ldots, P_n, \ldots$ (the terms of which represent, for example, the areas of the inscribed polygons considered in the proposition above) has been defined in the introduction to be the number C, such that, given any positive number ε, we can find a positive integer N, such that for $n > N$ it can be shown that $|C - P_n| < \varepsilon$. The spatial intuition of the method of exhaustion, with its application of areas, unlimited subdivision, and argumentation by a reductio ad absurdum, here gives way to definition in terms of formal logic and number, i. e., of infinite ordered aggregates of the positive integers. The method of exhaustion corresponds to an intuitional concept, described in terms of mental pictures of the world of sensory perception. The notion of a limit, on the other hand, may be regarded as a verbal concept, the explication of which is given in terms of words and symbols—such as number, infinite sequence, less than, greater than—with regard not to any mental visualization, but only to their definition in terms of the primary undefined elements. The limit concept is thus by no means to be considered ineffable; nor does it imply that there is other than

[84] Cf. Brunschvicg, *Les Étapes de la philosophie mathématique*, pp. 157–59.
[85] *The Works of Archimedes*, ed. by T. L. Heath, p. cxliii.

empirical experience. It simply makes no appeal to intuition or sensory perception. It resembles the method of exhaustion in that it allows our vague instinctive feeling for continuity to shift for itself in any effort that may be made to picture how the gap between the curvilinear and the rectilinear, or between the rational and the irrational, is bridged, for such an attempt is quite irrelevant to the logical reasoning involved. The limit C is not for this reason to be regarded as a sophistic or inconceivable quantity which somehow nevertheless enters into real relations with other similar quantities, nor is it to be visualized as the last term of the infinite sequence. It is to be considered merely as a number possessing the property stated in the definition. It is to be borne in mind that although adumbrations of the limit idea appear in the history of mathematics in ancient times, nevertheless the rigorous formulation of this concept does not appear in work before the nineteenth century—and certainly not in the Greek method of exhaustion.[86]

The apparent break, in the mathematical work of Eudoxus, from the metaphysical aspect of Platonism is seen equally clearly[87] in the philosophical thought of one who studied under Plato for twenty years and who was known as the "mind of the School."[88] Aristotle borrowed freely from the work of his predecessors, with the result that, although not primarily a mathematician, he was familiar with the difficulties and results of Greek mathematics, including the method of exhaustion. He wrote a work (now lost) *On the Pythagoreans*, discussed at some length the paradoxes of Zeno, mentioned Democritus frequently in mathematics and science (although always to refute him), was intimately familiar with Plato's thought, and was acquainted with the work of Eudoxus. In spite of his competence in mathematics and of his frequent use of geometry in his constructions,[89] Aristotle's approach to the problems involved was essentially scientific, in the inductively descriptive sense. Furthermore, for Plato's mathematical intellectualism he substituted a grammatical intuitionism.[90]

[86] Milhaud (*Les Philosophes géomètres*, p. 182) would have Eudoxus consider the circles as the limit of a polygon, but this could not have been in the sense in which the term limit is now employed.

[87] Cf., in this respect, Jaeger, *Aristoteles*.   [88] Ross, *Aristotle*, p. 2.

[89] Enriques, *Problems of Science*, p. 110.   See also Görland, *Aristoteles und die Mathematik*, and Heiberg, "Mathematisches zu Aristoteles."

[90] Brunschvicg, *Les Étapes de la philosophie mathématique*, p. 70.

Although he realized that the objects of mathematics are not those of sense experience and that the figures used in demonstrations are for illustration only[91] and in no way enter into the reasoning, Aristotle's whole attitude was governed by a strong dependence upon the evidence of the senses, as well as upon logic, and by an aversion to abstraction and extrapolation beyond the powers of sensory perception. As a consequence, he did not think of a geometrical line, as had Plato, as an idea which is prior to, and independent of, experience of the concrete. Neither did he regard it, as does modern mathematics, as an abstraction which is suggested, perhaps, though not in any way defined, by physical objects. He viewed it rather as a characteristic of natural objects which has merely been separated from its irrelevant context in the world of nature. "Geometry investigates physical lines but not *qua* physical," he said,[92] and added:

Necessity in mathematics is in a way similar to necessity in things which come to be through the operation of nature. Since a straight line is what it is, it is necessary that the angles of a triangle should equal two right angles.[93]

The decisions which Aristotle rendered on the indivisible, the infinite, and the continuous were consequently those dictated by common sense. In fact, with the exception of Plato's successors in the Academy and, perhaps, of Archimedes, they were those accepted by the body of Greek mathematicians after Eudoxus.

Only in the case of the indivisible, however, do Aristotle's views coincide with the present notions in mathematics. Modern science has opposed, modern mathematics upheld, Aristotle in his vigorous denial of the indivisible, physical and mathematical, of the atomic school. Recent physical and chemical theories of the atom have furnished a description of natural phenomena which offers a higher degree of consistency within itself and with sensory impressions than had the Peripatetic doctrine of continuous substantiality. Science has consequently, under Carneades' doctrine of truth, accepted the atom as a physical reality. Modern mathematics, on the other hand, agrees with Aristotle in his opposition to minimal indivisible line segments; not, however, because of any argument from experience, but because it has

[91] T. L. Heath, *History of Greek Mathematics*, I, 337.
[92] *Physica* II. 193b–194a.   [93] *Physica* II. 200a.

been unable to give a satisfactory definition and logical elaboration of the concept. However, the mathematical indivisible, in spite of the opposition of Aristotle's authority, was destined to play an important part in the development of the calculus, which in the end definitely excluded it. That the concept enjoyed an extensive popularity even in Aristotle's day, Greek logic notwithstanding, is seen by the fact that a Peripatetic treatise formerly ascribed to Aristotle (but now thought to have been written by Theophrastus, or Strato of Lampsacus, or perhaps by someone else), the *De lineis insecabilibus*,[94] was directed against it. This presents many arguments against the assumption of indivisible lines and concludes that "it conflicts with practically everything in mathematics."[95]

The work may have been composed as an answer to Xenocrates, the successor of Plato in the Academy, who apparently maintained the existence of mathematical indivisibles. It has been asserted[96] that neither Aristotle nor Plato's successors understood the infinitesimal concept of their master, and that only Archimedes rose to a correct appreciation of it. It is to be remarked, however, that such an assumption is wholly gratuitous. Plato in his extant works offered no clear definition of the infinitesimal. Archimedes, moreover, made no mention of any indebtedness to Plato in this matter, and, as will be seen, explicitly disclaimed any intention of regarding infinitesimal methods as constituting valid mathematical demonstrations.[97] The opposition of Aristotle to the doctrine of infinitesimals was wholly justified by considerations of logic, although from the point of view of the subsequent development of the calculus the uncritical use of the infinitely small was for a time most fruitful.

As in the case of the infinitesimal, so also with respect to the infinite the views of Aristotle constitute excellent illustrations of his abiding confidence in the ultimate interpretability of phenomena in terms of distinctly clear concepts derived from sensory experience.[98] The Pythagoreans had regarded space as infinitely divided, and Democritus had likewise spoken of the atoms as infinite in number. Plato, influ-

[94] *The Works of Aristotle*, Vol. VI, *Opuscula*.   [95] *De lineis insecabilibus* 970a.
[96] Hoppe, "Zur Geschichte der Infinitesimalrechnung," p. 152.
[97] *The Method of Archimedes*, ed. by T. L. Heath, p. 17.
[98] Aristotle's view of the infinite has been much discussed. For one of the most recent philosophical discussions of this subject, see Edel, *Aristotle's Theory of the Infinite*.

enced by these views and perhaps also by those of Anaximander and Anaxagoras, had not clearly distinguished the concrete from the abstract.[99] He had held that the infinite was located at the same time in ideas and in the sensible world,[100] the line as made up of points being one illustration of this fact. However, no one of Aristotle's predecessors had made quite clear his position with respect to the infinite. Anaxagoras, at least, seems to have realized that it is only the imagination which objects to the infinite and to an infinite subdivision.[101]

On the other hand, Aristotle, in adopting the inductive scientific attitude, did not go beyond what is clearly representable in the mind. In consequence he denied altogether the existence of the actual infinite and restricted the use of the term to indicate a potentiality only.[102] His clear distinction between an existent infinity and a potential infinity was the basis of much of the discussion of the Scholastics on the subject and of later controversies on the metaphysics of the calculus. His refusal to recognize the actual infinite was in keeping with his fundamental tenet that the unknowable exists only as a potentiality: that anything beyond the power of comprehension is beyond the realm of reality. Such a methodological definition of existence has led investigators in inductive science to continue to the present time the Aristotelian attitude of negation toward the infinite;[103] but such a view, if adopted in mathematics, would exclude the concepts of the derivative and the integral as extrapolations beyond the thinkable, and would, in fact, reduce mathematical thought to the intuitively reasonable.

That the Aristotelian doctrine of the infinite was abandoned in the mathematics of the nineteenth century was largely the result of a shift of emphasis from the infinite of geometry to that of arithmetic; for in the latter field assumptions appear to be less frequently dictated by experience. For Aristotle such a change of view would have been impossible, inasmuch as his conception of number was that of the Pythagoreans: a collection of units.[104] Zero was not included, of course; nor was "the generator of numbers," the integer one. "The

[99] Paul Tannery, *Pour l'histoire de la science hellène*, pp. 300–5.
[100] Brunschvicg, *Les Étapes de la philosophie mathématique*, p. 67.
[101] Paul Tannery, *Pour l'histoire de la science hellène*, pp. 293–94.
[102] *Physica* III. 206b.   [103] Barry, *The Scientific Habit of Thought*, p. 197.
[104] *Physica* III. 207b; cf. also Plato, *Republic* VII. 525e.

smallest number in the strict sense of the word 'number' is two,"
said Aristotle.[105] Such a view of number could not be reconciled with
the infinite divisibility of continuous magnitude which Aristotle
upheld so vigorously. When, then, Aristotle distinguished two kinds of
(potential) infinite—one in the direction of successive addition, or the
infinitely large, and the other in the direction of successive subdivision,
or the infinitely small—we find the behavior of number to be quite
different from that of magnitude:

Every assigned magnitude is surpassed in the direction of smallness, while
in the other direction there is no infinite magnitude .... Number on the
other hand is a plurality of "ones" and a certain quantity of them. Hence
number must stop at the indivisible. . . . But in the direction of largeness
it is always possible to think of a larger number. . . . Hence this infinite is
potential, ... and not a permanent actuality but consists in a process
of coming to be, like time . . . . With magnitudes the contrary holds.
What is continuous is divided ad infinitum, but there is no infinite in the
direction of increase. For the size which it can potentially be, it can also
actually be.[106]

In commenting on the view of mathematicians, Aristotle said,

In point of fact they do not need the infinite and do not use it. They postulate only that the finite straight line may be produced as far as they
wish. . . . Hence, for the purposes of proof, it will make no difference to
them to have such an infinite instead, while its existence will be in the
sphere of real magnitude.[107]

How well this characterizes Greek geometry can be seen in the method
of exhaustion as presented by Eudoxus slightly earlier than Aristotle
and by Euclid a little later. This method assumes in the proof only
that the bisection can be continued as far as one may wish, not carried out to infinity. How far it lies from the point of view of modern
analysis is indicated by the fact that the latter has been called "the
symphony of the infinite."

That Aristotle would be unable to cope with the problems of the
continuous is perhaps to be expected, both from his view of the
infinite and from the lack among the Greeks of an adequate arithmetical point of view. After having considered place and time in the
fourth book of the *Physica*, and change in the fifth, Aristotle turned
in the sixth book to the continuous. His account is based upon a

[105] *Physica* IV. 220a.   [106] *Physica* III. 207b.   [107] *Ibid.*

definition derived from the intuitive notion of the essence of continuous magnitude: "By continuous I mean that which is divisible into divisibles that are infinitely divisible."[108] This view was supplemented by a naïve appeal to the instinctive feeling of the necessity of a hang-togetherness—of the coincidence of the extremities of the component parts.[109] For this reason Aristotle denied that number can produce a continuum,[110] inasmuch as there is no contact in numbers.[111] Only a generation ago this Aristotelian view had not been entirely abandoned,[112] but the mathematical continuum accepted at the present time is defined precisely in terms of the concept of number, or of classes of elements, and of the notion of separation—as in the Dedekind Cut—rather than contact. The nature of continuous magnitude has been found to lie deeper than Aristotle believed, and it has been explained on the basis of concepts which require a broader definition of number than that held during the Greek period. The Aristotelian dictums on the subject were not unfruitful, however, for they led to speculations during the medieval period which in turn aided in the rise of the calculus and the modern doctrine of the continuum.

In connection with the study of continuous magnitude, Aristotle attempted also to clarify the nature of motion, criticizing the atomists for their neglect of the whence and the how of movement.[113] Although he was quite adept at detecting problems, he failed to make his formulation of these quantitative and was consequently infelicitous in their resolution. In the light of modern scientific method, this lack of mathematical expression gives to his treatment of motion and variability the appearance of a dialectical exercise, rather than of a serious effort to establish a sound basis for the science of dynamics.[114]

Aristotle's approach to the subject was qualitative and metaphysical. This is evidenced by his definition of motion as "the fulfillment of what exists potentially, in so far as it exists potentially," and by the further remark, "We can define motion as the fulfillment of the movable qua movable."[115] We shall find this qualitative explanation of motion—the result of the striving of a body to become actually what it is potentially—involved in less teleological forms in the idea

---

[108] *Physica* VI. 232b.   [109] *Physica* VI. 231a; cf. also *Categoriae* 5a.
[110] *Metaphysica* 1075b and 1020a; *Categoriae* 4b.   [111] *Metaphysica* 1085a.
[112] See Merz, *A History of European Thought in the Nineteenth Century*, II, 644.
[113] *Metaphysica* 985b and 1071b.
[114] Cf. Mach, *The Science of Mechanics*, p. 511.   [115] *Physica* III, 201a–202a.

of *impetus* developed by the Scholastics, in Hobbes' explanation of velocity and acceleration in terms of a *conatus*, and even in metaphysical interpretations of the infinitesimal of the calculus as an *intensive* quantity—that is, as a "becoming" rather than a "being." In this respect Aristotle's work may have encouraged the elaboration of notions leading toward the derivative. However, his influence was in another sense quite adverse to the development of this concept in that it centered attention upon the qualitative description of the change itself, rather than upon a quantitative interpretation of the vague instinctive feeling of a continuous state of change invoked by Zeno. The calculus has shown that the concept of continuous change is no more free from that of the discrete than is the numerical continuum, and that it is logically to be based upon the latter, as is also the idea of geometrical magnitude. As long as Aristotle and the Greeks considered motion continuous and number discontinuous, a rigorous mathematical analysis and a satisfactory science of dynamics were difficult of achievement.

The treatment of the infinite and of continuous magnitude found in the *Physica* of Aristotle has been regarded as presenting the appearance of a veritable introduction to a treatise on the differential calculus.[116] Such a view, however, is seen to be most unwarranted, inasmuch as Aristotle expressed his unqualified opposition to the fundamental idea of the calculus—that of an instantaneous rate of change. He asserted that "Nothing can be in motion in a present. . . . Nor can anything be at rest in a present."[117] This point of view necessarily operated against the mathematical representation of the phenomena of change and against the development of the calculus. Aristotle's denial of instantaneous velocity, as realized in the world described by science, is, to be sure, in conformity with the recognized limitations of sensory perception. Only average velocities, $\dfrac{\Delta s}{\Delta t}$, are recognizable in this sense. In the world of thought, on the other hand, it has been found possible—through the calculus and the limit concept—to give a rigorous quantitative definition of instantaneous velocity, $\dfrac{ds}{dt}$. Aristotle, however, in conformity with a view widely ac-

[116] Moritz Cantor, "Origines du calcul infinitésimal," p. 6.   [117] *Physica* VI. 234a.

cepted at the time, regarded mathematics as a pattern of the world known through the senses and consequently did not foresee such a possibility.

The failure of Aristotle to distinguish sharply between the worlds of experience and of mathematical thought resulted in his lack of clear recognition of a similar confusion in the paradoxes of Zeno. Aristotle refuted the arguments in the *stade* and the *arrow* by an appeal to sensory perception and the denial of an instantaneous velocity. Modern mathematics, on the other hand, has answered them in terms of thought alone, based on the concept of the derivative. In the same manner Aristotle resolved the paradoxes in the *dichotomy* and the *Achilles* by the curt assertion, suggested by experience, that although one cannot traverse an infinite space in finite time, it is possible to cover an infinitely divided space in finite time because of the infinite divisibility of the latter.[118]

Mathematics has, of course, given the solution of the difficulties in terms of the abstract concept of converging infinite series. In a certain metaphysical sense this notion of convergence does not answer Zeno's argument, in that it does not tell how one is to picture an infinite number of magnitudes as together making up only a finite magnitude; that is, it does not give an intuitively clear and satisfying picture, in terms of sense experience, of the relation subsisting between the infinite series and the limit of this series. If one demands that Zeno's paradoxes be answered in terms of our vague instinctive feeling for continuity—as essentially different from the discrete—no answers more satisfying than those of Aristotle (to whom we owe also the statement of the paradoxes, since we do not have Zeno's words) have been given. The unambiguous demonstration that the difficulties implied by the paradoxes are simply those of visualization and not those of logic was to require more precise and adequate definitions than any which Aristotle could furnish for such subtle notions as those of continuity, the infinite, and instantaneous velocity. Such definitions were to be given in the nineteenth century in terms of the concepts of the calculus; and modern analysis has, upon the basis of these, clearly dissented from the Aristotelian pronouncements in this field. The views of Aristotle are not on this account to be regarded—

[118] *Physica* VI. 239b–240a.

as is all too frequently and uncritically maintained[119]—as gross misconceptions which for two thousand years retarded the advancement of science and mathematics. They were, rather, matured judgments on the subject which furnished a satisfactory working basis for later investigations which were to result in the science of dynamics and in the mathematical continuum. Nevertheless, there is apparent in the work of Aristotle the cardinal weakness of Greek logic and geometry: a naïve realism which regarded thought as a true copy of the external world.[120] This caused him to place too ingenuous a confidence in certain instinctive feelings with respect to continuous magnitude and to seek, of all possible representations, that which presented the greatest plausibility in the light of sensory experience, rather than that which offered the widest consistency in thought.

It has been said[121] that the fifth book of Euclid's *Elements* and the logic of Aristotle are the two most unobjectionable and unassailable treatises ever written. The two men were roughly contemporaries: Aristotle lived from 384 to 322 B. C.; Euclid's birth has been placed at about 365 B. C. and the composition of the *Elements* may be accepted as between 330 and 320 B. C.[122] There is also a marked similarity between the Aristotelian apodictic and the mathematical method built up by Euclid.[123] Although Euclid was probably taught by the pupils of Plato,[124] the influence of the sober, hard-headed, scientific thought illustrated by Eudoxus and Aristotle must have predominated over the more abstract, speculative, and even mystical trend seen in the immediate successors of Plato to the leadership of the Academy and carried to excess by later Neoplatonists. There is in Euclid none of the metamathematics which played such a prominent part in Plato's thought, nor do metaphysical speculations on mathematical atomism enter. Mathematics was regarded by Euclid neither as a necessary form of cosmological intelligibility, nor as a mere tool of

[119] See, for example, Mayer, "Why the Social Sciences Lag behind the Physical and Biological Sciences"; cf. also Toeplitz, "Das Verhältnis von Mathematik und Ideenlehre bei Plato."
[120] Enriques, *The Historic Development of Logic*, p. 25.
[121] By Augustus De Morgan. See Hill, "Presidential Address on the Theory of Proportion."
[122] See Vogt, "Die Lebenzeit Euklids."
[123] Brunschvicg, *Les Étapes de la philosophie mathématique*, pp. 84–85.
[124] T. L. Heath, *History of Greek Mathematics*, I, 356.

pragmatic utilitarianism. For him it had entered the domain of logic, and in this connection Proclus tells us that Euclid subjected to rigorous proofs what had been negligently demonstrated by his predecessors.[125] Nevertheless, the *Elements* retained the realism which was so clearly apparent in Aristotelian logic.[126]

Although Aristotle had rejected Plato's doctrine of ideas, he had retained a belief in a natural order of science and in the necessary character of principles. This latter confidence was adopted likewise by Euclid. Greek geometry was not formal logic, made up of hypothetical propositions, as mathematics largely is today; but it was an idealized picture of the world of actuality. Just as Aristotle seems not to have clearly recognized the tentative character of scientific knowledge (thus leaving himself open to the attacks of the Skeptics), so also he failed to appreciate that although the conclusions drawn by mathematics are necessary inferences from the premises, nevertheless the latter are quite arbitrarily selected, subject only to an inner compatibility. Aristotle considered hypotheses and postulates as statements which are assumed without proof, but which are nevertheless capable of demonstration.[127] Although he admitted that "we must get to know the primary premises by induction" (rather than by pure intellection, as Plato had believed), he maintained that "since except intuition nothing can be truer than scientific knowledge, it will be intuition that apprehends the primary premises," and primary premises are therefore "more knowable than demonstrations."[128]

The Euclidean view was similar to the Peripatetic attitude in giving to geometry the characteristic form of logical conclusions from necessary postulates. As such, it excluded any notions the nature of which was not clearly and compellingly "felt" through intuition. The infinite was never invoked in the demonstrations, true to Aristotle's statement that it was unnecessary, its place being taken by the method of exhaustion which had been developed by Eudoxus. The limitation of the concept of number to that of positive integers apparently was

[125] Cf. Proclus Diadochus, *In primum Euclides elementorum librum commentariorum . . . ibri IIII*, p. 43.

[126] Enriques, *The Historic Development of Logic*, p. 25; cf. also Burtt, *Metaphysical Foundations of Modern Physical Science*, p. 31.

[127] *Analytica posteriora* I. 76b.

[128] *Analytica posteriora* II. 100b; cf. also Brunschvicg, *Les Étapes de la philosophie mathématique*, pp. 86–93.

continued, a broader view being made unnecessary by the Eudoxian theory of proportion.[129]

For Euclid ratio was not a number in the abstract arithmetical sense (and in fact it did not become so until the time of Newton),[130] and the treatment of the irrational in the *Elements* is completely geometrical. Furthermore, the axioms, postulates, and definitions of Euclid are those suggested by common sense, and his geometry never loses contact with spatial intuition.[131] His premises are the dictates of sensory experience, much as Aristotle's science may be characterized as a glorification of common sense. Such purely formal, logical concepts as those of the infinitesimal and of instantaneous velocity, of infinite aggregates and the mathematical continuum, are not elaborated in either Euclidean geometry or in Aristotelian physics, for common sense has no immediate need for them. The ideas which were to lead to the calculus had not in Euclid's time reached a stage at which a logical basis could have been afforded; mathematics had not attained the degree of abstraction demanded for symbolic logic. Although the origin of the notions of the derivative and the integral are undoubtedly to be found in our confused thought about variability and multiplicity, the rigorous formulation of the concepts involved, as we shall find, demanded an arithmetical abstraction which Euclid was far from possessing. Even Newton and Leibniz, the inventors of the algorithmic calculus, did not fully recognize the need for it. The logical foundations of the calculus are much further removed from the vague suggestions of experience—much more subtle—than those of Euclidean geometry. Since, therefore, the ideas of variability, continuity, and infinity could not be rigorously established, Euclid omitted them from his geometry. The *Elements* are based on "refined intuition,"[132] and do not allow free scope to the "naïve intuition" which was to be especially active in the genesis of the calculus in the seventeenth century.[133]

From the point of view of the development of the calculus, there-

[129] See Stolz, *Vorlesungen über allgemeine Arithmetik*, I, 94; cf. also Schubert, "Principes fondamentaux de l'arithmétique," pp. 8–9.

[130] See Newton, *Opera omnia*, I, 2.

[131] Cf. Barry, *The Scientific Habit of Thought*, pp. 215–17.

[132] As Felix Klein (*The Evanston Colloquium Lectures on Mathematics*, pp. 41–42) so aptly puts it.

[133] *Ibid.*

fore, the *Elements* of Euclid show an uninteresting inflexibility of rigor, discouraging to the growth of such new speculations and discoveries. The work of Euclid represents the final synthetic form of all mathematical thought—the elaboration by deductive reasoning of the logical implications of a set of premises. Back of his geometry, however, stood several centuries of analytical investigation, carried out often on the basis of empirical research, or on uncritical intuition, or, not infrequently, on transcendental speculation. It was to be largely from indagation of a similar type, rather than from the rigorously precise thought of Euclid, that the development of the concepts of the calculus was to proceed. This in its turn was necessarily to give way, in the nineteenth century, to a formulation as eminently deductive—albeit arithmetic rather than geometric—as that found in the *Elements*.

The greatest mathematician of antiquity, Archimedes of Syracuse, displayed two natures, for he tempered the strong transcendental imagination of Plato with the meticulously correct procedure of Euclid. He "gave birth to the calculus of the infinite conceived and brought to perfection successively by Kepler, Cavalieri, Fermat, Leibniz, and Newton,"[134] and so made the concepts of the derivative and the integral possible. In the demonstration of his results, however, he adhered to the clearly visualized details of the Eudoxian procedure, modifying the method of exhaustion by considering not only the inscribed figure but the circumscribed figure as well. The deductive method of exhaustion was not a tool well adapted to the discovery of new results, but Archimedes combined it with infinitesimal considerations toward which Democritus and the Platonic school had groped. The freedom with which he handled these is shown most clearly in the treatise to which we have already referred, the *Method*.[135]

This work, addressed to Eratosthenes the geographer, astronomer, and mathematician of Alexandria, was lost and remained largely un-

[134] Chasles, *Aperçu historique sur l'origine et le développement des méthodes en géométrie*, p. 22.

[135] For the works of Archimedes in general, see Heiberg, *Archimedis opera omnia* and T. L. Heath, *The Works of Archimedes*. For Archimedes' *Method*, see T. L. Heath, *The Method of Archimedes, Recently Discovered by Heiberg*; Heiberg and Zeuthen, "Eine neue Schrift des Archimedes"; and Smith, "A Newly Discovered Treatise of Archimedes."

known until rediscovered in 1906. In it Archimedes disclosed the method which is presumably that which he employed in reaching many of his conclusions in problems involving areas and volumes. Realizing that it is advantageous to have a preliminary notion of the result before carrying through a deductive geometrical demonstration, Archimedes employed for this purpose, in conjunction with his law of the lever, the idea of a surface as made up of lines. For example, he showed that the truth of the proposition that a parabolic segment is ⅓ the triangle having the same base and vertex (the vertex of the segment being taken as the point from which the perpendicular to the base is greatest) is indicated by the following considerations from

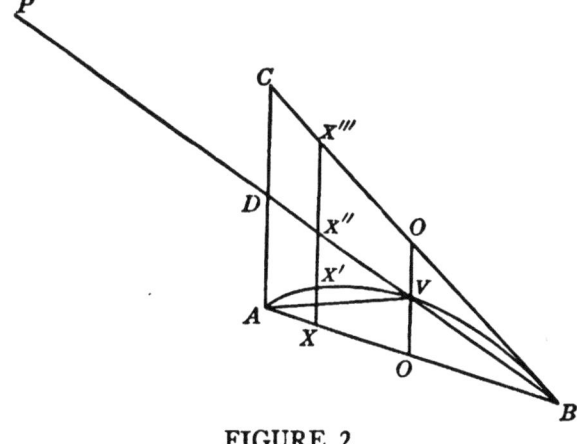

FIGURE 2

mechanics.[136] In the diagram given in Figure 2, in which V is the vertex of the parabola, BC is tangent at B, $BD = DP$, and X is any point on AB, we know from the properties of the parabola that for any position of X we have the ratio $\dfrac{XX'''}{XX'} = \dfrac{AB}{AX} = \dfrac{BD}{DX''} = \dfrac{DP}{DX''}$. But X'' is the center of gravity of XX''', so that from the law of the lever we see that XX', if brought to P as its midpoint, will balance XX''' in its present position. This will be true for all positions of X on AB. Inasmuch as the triangle ABC consists of the straight lines XX''' in this triangle, and since the parabolic segment AVB is likewise made

[136] See T. L. Heath, *The Method of Archimedes*, Proposition I, pp. 15–18.

up of the lines XX', we can conclude that the triangle ABC in its present position will be in equilibrium at D with the parabolic segment when this is transferred to P as its center of gravity. But the center of gravity of ABC is on BD and is $\frac{1}{3}$ the distance from D to B, so that the segment AVB is $\frac{1}{3}$ the triangle ABC, or $\frac{4}{3}$ the triangle AVB.

This method of Archimedes indicates an anticipation of the use of the concept of the indivisible which was to be made in the fourteenth century and which, when developed again more freely in the seventeenth century, was to lead directly to the procedures of the calculus. The basis of the method is to be found in the assumption of Archimedes that surfaces may be regarded as consisting of lines. We do not know in precisely what sense he intended this to be understood, for he did not speak of the number of elements in each figure as infinite, but said rather that the figure is *made up of all* the elements in it. That he probably thought of them as mathematical atoms is indicated not only by this manner of expression, but also by the highly suggestive fact that he was led to many new results by a process of balancing, in thought, elements of dissimilar figures, using the principle of the lever precisely as one would in weighing mechanically a collection of thin laminae or material strips.

Using this heuristic method, Archimedes was able to anticipate the integral calculus in achieving a number of remarkable results. He discovered, among other things, the volumes of segments of conoids and cylindrical wedges and the centers of gravity of the semicircle, of parabolic segments, and of segments of a sphere and a paraboloid.[137] However, to assert that here "for the first time one can correctly speak of an integration"[138] is to misinterpret the mathematical process known by this name. The definite integral is defined in mathematics as the limit of an infinite sequence and not as the sum of an infinite number of points, lines, or surfaces.[139] Infinitesimal considerations, similar to those in the *Method*, were at a later period to furnish perhaps the strongest incentive to the development of the calculus, but,

[137] See *The Works of Archimedes*, Chap. VII, "Anticipations by Archimedes of the Integral Calculus"; also *Method*.

[138] Hoppe, "Zur Geschichte der Infinitesimalrechnung," p. 154; cf. also p. 155.

[139] T. L. Heath (*The Method of Archimedes*, pp. 8–9) correctly points out that the method here used is not integration; but he gratuitously imputes to Archimedes the concept of the differential of area.

as Archimedes realized, they lacked in his time all basis in rigorous thought. This they continued to do until the concepts of variability and limit had been carefully analyzed. For this reason Archimedes considered that this method merely indicated, but did not prove, that the result is correct.[140]

Archimedes employed his heuristic method, therefore, simply as an investigation preliminary to the rigorous demonstration by the method of exhaustion. It was not a generous gesture that led Archimedes to supplement his "mechanical method" by a proof of the results in the rigorous manner of the method of exhaustion; it was, rather, a mathematical necessity. It has been asserted that Archimedes' method "would be quite rigorous enough for us today, although it did not satisfy Archimedes himself."[141] Such an assertion is strictly correct only if we ascribe to him our modern doctrines on number, limit, and continuity. This ascription is hardly warranted, inasmuch as Greek geometry was concerned with *form* rather than with *variation*. It was, as a result, necessarily unable to frame a satisfactory definition of the infinitesimal, which of necessity was to be regarded as a fixed quantity rather than as an auxiliary variable. Archimedes was probably well aware of the lack of any sound basis for his method and for this reason recast all of his analysis by infinitesimals in the orthodox synthetic form, much as Newton was to do almost nineteen hundred years later after the methods of the calculus had been discovered but still lacked adequate foundation.

The suggestive analysis of the problem of determining the area of a parabolic segment had been given by Archimedes in the *Method*. However, formal proofs (both mechanical and geometrical) of the proposition were carried out by the method of exhaustion in another treatise, the *Quadrature of the Parabola*.[142] In these proofs Archimedes followed his illustrious predecessors in omitting all reference to the infinite and the infinitesimal. In the geometrical demonstration, for example, he inscribed within the parabolic segment a triangle of area A, having the same base and vertex as the segment. Then within each

[140] T. L. Heath, *History of Greek Mathematics*, II, 29.

[141] T. L. Heath, *The Method of Archimedes*, p. 10.

[142] See *The Works of Archimedes*, and T. L. Heath, *History of Greek Mathematics*, II, 85–91. A good adaptation of the geometrical proof is given in Smith, *History of Mathematics*, II, 680–83.

of the two smaller segments having the sides of the triangle as bases, he similarly inscribed triangles. Continuing this process, he obtained a series of polygons with an ever-greater number of sides, as illustrated (fig. 3). He then demonstrated that the area of the nth such polygon was given by the series $A\left(1 + \frac{1}{4} + \frac{1}{4^2} + \ldots + \frac{1}{4^{n-1}}\right)$, where A is the area of the inscribed triangle having the same vertex and base as the segment. The sum to infinity of this series is $\frac{4}{3}A$, and it was probably from this fact that Archimedes inferred that the area of the parabolic segment was also $\frac{4}{3}A$.[143]

However, he did not state the argument in this manner. Instead of finding the limit of the infinite series, he found the sum of n terms and added the remainder, using the equality

$$A\left(1 + \frac{1}{4} + \frac{1}{4^2} + \ldots + \frac{1}{4^{n-1}} + \frac{1}{3}\cdot\frac{1}{4^{n-1}}\right) = \frac{4}{3}A.$$

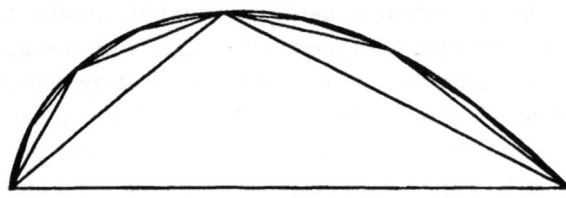

FIGURE 3

As the number of terms becomes greater, the series thus "exhausts" $\frac{4}{3}A$ only in the Greek sense that the remainder, $\frac{1}{3}\left(\frac{1}{4^{n-1}}\right)A$, can be made as small as desired. This is, of course, exactly the method of proof for the existence of a limit,[144] but Archimedes did not so interpret the argument. He did not express the idea that there is no remainder in the limit, or that the infinite series is rigorously equal to $\frac{4}{3}A$.[145] Instead, he proved, by the double reductio ad absurdum of the method of exhaustion, that the area of the parabolic segment could be neither greater nor less than $\frac{4}{3}A$. In order to be able to define $\frac{4}{3}A$ as the sum of the infinite series, it would have been necessary to develop the general concept of real number. Greek mathematicians did not possess this, so that for them there was always a gap between the real (finite) and the ideal (infinite).

[143] T. L. Heath, "Greek Geometry with Special Reference to Infinitesimals."
[144] As Miller pointed out in "Some Fundamental Discoveries in Mathematics."
[145] *The Works of Archimedes*, p. cxliii.

It is not strictly correct, therefore, to speak of Archimedes' geometrical procedure as a passage to the limit, for the essential part of the definition of a limit is the infinite sequence.[146] Inasmuch as he did not invoke the limit concept, it is hardly correct to say that in finding the sum of such series Archimedes answered in a very explicit and definite manner some of the difficult questions raised by Zeno, and that "These difficulties were completely solved by the Greek mathematicians, and further serious arguments along this line seem to be based upon ignorance or perversity."[147] The notion of the limit of an infinite series is essential for the clarification of the paradoxes; but Greek mathematicians (including Archimedes) excluded the infinite from their reasoning. The reasons for this ban are obvious: intuition could at the time afford no clear picture of it, and it had as yet no logical basis. The latter difficulty having been removed in the nineteenth century and the former being now considered irrelevant, the concept of infinity has been admitted freely into mathematics. The related limit concept is now invoked in the explication of the paradoxes, as well as in a simplification of Archimedes' long indirect demonstrations.

The series given above is not the only one found in Archimedes' work. In determining, by the method of exhaustion, the volume of a segment of a paraboloid of revolution, he was led to investigations of a similar nature. A consideration, in some detail, of the application of the method which Archimedes here made[148] may be desirable at this point, in order to bring out clearly the general character of his procedure and to point out to what a remarkable extent it resembles that used in the integral calculus, even though Archimedes did not explicitly employ the limit concept.

Archimedes first circumscribed about the solid ABC (which he called a conoid) the cylinder $ABEF$ (fig. 4) having the same axis, CD, as has the paraboloidal segment. He then divided the axis into n equal parts of length h and through the points of division passed planes parallel to the base. On the sections of the paraboloid thus formed he constructed inscribed and circumscribed cylinder frusta, as

[146] C. R. Wallner, "Über die Entstehung des Grenzbegriffes," p. 250. See also Hankel, "Grenze"; and Wieleitner, *Die Geburt der Modernen Mathematik*, II, 12.

[147] Miller, "Some Fundamental Discoveries," p. 498.

[148] *On Conoids and Spheroids*, Propositions 21, 22; *The Works of Archimedes*, pp. 131–33.

shown in the figure. He was then able to establish the equivalent of the proportions:

$$\frac{\text{Cylinder } ABEF}{\text{Inscribed figure}} = \frac{n^2h}{h + 2h + \ldots + (n-1)h}.$$

and

$$\frac{\text{Cylinder } ABEF}{\text{Circumscribed figure}} = \frac{n^2h}{h + 2h + \ldots + nh}.$$

Archimedes had previously shown[149] (using a method cast in geometrical form but otherwise much like that ordinarily employed in elementary algebra to determine the sum of an arithmetic progression) that $h + 2h + \ldots + (n-1)h < \frac{1}{2}n^2h$ and that $h + 2h + \ldots + nh > \frac{1}{2}n^2h$. At this point modern mathematics would employ the limit

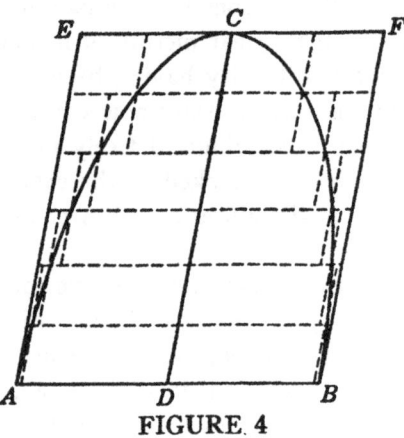

FIGURE. 4

concept and, allowing the series $h + 2h + 3h + \ldots$ to become infinite, would conclude that $\lim\limits_{n \to \infty} \left(\dfrac{n^2h}{h + 2h + 3h + \ldots + nh} \right) = 2$. Not so Archimedes. Instead of doing this, he showed that the proportions above may be written, as a result of these inequalities,

$$\frac{\text{Cylinder } ABEF}{\text{inscribed figure}} > \frac{2}{1} \text{ and } \frac{\text{Cylinder } ABEF}{\text{circumscribed figure}} < \frac{2}{1}.$$

Now by the principle of exhaustion and the usual reductio ad absurdum, he concluded that the paraboloidal segment can be neither greater than, nor less than, half the cylinder $ABEF$.

In the argument of Archimedes, the logical validity of a conclusion

[149] In Proposition 10 of the work *On Spirals*. See *Works of Archimedes*, pp. 163–65.

based on the numerical concept of the limit of an infinite series was not admitted, but was replaced by one based on the rigorous geometrical method of exhaustion. It has been said that the differences in the methods of infinitesimals, of exhaustion, and of limits are felt to be more in the words than in the ideas;[150] but such an assertion may lead to serious misinterpretation. The methods are, of course, interrelated, and in consequence they lead to identical results; but the points of view are distinctly different, as we have seen. Although in a broad sense the procedures of Archimedes may be considered as "practically integrations,"[151] or even as representing in a general way "an integration process,"[152] it is indeed far from correct to speak of any one of them as a "veritable integration"[153] or as "the equivalent of genuine integration."[154] The definite integral requires for its correct formulation an appreciation of the notions of variability and functionality, the formation of the characteristic sum $\sum_{i=1}^{n} f(x_i) \triangle x_i$, and the application of the concept of the limit of the infinite sequence obtained from this sum by allowing n to increase indefinitely as $\triangle x_i$ becomes indefinitely small. These essential aspects of the integral are, of course, at no place indicated in the work of Archimedes, for they were extrinsic to the whole of Greek mathematical thought.

The series $h + 2h + 3h + \ldots$, employed by Archimedes in the proposition above, was to figure prominently in the gropings toward the calculus in the seventeenth century, so that it may be well to point out that this geometrical demonstration is in broad outline equivalent to performing the integration indicated by $\int_0^a x\,dx$. In determining that the area bounded by the polar axis and by one turn of the spiral $\rho = \dfrac{a\theta}{2\pi}$ is $\frac{1}{3}$ that of the circle of radius a,[155] Archimedes had occasion to make a similar calculation, equivalent to evaluating $\int_0^a x^2\,dx$. This was not done directly and arithmetically by determining

$$\lim_{n \to \infty} \left(\frac{h^2 + (2h)^2 + \ldots + (nh)^2}{n(nh)^2} \right)$$ to be $\frac{1}{3}$, as he might easily

[150] See Milhaud, *Nouvelles études*, p. 149.
[151] T. L. Heath, *History of Greek Mathematics*, II, 3.
[152] Karpinski, "Is There Progress in Mathematical Discovery?" p. 48.
[153] Zeuthen, *Geschichte der Mathematik im Altertum und Mittelalter*, p. 181.
[154] *The Works of Archimedes*, p. cliii; cf. also p. cxliii.
[155] *On Spirals*, Proposition 24, *The Works of Archimedes*, pp. 178–82.

have done had the Greeks not interdicted the infinite, but indirectly and geometrically by coupling the inequalities

$$\frac{h^2 + (2h)^2 + \ldots + (nh)^2}{n(nh)^2} > \tfrac{1}{3} \text{ and } \frac{h^2 + (2h)^2 + \ldots + [(n-1)h]^2}{n(nh)^2} < \tfrac{1}{3}$$

with the proof by the method of exhaustion, in a manner similar to that employed in the proposition on the paraboloidal segment. Archimedes may have known the equivalent result for the sum of cubes as well, and much later the Arabs extended his work to include also the fourth powers. In the seventeenth century at least a half dozen mathematicians —Cavalieri, Torricelli, Roberval, Fermat, Pascal, and Wallis—were to extend (all more or less independently) this work of Archimedes still further by the determination of $\int_0^a x^n dx$ for yet other values of n, and in this way to point immediately to the algorithm of the calculus. The methods which these men were to use were not, in general, the careful geometric procedures found in the propositions of Archimedes, but were to be based on ideas of indivisibles and of infinite series—notions which the decline of the idea of mathematical rigor during the intervening years was to make more acceptable to mathematicians, even though they were to be at that time no less impeachable than in Archimedes' day.

We have seen that Greek geometry was concerned largely with form rather than with variation, so that the function concept was not developed. That motion was nevertheless occasionally invoked in mathematics is indicated by Plato's suggestion that a line is generated by a moving point, as well as by the fact that certain special curves, described by a double motion, had been discussed even before Plato's time. The most famous of such curves was perhaps the quadratrix of Hippias, the Sophist. Archimedes may have been influenced by Hippias' idea when he defined his spiral as the locus of a point which moved with uniform radial velocity along a line, while the line in turn revolved uniformly about one of its end points which is kept fixed.[155]

In the further study of this curve Archimedes was led, in attempting to determine its tangent, to one of the few considerations corresponding to the differential calculus to be found in Greek geometry. In conformity with the static geometry of design, a tangent to a circle

[154] *The Works of Archimedes*, p. 165.

had been defined by Euclid[157] as a line touching the circle at only one point, and this definition was extended by Greek geometers to apply to other curves as well. There is also ascribed to the ancients[158] the definition, following the suggestion contained in a proposition of Euclid,[159] of a tangent as a line touching a curve and such that in the space between the straight line and the curve no other straight line can be interposed. These definitions were, of course, of restricted applicability and did not, in general, suggest a method of procedure for drawing tangents. Although Archimedes did not offer a more satisfactory definition, he appears, nevertheless, to have employed, for the determination of the tangent to his spiral, a method suggestive of a more general point of view. As in quadratures he had used considerations from the science of statics, so here in the problem of tangents Archimedes appears to have had recourse to a representation derived from kinematics. It seems likely, although Archimedes did not thus express the idea, that he found the tangent to the spiral $\rho = a\theta$ through a determination of the instantaneous direction of motion of the point P, by which it is traced.[160] This he probably did by applying to the motions by which the spiral may be generated the parallelogram of velocities, the principle of which had been perceived by the Peripatetics.[161] The motion of P may be regarded as compounded of two resultant motions: one with a radial velocity of constant magnitude V_r directed along the line OP (see fig. 5), and the other in a direction perpendicular to this and having a magnitude which is given by the variable product, V_a, of the distance OP and the uniform speed of rotation. Inasmuch as the distance OP and the speeds are given, the parallelogram (in this case a rectangle) of velocities can be constructed and the direction of the resultant velocity, and therefore also the tangent PT, determined.

The determination by Archimedes of the tangent to the spiral has been characterized as "a differentiation,"[162] or as "corresponding to

---

[157] Book III, Definition 2, in T. L. Heath ed., II, 1.

[158] See, for example, Comte, *The Philosophy of Mathematics*, pp. 108–10.

[159] Book III, Proposition 16, in T. L. Heath ed., II, 37.

[160] T. L. Heath, *History of Greek Mathematics*, II, 556–61.

[161] See *Mechanica* XXXIII. 854b–855a in *The Works of Aristotle*, Ross ed., Vol. VI, *Opuscula;* cf. also Duhem, *Les Origines de la statique*, II, 245; and Mach, *The Science of Mechanics*, p. 511.

[162] Simon, "Zur Geschichte und Philosophie der Differentialrechnung," p. 116.

our use of the differential calculus."[143] Such a designation, however, is hardly justified. Apparently, Archimedes made no effort to develop the idea here involved into a uniform method of attack upon the problem of tangents to other curves. When, in the seventeenth century, the method again appeared in the work of Torricelli, Roberval, Descartes, and Barrow, the scope of its applicability was somewhat extended; but only with the method of fluxions of Newton was there presented an algorithmic procedure for determining from the equation of any curve a pair of generating motions from which the tangent might be found. There was in Greek geometry no idea of a curve as

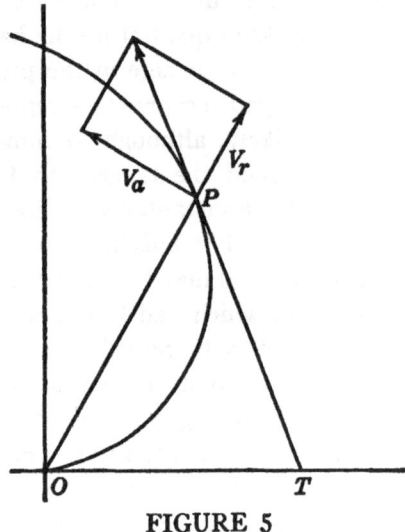

FIGURE 5

corresponding to a function, nor was there a satisfactory definition of a tangent in terms of the limit concept. There was therefore in the thought of Archimedes no anticipation of the realization that the geometrical notion of tangency is to be based upon the function concept and upon the numerical idea of a limit, i. e., upon the expression $\lim_{h \to 0} \left(\dfrac{f(x + h) - f(x)}{h} \right)$, which furnishes the basis for the differential calculus. There is in the whole of Greek mathematics no clear recognition of the need for the limit concept, both for the determina-

[143] T. L. Heath, *History of Greek Mathematics*, II, 557.

tion of curvilinear areas and of tangents to curves, and even for the very definition of these ideas which intuition vaguely suggests.

It is therefore incorrect to impute to Archimedes the ideas expressed in the integral and the derivative. These notions were not part of Greek geometry. Nevertheless, the problems and methods of Archimedes furnished probably the strongest incentive to the later development of such ideas, for they were leading in precisely this direction. The work of Archimedes so strongly suggests the newer methods of analysis that in the seventeenth century Torricelli and Wallis hazarded the opinion that the ancient Greek mathematicians had deliberately concealed under their synthetic demonstrations the analytic devices by which they had been led to their discoveries.[144] Through the discovery of the *Method* of Archimedes, the assumption of the existence of such methods has been proved correct, but the failure of ancient geometers further to elaborate these in their works is not to be regarded as indicating an intent to deceive. Although the notions of instantaneous velocity and of the infinitely small were accepted—all too uncritically—by Torricelli, Wallis, and their contemporaries, these ideas were not considered by Greek thinkers as admissible in mathematics. In the seventeenth century, however, the infinitesimal and kinematic methods of Archimedes were made the basis of the differential and the fluxionary forms of the calculus. Although Leibniz and Newton were thus to admit into the calculus the notions of the infinitely small and of instantaneous velocity, these were to remain open to criticism even after that—in fact, until, in the nineteenth century, the basic concept, that of the derivative, had been carefully defined.

After the time of Archimedes, the trend of Greek geometry was toward *applications*, rather than toward new theoretical developments.[165] No Greek mathematician approached nearer to the calculus than had Archimedes. His successors—Hipparchus, Heron, Ptolemy, and others—turned to various mathematical sciences, such as astronomy, mechanics, and optics. Nevertheless, the infinitesimal considerations of Archimedes were not forgotten by the Greek geometers of later times. Toward the end of the third century of our era, in a

[144] T. L. Heath, *History of Greek Mathematics*, II, 21, 557; Torricelli, *Opere*, I (Part 1), 140.
[165] T. L. Heath, *History of Greek Mathematics*, II, 198.

period which is regarded as one of decline in mathematics, the geometer Pappus not only displayed a familiarity with these methods, but was able also to add a new result—known as the Pappus theorem—to the work of Archimedes on centers of gravity.[166]

However, further significant advances in geometrical method were to be dependent upon certain broad changes in other branches of mathematics. One of these was the development of a more highly elaborated abstract symbolic algebra; another was the introduction into algebra and geometry of the notion of variation—of variables and functionality. The *Arithmetic* of Diophantus, which represents the highest development of Greek algebraical thought, was really theoretical logistic,[167] rather than generalized arithmetic or the study of certain functions of variables. In it the Peripatetic basis in logic is stronger than the Platonic ontological conception of mathematics. Only one unknown was introduced in this work, and the only solutions which were accepted as having meaning were those which, in accordance with Aristotelian tradition, could be expressed as the quotient of integers. Irrational and imaginary numbers were not recognized.[168]

The *Arithmetic* of Diophantus represents a "synocopated algebra," in that abbreviations for certain recurring quantities and operations are systematically introduced in it. However, in order that his work might be associated with geometrical results and later serve as the suitable basis for the calculus, it had to be made more completely symbolic, the concept of number had to be generalized, and the ideas of variable and function had to be introduced. During the Middle Ages the interest of the Hindus and the Arabs in algebraic development, and the attack by the Scholastic philosophers upon the problems offered by the continuum, were to supply in some measure the background required for these changes. When, therefore, in the sixteenth and seventeenth centuries, the classical work of Archimedes was developed into the methods constituting the calculus, the advance was made along lines suggested by the traditions built up during the medieval period.

[166] We cannot be sure, however, that the theorem was an original contribution on the part of Pappus or that he possessed a proof of it. See Pappus of Alexandria, *La Collection mathématique*, trans. by Ver Eecke; cf. also Weaver, "Pappus."

[167] Jakob Klein, "Die griechische Logistik und die Entstehung der Algebra"; see also Tannery, *Pour la science hellène*, p. 405.

[168] T. L. Heath, *History of Greek Mathematics*, I, 462.

III. Medieval Contributions

THE MOOT points as to the origin and antiquity of Hindu mathematics are not immediately pertinent in a *précis* of the development of the derivative and the integral; for these concepts depend on certain logical subtleties, the significance of which appears to have surpassed the appreciation, or at least to have escaped the interest, of the early Indian mathematicians.

The Hindus apparently were attracted by the arithmetical and computational aspects of mathematics,[1] rather than by the geometrical and rational features of the subject which had appealed so strongly to the Hellenic mind. Their name for mathematics, *gaṇita*, meaning literally the "science of calculation,"[2] well characterizes this preference. They delighted more in the tricks that could be played with numbers than in the thoughts the mind could produce, so that neither Euclidean geometry nor Aristotelian logic made a strong impression upon them. The Pythagorean problem of the incommensurable, which was of intense interest to Greek geometers, was of little import to Hindu mathematicians, who treated rational and irrational quantities, curvilinear and rectilinear magnitudes indiscriminately.[3] With respect to the development of algebra, this attitude occasioned perhaps an incidental advance, since by the Hindus the irrational roots of quadratics were no longer disregarded, as they had been by the Greeks, and since to the Hindus we owe also the immensely convenient concept of the absolute negative.[4] These generalizations of the number system and the consequent freedom of arithmetic from geometrical representation were to be essential in the development of the concepts of the calculus, but the Hindus could hardly have appreciated the theoretical significance of the change.

Similarly, another consequence of the lack of nice distinction in Hindu thought happens to have been responsible for a change which corresponds to the modern view. We have seen that the crisis of

[1] Cf. Karpinski, *The History of Arithmetic*, p. 46.
[2] Datta and Singh, *History of Hindu Mathematics*, Part I, p. 7.
[3] Lasswitz, *Geschichte der Atomistik*, I, 185.
[4] Fine, *The Number System of Algebra*, p. 105.

the incommensurable led to the abandonment by the Greeks of the attempt to associate numbers with all geometrical magnitudes. The Pythagorean problem of the application of areas was in general insoluble in terms of the conceptions of number and geometrical magnitude then current. The area of the circle could not literally be exhausted by applying rectilinear configurations to it, inasmuch as curvilinear magnitudes were fundamentally different. The area of a circle was not to be compared with that of a square, for area was not a numerical concept, and equality in general—as distinct from equivalence— meant congruence. The number π would have had no meaning in Greek mathematics. With the Hindus the view was different. They saw no essential unlikeness between rectilinear and curvilinear figures, for each could be measured in terms of numbers; arithmetic and mensuration, rather than geometry and considerations of congruence, were fundamental. The strong Greek distinction between the discreteness of number and the continuity of geometrical magnitude was not recognized, for it was superfluous to men who were not bothered by the paradoxes of Zeno or his dialectic. Questions concerning incommensurability, the infinitesimal, infinity, the process of exhaustion, and other inquiries leading toward the conceptions and methods of the calculus were neglected.

Operational difficulties were felt in dealing with the number zero— difficulties which led Brahmagupta to regard zero as an infinitesimal quantity which ultimately reduces to nought; and which caused Bhaskara to say that the product of a number and zero is zero, but that the number must be retained as a multiple of zero if any further operations impend.[5] Yet these difficulties do not appear to have been considered with the intention of resolving the logical questions, implicit in the use of indeterminate forms, which were later to puzzle the early users of the calculus. Questions of limits, although implied in their work, were not expressly stated.[6] The emphasis which Hindu mathematicians placed on the numerical aspect of the subject, together with the use of the Hindu numerals and of the principle of positional notation (the latter having been employed also by the Babylonians) did, of course, make more easily possible the develop_

[5] Datta and Singh, *op. cit.*, p. 242.
[6] Sengupta, "History of the Infinitesimal Calculus in Ancient and Mediaeval India," p. 224.

ment of algebra, and, subsequently, that of the *algorithmic procedures* of the calculus. However, the *logical concepts* of the derivative and the integral are just as easily defined in terms of Greek numeration as of our own, so that Hindu mathematics added no thought essential to the development of these ideas.

The Hindu numerals reached Europe through the medium of the Arabic civilization. This was preëminently eclectic, so that in Arabian mathematics we find both Greek and Hindu elements. The Arabic reckoning is based on that of Hindus, and Arabic trigonometry is Indian also in its use of the sine and the arithmetic form, rather than of the Hipparchan chord and geometric representation. Arabic geometry, however, shows the influence of Euclid and Archimedes, and Arabic algebra indicates a return to Greek geometric demonstration and the Diophantine avoidance of negative numbers.[7] The general character of Arabic algebra, however, is somewhat different from that of Diophantus' *Arithmetic*, for it is rhetorical rather than syncopated and deals mostly with problems of practical life, rather than with the abstract properties of numbers.[8] The whole trend of Arabic mathematics was, like that of the Hindus, directed away from the speculations on the incommensurable, continuity, the indivisible, and infinity—ideas which in Greek geometry were leading toward the calculus. Furthermore, additions to the classic Greek works—such as the treatise we have by Alhazen (or Ibn al-Haitham) on the measurement by infinitesimals of the paraboloid and on the summations of the cubes and the fourth powers of the positive integers[9]—were slight, for Arabic thought lacked the interest which was necessary to pursue further such fecund ideas. To the Arabs, however, we owe the preservation and transmission to Europe of much of the Greek work which would otherwise have been lost.

Christian Europe had, since the time of Pappus, added practically nothing to traditional mathematical theory, and was, in fact, largely unfamiliar with the ancient treatises until, in the twelfth century, Latin translations were made from Arabic, Hebrew, and Greek

[7] Fine, *Number System;* cf. also Paul Tannery, *Notions historiques*, p. 333, and Karpinski, *Robert of Chester's Latin Translation of the Algebra of Al-Khowarizmi*, p. 21.

[8] Gandz, "The Sources of al-Khowarizmi's Algebra," pp. 263–77.

[9] See Ibn al-Haitham (Suter), "Die Abhandlung über die Ausmessung des Paraboloids." Cf. also Wieleitner, "Das Fortleben der archimedischen Infinitesmalmethoden bis zum Beginn der 17. Jahr., insbesondere ueber Schwerpunkt bestimmungen."

manuscripts. Before this time the work of Euclid was known chiefly through the enunciation, largely without proofs, of selected propositions given in the early sixth century by Boethius, in his *Geometry*.[10] The work of Archimedes fared no better, for by the sixth century the only works of his generally known (and commented upon by Eutocius) were those on the sphere and cylinder, on the measurement of a circle, and on the equilibrium of planes (that is, on the law of the lever).[11]

When in the twelfth and thirteenth centuries the Greek works began to appear in Latin translations, they did not meet with an enthusiastic reception on the part of European scholars, interested as these men were in theology and metaphysics. The interest of Roger Bacon and the appearance of works such as those of Jordanus Nemorarius show that in the thirteenth century there was no apathetical lack of mathematical activity; but the knowledge displayed at this time indicates an inadequate familiarity with the classic works of Greek geometry[12]—a nescience which had led to the designation *fuga miserorum*[13]—for the fifth proposition of Euclid's *Elements*.[14]

The inadequate attention paid to Greek geometry during the later medieval period was paralleled by a similar lack of zeal for Greek and Arabic algebraic methods. The thirteenth century opened auspiciously with the appearance, in 1202, of the *Liber abaci* of Leonardo of Pisa. This, however, was not followed by a comparable work for almost 300 years—that is, until the appearance in 1494 of the *Summa de arithmetica* of Luca Pacioli. The dearth of advances in the mathematical tradition during this period has been made the basis for very severe strictures on the mathematical work done in this interval.[15] Such condemnation is justified by thinking in terms of contri-

---

[10] Ball, *A Short Account of the History of Mathematics*, p. 107.
[11] See *The Works of Archimedes*, p. xxxv.
[12] Ginsburg, "Duhem and Jordanus Nemorarius," p. 361.
[13] Later *pons asinorum*.
[14] Smith, "The Place of Roger Bacon in the History of Mathematics," pp. 162–67.
[15] Hankel (*Zur Geschichte der Mathematik im Alterthum und Mittelalter*, p. 349) says: "Mit Erstaunen nimmt man wahr, dass das Pfund, welches einst Leonardo der lateinischen Welt übergeben, in diesen drei Jahrhunderten durchaus keine Zinsen getragen hatte; wir finden, von Kleinigkeiten abgesehen, keinen Gedanken, keine Methode, welche nicht aus dem liber abaci oder der practica geometriae bereits wohl bekannt oder ohne Weiteres abzuleiten wäre." See also pp. 357–58 for a similar complaint. Cajori (*A History of Mathematics*, p. 125) says that in this period "the only noticeable advance is a simplification of numerical operations and a more extended application of them." Similarly, Archibald (*Outline of the History of Mathematics*, p. 27) characterizes the interval from

butions to Greek geometry[16] and to Arabic algebra (to which, however, Nicole Oresme did add in the fourteenth century the conception of a fractional power[17] which was later to add to the facility of application of the methods of the calculus). During the interval from 1202 to 1494 there may, indeed, have appeared no successor to either Archimedes or to Leonardo of Pisa.[18] If, on the other hand, we regard the broader aspects of mathematics—the speculations and investigations which lead up to the propositions which are in the end deductively demonstrated—it will appear that this so-called barren period furnished points of view of significance in the development of the calculus. In this respect, as in others, there was perhaps as much originality in medieval times as there is now.[19]

Throughout the early period of the Middle Ages, Aristotle had been known in Europe largely through his logical works. During the thirteenth century, however, his scientific treatises circulated freely and, although these were condemned at Paris in 1210,[20] their study was again established in the university by 1255, at which time nearly all of Aristotle was prescribed for candidates for the master's degree.[21] In the *Physica*, Aristotle had considered in some detail the infinite, the infinitesimal, continuity, and other topics related to mathematical analysis. These became, particularly in the next century, the center of a lively discussion on the part of scholastic philosophers. They were studied in the light of Peripatetic philosophy, rather than in terms of mathematical postulational thought, but the resulting speculations were of service in sustaining an interest in such conceptions until, at a later date, they became a part of mathematics.

Leonardo of Pisa to Regiomontanus as "a period of about 250 barren years." Björnbo ("Über ein bibliographisches Repertorium der handschriftlichen mathematischen Literatur des Mittelalters," p. 326) says "Von einer Entwickelung in dieser Epoche ist kaum zu reden; bedeutende mathematische Fortschritte wird man hier vergebens suchen."

[16] It has been well said by Sarton (*Introduction to the History of Science*, I, 19–20) that we do not do justice to the medievals because we judge their first steps by the Greek last steps.

[17] Cf. *Algorismus proportionum*, pp. 9–10. Fine (*Number System*, p. 113) goes so far as to say that this is the only contribution of the period to algebra.

[18] Eneström, "Zwei mathematische Schulen im christlichen Mittelalter."

[19] See Sarton, *op. cit.*, I, 16.

[20] Denifle and Chatelain, *Chartularium Universitatis Parisiensis*, Vol. I, p. 70, No. 11. Cf. also Vol. I, pp. 78–79, No. 20.

[21] *Ibid.*, Vol. I, pp. 277–79, No. 246. Cf. also Rashdall, *The Universities of Europe in the Middle Ages*, I, 357–58.

One of the best examples of Scholastic thought on these topics is that found in the work of a man who exerted a great influence upon medieval thought,[22] Thomas Bradwardine, "doctor profundus" and Archbishop of Canterbury, and perhaps the greatest English mathematician of the fourteenth century. In his *Geometria speculativa*[23] and in the *Tractatus de continuo*[24] Bradwardine discussed, among other things, the nature of continuous magnitude, his view being dominated by the Peripatetic opposition to any atomism. The doctrine of Leucippus and Democritus, which had denied divisibility to infinity, has at all times had partisans and adversaries, and the Scholastic period was far from exceptional in this respect. The idea of the indivisible, during the earlier Middle Ages, was often more elementary than that held by Democritus long before. It seems to have been believed by Capella, Isidore of Seville, Bede, and others, that time is composed of indivisibles, an hour being made up of 22,560 such instants.[25] It seems probable, or at least possible, that these instants were regarded as atoms of time. During the later medieval period the idea of indivisibles under various forms and modifications was upheld by Robert Grosseteste, Walter Burley, and Henry Goethals, among others.[26] On the other hand, Roger Bacon protested, in his *Opus majus*, that the doctrine of indivisibles was inconsistent with that of incommensurability,[27] an argument developed further by Duns Scotus, William of Occam, Albert of Saxony, Gregory of Rimini, and others.[28] Bradwardine considered, in the light of the problem of the continuum, the divers points of view represented by proponents of the doctrine of indivisibles. Some interpreted the question in terms of physical atomism, others of mathematical points; some assumed a finite, others an infinite, number of points; some postulated immediate contiguity, others a discrete set of indivisibles.[29] Bradwardine himself maintained

[22] Duhem, *Les Origines de la statique*, II, 323.

[23] For brief indications of the contents of this work, see Moritz Cantor, *Vorlesungen*, II, 103–6; and Hoppe, "Zur Geschichte der Infinitesimalrechnung," pp. 158–60.

[24] For an analysis of this see Stamm, "Tractatus de continuo von Thomas Bradwardina"; see also Cantor, *Vorlesungen*, II, 107–9.

[25] Paul Tannery, "Sur la division du temps en instants au moyen âge," p. 111.

[26] Stamm, "Tractatus de continuo," pp. 16–17; cf. also Duhem, *Études sur Léonard de Vinci*, II, 10–18.

[27] Smith, *The Place of Roger Bacon in the History of Mathematics*, p. 180, n.

[28] Duhem, *Études sur Léonard de Vinci*, II, 8.

[29] Stamm, "Tractatus de continuo," p. 16.

that continuous magnitudes, although including an infinite number of indivisibles, are not made up of such atoms.[30] "Nullum continuum ex indivisibilibus infinitis integrari vel componi," said Bradwardine,[31] using perhaps for the first time in this connection the word which Leibniz was to adopt (upon the suggestion of the Bernoulli brothers[32]) to designate in his calculus the sum of an infinite number of infinitesimals—the integral. Bradwardine asserted, on the contrary, that a continuous magnitude is composed of an infinite number of continua of the same kind. The infinitesimal, therefore, evidently possessed for him, as for Aristotle, only potential existence.[33]

William of Occam seems to have occupied a position intermediate between that of Bradwardine and that held by the supporters of indivisible lines. While admitting that no part of any continuum is indivisible, he maintained that, contrary to the teachings of Aristotle, the straight line does actually (not only potentially) consist of points.[34] In another connection, however, Occam said that points, lines, and surfaces are pure negations, having no reality in the sense that a solid is real.[35] One must not, moreover, read too much into the Scholastic views on the nature of the continuum. The opinion of Bradwardine has been rather freely identified with that of Brouwer and the modern intuitionists, who conceive of the continuum as made up of an infinite number of infinitely divided continua;[36] the view of Occam has been said to correspond to that held by Russell and the formalists, who regard the continuum as a perfect set of points everywhere dense.[37] Such comparisons are justifiable only in a very general sense, for the Scholastic speculations invariably centered upon the metaphysical question of the reality of indivisibles, rather than upon the search for a representation which should be consistent with the premises of mathematics. There was in the medieval views no conception of the rigorous axiomatic foundation of arithmetic which has

[30] Moritz Cantor, *Vorlesungen*, II, 108.

[31] Moritz Cantor, *Vorlesungen*, II, 109, n.; Hoppe, "Zur Geschichte der Infinitesimalrechnung," p. 159; cf. also Stamm, *op. cit.*, p. 17.

[32] James Bernoulli, "Analysis problematis antehac propositi," p. 218.

[33] Lasswitz, *Geschichte der Atomistik*, II, 201; Stamm, "Tractatus de continuo," p. 17.

[34] Burns, "William of Ockham on Continuity."

[35] Duhem, *Études sur Léonard de Vinci*, II, 16–17; III, 26.

[36] Stamm, "Tractatus de continuo," p. 20.

[37] See Birch, "The Theory of Continuity of William of Ockham," p. 496.

characterized modern thought upon this subject. Nevertheless, the fourteenth century disputations on the indivisible represent a keen appreciation of the difficulties involved and a clarity of thought which was, several centuries later, to lend an air of respectability to the infinitesimal methods leading to the calculus.

In discussions of indivisibles, the question of infinite division and the nature of the infinite arose, of necessity. In fact, the medieval philosophers discussed the question more from the point of view of infinite divisibility and infinite aggregates than from that of infinitely great magnitudes. Aristotle, it will be recalled, had distinguished two kinds of infinity—the potential and the actual. The existence of the latter he had categorically denied, and the former he had admitted as realized only in cases of infinitely small continuous magnitudes and of infinitely large numbers.[38] The Roman poet Lucretius had, with keen imagination, upheld the notion of the infinite as indicating more than the potentiality of indefinite increase. In focusing attention upon infinite multitudes rather than magnitude, he adumbrated a number of properties of infinite aggregates, such as that a part may in this case be equal to the whole.[39] The work of Lucretius was not, however, familiar to the European scholars of the Middle Ages. The Aristotelian distinction, on the other hand, was continued by the Scholastic philosophers, although with modifications resulting, perhaps, from the fact that Christianity recognized an infinite God. In the thirteenth century Petrus Hispanus, who became Pope John XXI, recognized in his *Summulae logicales*[40] two kinds of infinity: a categorematic infinity, in which all terms are actually realized, and a syncategorematic infinity, which is bound up always with potentiality.[41] This distinction is not greatly different from that which has been suggested recently by a mathematician and scientist who would discriminate between saying that an infinite aggregate is conceivable and saying it is actually conceived.[42]

The discussion of the two infinities, which was begun in the thirteenth century, was continued throughout the fourteenth also. Albert

[38] *Physica*, Book III.
[39] See Keyser, "The Rôle of the Concept of Infinity in the Work of Lucretius."
[40] Duhem, *Études sur Léonard de Vinci*, II, 22.
[41] Duhem, *op. cit.*, Vol. II, *Léonard de Vinci et les deux infinis*, pp. 1–53, gives an extensive account of this question.
[42] Enriques, *Problems of Science*, pp. 127–28.

of Saxony, for example, brought out the distinction nicely by a mere transposition of words, saying that the two views were illustrated respectively by the sentences: "In infinitum continuum est divisible," and "Continuum est divisible in infinitum."[43] Bradwardine brought out the difference, perhaps with less subtlety but certainly more clearly, in saying that the categorematic infinity is a quantity without end, whereas the syncategorematic infinity is a quantity which is not so great but that it can be made greater.[44]

Although the distinction between the two infinities was generally recognized by the philosophers of the Scholastic period, there were significant differences, then as now, on the question of their existence. William of Occam, in conformity with his nominalistic attitude and with the principle of economy enunciated in his well-known "razor," agreed with Aristotle in denying that the categorematic infinity is ever realized. Gregory of Rimini, on the other hand, maintained[45] what mathematics in the nineteenth century was to demonstrate: that there is in thought no self-contradiction involved in the idea of an actual infinity—the so-called completed infinite.

More interesting from the mathematical point of view than these philosophical and discursive discussions are the remarks made on the subject of the infinite by Richard Suiseth, popularly known as the Calculator, in his *Liber calculationum*. This was composed later than 1328, inasmuch as it refers to Bradwardine's[46] *Liber de proportionibus* of that year,[47] and probably dates from the second quarter of the fourteenth century.[48] Although more interested in dialectical argu-

[43] Duhem, *Études sur Léonard de Vinci*, II, 23.
[44] Stamm, "Tractatus de continuo," pp. 19–20.
[45] See Duhem, *Études sur Léonard de Vinci*, II, 399–401.
[46] Duhem (*Études sur Léonard de Vinci*, III, 429) has incorrectly asserted that Calculator names in his work only Bradwardine, Aristotle, and Averroes, but there are in the *Liber calculationum* references also to ancient mathematicians, for example, Euclid (fol. 29ᵛ, cols. 1–2) and Boethius (fol. 43ᵛ, col. 1).
[47] "Ut venerabilis magister Thomas de berduerdino in suo libro de proportionibus liquide declarat." *Liber calculationum*, fol. 3ᵛ, col. 1. Professor Lynn Thorndike has kindly allowed me to make use of his rotograph of a copy in the British Museum, I B. 29, 968, fol. 1ʳ—83.ᵛ This is a copy of the undated *editio princeps* at Padua, placed by Thorndike (*A History of Magic and Experimental Science*, III, 372) in the year 1477. Inasmuch as this work is neither well known nor readily available, passages from it will be cited at some length. The printed text contains so many abbreviations that it is difficult to read, but transcriptions from it will be given in full.
[48] See Thorndike, *History of Magic*, III, 375. Stamm ("Tractatus de continuo," p. 24) would place its appearance after 1350.

ments and subtle sophisms concerning infinity than in its adequate definition, Suiseth made several comments on the subject which are of particular mathematical significance. In the second chapter he said that all sophisms regarding the infinite could be easily resolved by recognizing that a finite part can have no ratio to an infinite whole.[49] This conclusion, he said, would be conceded by the imagination, for the contrary would imply that any part, when added to the whole, would not change it in magnitude. Arguments with respect to the infinite do not proceed, therefore, as do those concerning finite quantities.[50]

About two hundred years later Galileo remarked still more clearly the essential difference between the rules for the finite and those for the infinite, but he focused attention upon the *correspondence* between infinite *aggregates*, rather than upon the *ratio* of finite to infinite *magnitudes*—a change of view which led to the final formulation of the calculus in the nineteenth century. There is in the statements of Calculator a cogency and a warning which might well have been observed in later centuries, but they display the Peripatetic propensity to regard the infinite as a magnitude, rather than as an aggregation of terms. The fruitfulness of the infinite in the work of Archimedes arose out of the infinite collections of lines or of terms in a series. Reference will be made below, to be sure, to the study by Calculator of an infinite series, but in this connection it will be seen that it was not the endlessness of the sequence of terms which most interested him, but a certain infinite magnitude. A consideration of this series will require, however, the study of a larger problem with which it was associated, and to which we shall now turn.

The blending of theological, philosophical, mathematical, and scientific considerations which has so far been evident in Scholastic thought is seen to even better advantage in a study of what was per-

<hr/>

[49] "Infinita quasi sophismata possunt fieri de infinito que omnia si diligenter inspexeris quod nullius partis ad totum infinitum est aliqua proportio faciliter dissoluere poteris per predicta." *Liber calculationum*, fol. 8ᵛ, col. 2.

[50] "Que potest concedi de inmaginatione et causa est quia nulla pars finita finite intensa respectu tocius infiniti aliquid confert quia nullam habet proportionem ad illud infinitum si tamen subiectum esset finitum conclusio non foret inmaginabilis quia tunc conclusio inmediate repugnaret illi positioni quia tunc quelibet pars in comparatione ad totum conferet aliquantum et sic non est nunc ideo nullum argumentum proceditur de infinito sicut faceret de finito." *Ibid.*, fol. 8ʳ, col. 2, fol. 8ᵛ, col. 1.

haps the most significant contribution of the fourteenth century to the development of mathematical physics. It has commonly been protested that additions, if any, which were made to scientific knowledge in the medieval period lay solely in the field of practical discoveries and applications; and that the only mathematical achievement during this time was the simplification of the rules of operation for the Hindu-Arabic numerals, the latter having been made known in Europe by Leonardo of Pisa and other men of the thirteenth century. There is at least one exception to such an assertion, for it was precisely during this interval, and particularly in the fourteenth century, that a theoretical advance was made which was destined to be remarkably fruitful in both science and mathematics, and to lead in the end to the concept of the derivative. This consisted in the idea— often expressed, to be sure, in terms of dialectical rather than mathematical method—of studying change quantitatively, and thus admitting into mathematics the concept of variation.[51]

Heraclitus, Democritus, and Aristotle had made some qualitative metaphysical speculations on the subject of motion, and occasionally Greek geometers (Hippias, Archimedes, Nicomedes, Diocles) had allowed this notion to enter their thoughts (though not their proofs); but the idea of representing continuous variation by means of geometrical magnitude or of studying it in terms of the discreteness of number does not seem to have arisen with them. The Greek sciences of astronomy, optics, and statics had all been elaborated geometrically, but there was no such representation of the phenomena of change. Archimedes' famous work in statics was not paralleled by any equivalent kinematic system which admitted of representation in the form of mathematical propositions.

Perhaps the demand for rigor in Greek thought, which made congruence fundamental in geometry and which allowed no confusion of

[51] It is preposterous to assert—as does Tobias Dantzig (*Aspects of Science*, p. 45)— that, with reference to Aristotle and the Schoolmen, "his casuistry, his predilection for the static, his aversion for everything that moved, changed, flowed, or evolved admirably suited their purposes." A brief examination of the works of Aristotle on physical science and a glance at *A Catalogue of Incipits of Mediaeval Scientific Writings in Latin* by Thorndike and Kibre will show how untrue is such a statement. Aristotle made extensive investigations, from the point of view of physics, into the phenomena of motion. Scholastic philosophers not only continued his work, but also added the quantitative form of expression which was so successfully developed later in the seventeenth century.

the discrete with the continuous, could not be met by any attempt to establish a science of dynamics. Moreover, Greek astronomy lacked the concept of acceleration; the motions involved were all uniform (and hence eternal) and circular, and could, in this case, be represented by the geometry of the circle. No such uniformity was apparent for local motion—that is, for terrestrial changes of position. Motion was, it appeared, a quality rather than a quantity; and there was among the ancients no systematic quantitative study of such qualities.[52] Aristotle had spoken of mathematics as concerned with "things which do not involve motion," and had held that mathematics studies objects *qua* continuous, physics *qua* moving, and philosophy *qua* being.[53] In general Greek mathematics was the study of form, rather than variability. The quantities entering into Diophantine algebraic equations are constants, rather than variables, and this is true also of Hindu and Arabic algebra. In the Scholastic period, however, there arose a problem which was ultimately to change this view. Aristotle had considered motion a quality that does not increase and decrease through the joining together of parts, as does a quantity,[54] and this idea dominated most thought until toward the end of the thirteenth century,[55] at which time a reaction against Peripateticism arose, which was to lead, at Paris, to new views on the subject of motion.[56]

Bacon, in the second half of the thirteenth century, still followed the Aristotelian discussion of motion,[57] but a new approach to the problem was evidenced early in the fourteenth century by the introduction of the idea of impetus—the notion that a body, once set in motion, will continue to move because of an internal tendency which it then possesses, rather than, as Peripatetic doctrine had taught, because of the application of some external force, such as that of the air, which continues to impel it. This doctrine, which has been ascribed to Jean Buridan,[58] was particularly significant as an adumbration of the famous work in dynamics of Galileo almost three centuries later. At

[52] Brunschvicg, *Les Étapes de la philosophie mathématique*, p. 97.

[53] *Physica* II. 198a; *Metaphysica* 1061.

[54] Duhem, *Études sur Léonard de Vinci*, III, 314–16; cf. also Wieleitner, "Ueber den Funktionsbegriff und die graphische Darstellung bei Oresme," pp. 196–97.

[55] Wieleitner, "Ueber den Funktionsbegriff," p. 197.

[56] Duhem, *Études sur Léonard de Vinci*, II, pp. iii–iv.

[57] Thomson, "An Unnoticed Treatise of Roger Bacon on Time and Motion."

[58] See Duhem, *Études sur Léonard de Vinci, passim.*

the time of its inception it served also to make more acceptable the intuitive notion of instantaneous velocity, an idea excluded by Aristotle from his science, but implied by the quantitative study of variation of the fourteenth century. At that stage no precise definition of instantaneous rate of change could, of course, be given—nor was one given by Galileo—but there appeared at the time a large number of works, more philosophical than mathematical, all based on the intuition of this concept which everyone thinks he possesses. These tractates were devoted to a discussion of the latitude of forms, that is, of the variability of qualities.

There seems to be no scientific term which correctly expresses the equivalent of the word *form* as here used. It refers in general to any quality which admits of variation and which involves the intuitive idea of intensity—that is, to such notions as velocity, acceleration, density. These concepts are now expressed quantitatively in terms of limits of ratios—that is, simply as numbers—so that no need is now felt for a word to express the medieval idea of a form. In general, the latitude of a form was the degree to which the latter possessed a certain quality, and the discussion centered about the *intensio* and the *remissio* of the form, or the alterations by which this quality is acquired or lost. Aristotle had distinguished between uniform and non-uniform velocity, but the critical dialectical treatment of the Scholastics went much further. In the first place, the time rates of change which they considered were not necessarily those of distance, but included many others as well, such as those of intensity of illumination, of thermal content, of density. Secondly—and more significantly—they distinguished not only between *latitudo uniformis* and *latitudo difformis* (that is, between uniform and nonuniform rates of change), but proceeded further to classify the latter as either *latitudo uniformiter difformis* or *latitudo difformiter difformis* (that is, according as the instantaneous rate of change of the rate of change was uniform or not); and the last-mentioned were sometimes in turn further divided into either *latitudo uniformiter difformiter difformis* or *latitudo difformiter difformiter difformis*.

These attempts to introduce order, by means of verbal arguments, into the disconcerting problem of variability were destined to be replaced centuries later by equivalent statements, expressed in mathematics and science by the remarkably concise terminology and symbolism of algebra and the differential calculus. At the time at which they were made, however, they represented the first serious and careful effort to make quantitative the idea of variability.

The origin of the question of the latitude of forms is shrouded in doubt. Duns Scotus appears to have been among the first to consider the increase and the decrease (*intensio* and *remissio*) of forms,[59] although the loose idea of latitude of forms apparently goes back to some time before this, inasmuch as Henry Goethals used the word *latitudo* in this connection.[60] In the early part of the fourteenth century there appeared works on variability and the latitude of forms by James of Forli, Walter Burley, and Albert of Saxony.[61] Similar ideas appeared also in 1328, at Oxford, in the treatise on proportions of Bradwardine.[62] This work is devoted more particularly to mechanics than to arithmetic, but the archbishop did not make a special investigation into the theory of the latitude of forms.[63]

"In the fourteenth century the study of the mathematical sciences flourished greatly in Oxford,"[64] and it was here that appeared not only the work of Bradwardine, but also the "leading model"[65] of such treatises on the latitude of forms—the *Liber calculationum* of Suiseth to which reference has already been made.[66]

That the doctrine of the latitude of forms was well known by the middle of the fourteenth century is indicated by the fact that Calculator begins *in medias res* by a general consideration, in the first chapter, of the intension and remission of forms and the question as to whether a *latitudo difformis* corresponds to its maximum or minimum *gradus* or intensity. The sense in which a form could correspond to any *gradus* is not clear, but Calculator appears here to be striving

[59] Von Prantl, *Geschichte der Logik im Abendlande*, III, 222–23. Cf. Stamm, "Tractatus de continuo," p. 23.

[60] Duhem, *Études sur Léonard de Vinci*, III, 314–42.

[61] Wieleitner, "Der 'Tractatus de latitudinibus formarum des Oresme,'" pp. 123–26.

[62] Duhem, *Études sur Léonard de Vinci*, III, 290–301.

[63] Stamm, "Tractatus de continuo," p. 24.

[64] Gunther, *Early Science in Oxford*, II, 10.

[65] Thorndike, *History of Magic*, III, 370.

[66] For a general description of the contents of this work, see Thorndike, *History of Magic*, Vol. III, Chap. XXIII; and Duhem, *Études sur Léonard de Vinci*, III, 477–81. No analysis of the *Liber calculationum* from the mathematical point of view appears to be available.

toward the idea of average intensity, an idea which could not be made precise without the use of the concepts of the calculus. In Chapter II, however, he arrived, in this connection, at a result which was to be of particular significance in the later development of science and mathematics. Here, in considering the intension in difform things, he reached the conclusion, in connection with problems on thermal content, that the average intensity of a form whose rate of change over an interval is constant, or of a form which is such that it is uniform throughout each half of the interval, is the mean of its first and last intensities.[67] The rigorous proof of this requires the use of the limit concept, but Calculator had resort to dialectical reasoning, based on physical experience of rate of change. He argued, in connection with such a form, that if the greater intensity is allowed to decrease uniformly to the mean while the lesser is increased at the same rate to this mean, then the whole is neither increased nor decreased.[68] The argument is pursued at further length here, and the author adds, as well, numerical illustrations such as the following: If the intensity increases uniformly from four to eight, or if for the first half of the time it is four, and for the last half it is eight, then the effect is that which would result from a uniform intensity of six, operating throughout the whole time.[69]

The methods presented here are amplified and applied, in other chapters of the *Liber calculationum*, to questions dealing with density, velocity, and the intensity of illumination. In these Suiseth again made use of the law that if the rate of change of one of these is constant, or if the form is such that in each of the two halves of the

[67] "Primo arguitur latitudinem caliditatis uniformiter difformem seu etiam difformem cuius utraque medietas est uniformis, et calidum uniformiter difforme suo gradui medio correspondere."¹ *Liber calculationum*, fol. 4ᵛ, col. 2.

[68] "Capiatur talis caliditas seu tale calidum et remittatur una medietas ad medium et intendatur alia ad medium equeuelociter: et sequitur totum non intendi nec remitti; eo quod totam latitudinem acquiret. Secundum unam medietatem seu partem sicut deperdet secundum aliam partem equalem et in fine erit uniforme sub tali gradu medio. Igitur nunc correspondet tali gradui." *Ibid.*

[69] "Sit enim tale uniformiter difforme seu difforme cuius utraque medietas est uniformis una ut. VIII. et alia ut .IIII. gratia argumenti. Tunc prima qualitas ut .VIII. extenditur per medietatem totius per predicta. Ergo solum denominat totum ut quatuor per idem prima qualitas ut quatuor per aliam medietatem extensa solum facit ut duo ad totius denominatorem. Igitur ille due qualitates totum precise denominabunt ut .VI. qui est gradus medius inter illas medietates. Sequitur igitur positio sic in speciali." *Ibid.*, fol. 5ᵛ, col. 2.

interval the rate of change is zero, then the average intensity is the mean of the first and last values. As was the case in the study of variation of the intensity of heat, no definition is given of the terms employed, inasmuch as this would presuppose an appreciation of the limit concept. The lack of such precise definitions led Calculator into difficulties involving the infinite. He considered, for example, a rarity of degree zero as a density of infinite degree, and conversely,[70] and consequently became involved unnecessarily often in the paradoxes of the infinite.

The tendency displayed by Calculator to consider the infinite from the point of view of intensity or magnitude, rather than aggregation, is brought out again in the discussion of an example of nonuniform variation, which might have had a significant influence upon the development of the calculus, had Suiseth's purpose been less that of introducing mathematics into dialectical discussions of change and more that of bringing the problem of variation into the realm of mathematics. Calculator, in the second book of his *Liber calculationum*, had occasion to consider the following problem: if throughout half of a given time interval a variation continues at a certain intensity, throughout the next quarter of the interval at double this intensity, throughout the following eighth at triple this, and so ad infinitum; then the average intensity for the whole interval will be the intensity of the variation during the second subinterval (or double the initial intensity).[71]

This is equivalent to a summation of the infinite series $\frac{1}{2} + \frac{2}{4} + \frac{3}{8} + \frac{4}{16} + \ldots + \frac{n}{2^n} + \ldots = 2$. It will be recalled that Archimedes

---

[70] "Ergo .a. est infinite densum et per consequens non est rarum. Ex istis sequitur ista conclusio quod aliquid est rarum quod non est rarum quia .a. est rarum quia est uniformiter difforme rarum et non est rarum quia est infinite densum. Pro istis negatur utraque conclusio et tunc ad casum positum quod .a. sit unum uniformiter difformiter rarum terminatur ad non gradum raritatis negatur casus nam ex quo raritas se habet privative sequitur ut argutum est quod ab omni gradu raritatis usque ad non gradum raritatis est latitudo infinita quia non gradus raritatis est infinitus gradus densitatis sed impossibile est quod aliquid sit uniformiter difforme aliquale mediante latitudine infinita. Ideo casus est impossibilis sicut est impossibile quod aliquid sit uniformiter difforme remissum ad non gradum remissionis terminatum." *Ibid.*, fol. 19ᵛ, cols. 1–2.

[71] "Contra quam positionem et eius fundamentum arguitur sic quia sequitur quod si prima pars proportionalis alicuius esset aliqualiter intensa, et secunda in duplo intensior, et tertia in triplo intensior, et sic in infinitum totum esset equale intensum precise sicut est secunda pars proportionalis quod tamen non uidetur visum." *Ibid.*, fol. 5 , col. 2.

had made use of certain simple series in connection with his geometry, but that the characteristic Greek interdiction of the infinite had led him to consider the sum to n terms only. It was apparent to him that the series $1 + \frac{1}{4} + \frac{1}{16} + \frac{1}{64} + \ldots + \frac{1}{4^{n-1}}$ approached $\frac{4}{3}$ in such a way that the difference could be made, by taking a sufficiently large number of terms, less than any specified quantity. He did not, however, go so far as to define $\frac{4}{3}$ as the "sum" of the infinite series, for this would have exposed his thought to the paradoxes of Zeno, unless he had invoked the precisely formulated concept of a limit as given in the nineteenth century. The Scholastic discussions of the fourteenth century, on the other hand, referred frequently to the infinite, both as actuality and as potentiality, with the result that Suiseth, with perfect confidence, invoked an infinite subdivision of the time interval to obtain the equivalent of an infinite series. He did not resolve the aporias of Zeno, to show in what sense an infinite series may be said to have a sum—a problem which future mathematicians were to consider at length. Calculator, instead, was more particularly interested in infinite magnitudes than in infinite series. Not only is the time interval in his problem infinitely divided, but the intensity itself becomes infinite. Now how can a quantity, whose rate of change becomes infinite, have a finite average rate of change? Suiseth admitted that this paradoxical result was in need of demonstration and so furnished at great length the equivalent of a proof of the convergence of the infinite series. This he did as follows.

Consider two uniform and equal rates of change, a and b, operating throughout a given time interval, which has been subdivided in the ratios $\frac{1}{2}$, $\frac{1}{4}$, $\frac{1}{8}$, ... Now let the rate of change b be doubled throughout the interval; but in the case of a, let it be doubled in the second subinterval; tripled in the third; and so to infinity, as given in the problem above. Now the increase in a in the second subinterval, if continued constantly throughout this and all following subintervals, would result in an increase in the effect equal to that brought about by the change in b during the first half of the time. The tripling of a in the third subinterval, if continued constantly throughout this and the ensuing subintervals, would in turn result in a further increase in the effect of a equal to that brought about by the change in b in the second subinterval, and so to infinity. Hence the increase resulting from the doubling, tripling, and so forth of a is equal to that caused by the doubling of b; i. e., the average rate of change in the problem considered above is the rate of change during the second subinterval, which was to be proved.[72]

The tediously verbose proof given by Calculator is, of course, based entirely on arguments appealing to our intuition of uniform rate of change. Because Suiseth gave no unambiguously clear definitions of velocity, density, intensity of illumination, and other terms used freely in his dissertation, his work not infrequently presents—as had also, to a certain extent, the *Physica* of Aristotle[73]—the appearance of an effort to propound sophistical questions on the subject of change, rather than that of a serious effort to establish a scientific basis for the study of the phenomena of motion and variability.[74]

[72] "Nam apparet quod illa qualitas est infinita ergo si sit sine contrario infinite denominabit suum subiectum. Et quod conclusio sequatur arguitur sic: sint .a. .b. duo uniforma eodem gradu, et dividatur .b. in partes proportionales et est illa hora ita quod partes maiores terminentur seu incipiant ab hoc instanti, et ponatur quod in prima parte proportionali illius hore intendatur prima pars .b. ad duplum, et in secunda parte proportionali intendatur secunda pars proportionalis illius ad duplum, et sic in infinitum, ita quod in fine erit .b. uniforme sub gradu duplo ad gradum nunc habitum. Et ponatur quod .a. in prima parte proportionali illius hore intendatur totum residuum .a. prima parte proportionali .a. acquirendo totam latitudinem sicut tunc acquireret prima pars proportionalis .b. et in secunda parte proportionali euisdem hore intendatur totum residuum .a. a prima parte proportionali et secunda illius .a. acquirendo tantam latitudinem sicut tunc acquiret pars proportionalis secunda .b. et in tertia parte proportionali indendatur residuum a prima parte proportionali et secunda et tertia acquirendo tantam latitudinem sicut tunc acquiret tertia pars proportionalis .b. et sic in infinitum scilicet quod quandocumque aliqua pars proportionalis .b. intendetur pro tunc intendatur .a. secundum partes proportionales subsequentes partem correspondentem in .a. acquirendo tantam latitudinem sicut acquiret pars prima in .b. et sint .a.b. consimilia quantitatiue continue quo posito sequitur quod .a. et .b. continue equeuelociter intendentur quia .a. continue per partes proportionales similiter intendetur sicut .b. quia residuum a prima parte proportionali .a. est equale prime parti proportionali eidem. Cum igitur .b. in prima parte proportionali illius hore continue intendetur per primam partem proportionalem et .a. per totum residuum a prima sua parte proportionali patet quod .a. in prima parte proportionali equeuelociter intendetur cum .b. et sic de omni alia parte eo quod quandocumque .b. indendetur per aliquam partem proportionalem .a. intendetur per totum interceptum inter partes correspondentes sibi et extremum ubi partes terminantur scilicet minores. Cum ergo quelibet pars proportionalis cuiuslibet continui sit equalis toti intercepto inter eandem et extremum ubi partes minores terminantur. Igitur patet quod .a. continue equeuelociter intendetur cum .b. et nunc est eque intensum cum .b. ut ponitur in casu. Ergo in fine .a. erit .a. eque intensum cum .b. et .a. tunc est tale cuius prima pars proportionalis erit aliqualiter intensa, et secunda pars proportionalis in duplo intensior, et tertia in triplo intensior et sic in infinitum. Et .b. erit uniforme gradu sub quo erit secunda pars proportionalis .a. ergo sequitur conclusio." *Ibid.*, fol. 5ᵛ, col. 2—fol. 6ᵛ, col. 1.

[73] Cf. Mach, *The Science of Mechanics*, p. 511.

[74] "Multa alia possent fieri sophismata per rarefactionem subiecti, et per *fluxum* qualitatis et alterationis qualitatis si subiectum debet intendi et remitti per huiusmodi rare-

Suiseth's unfortunate adoption of the Peripatetic attitude that such qualities as dryness, coldness, and rarity are the opposites of moistness, warmth, density—rather than simply degrees of the latter[7a]—complicated his consideration of these through the unnecessarily frequent introduction of the paradoxes of the infinite. Nevertheless, to Calculator we owe perhaps the first serious effort to make quantitatively understandable these concepts of mathematical physics. His bold study of the change of such quantities anticipated not only the scientific elaboration of these, but also adumbrated the introduction into mathematics of the notions of variable quantity and derivative. In fact, the very words *fluxus* and *fluens*, which Calculator used in this connection,[76] were to be employed by Newton some three hundred years later, when in his calculus he spoke of such a variable mathematical quantity as a fluent and called its rate of change a fluxion. Newton apparently felt as little need as Suiseth for a definition of this notion of fluxion, and was satisfied to make a tacit appeal to our intuitions of motion. Our definitions of uniform and nonuniform rate of change, are, as Suiseth anticipated, numerically expressed; but their rigorous definition could be given only after the development, to which Newton contributed, of the limit concept. This latter arose out of the notions of the calculus, which, in their turn, had evolved from the intuitions of geometry. The prolix dialectic of Calculator made no appeal to the geometrical intuition, which was to act as an intermediary between his early attempts to study the problem of variation and the final formulation given by the calculus. This link between the interminable discursiveness of Suiseth and the concise symbolism of algebra was supplied by others of the fourteenth century who studied the latitude of forms. Of these the most famous was Oresme.

Nicole Oresme was born about 1323 and died in 1382. He was thus somewhat younger than Calculator, from whom we have a manuscript

---

factionem, et *fluxum* et alterationem, ad que omnia considerando proportionem totius ad partem responsionem alicere poteris ex predictis faciliter." *Liber calculationum*, fol. 9ʳ, col. 2. The word *fluxum* has been italicized by me for emphasis.

[7a] It should be remarked, in this connection, that several centuries later Galileo entered into an extensive discussion as to whether the Peripatetic qualities are to be regarded as positive (*Le Opere di Galileo Galilei*, I, 160 ff.), and that even in the eighteenth and nineteenth centuries the question as to the existence of "frigerific" rays was still argued. See, for example, Neave, "Joseph Black's Lectures on the Elements of Chemistry," p. 374.

[76] *Liber calculationum*, fol. 9ʳ, col. 2; fol. 75ᵛ, col. 1.

dated 1337,[77] and his doctrines were in all probability derived from those of Suiseth and others of the Oxford School.[78] Furthermore, from the very character of their work it would seem reasonable to place the composition of the *Liber calculationum* at Oxford before the appearance of the work of Oresme at Paris, on the grounds that the latter lacks the chief defects of the former—the prolixity of the complicated dialectical demonstrations and the author's propensity for subtle sophisms. There is in the entire *Liber calculationum* no diagram or reference to geometrical intuition, the reasoning being purely verbal and arithmetical.[79] On the other hand, Oresme felt that the multiplicity of types of variation involved in the latitude of forms is discerned with difficulty, unless reference is made to geometrical figures.[80] The work of Oresme therefore makes most effective use of geometrical diagrams and intuition, and of a coördinate system, to give his demonstrations a convincing simplicity. This graphical representation given by Oresme to the latitude of forms marked a step toward the development of the calculus, for although the logical bases of modern analysis have recently been divorced as far as possible from the intuitions of geometry, it was the study of geometrical problems and the attempt to express these in terms of number which suggested the derivative and the integral and made the elaboration of these concepts possible.

The *Tractatus de latitudinibus formarum*[81] has been generally

[77] See Thorndike, *History of Magic*, III, 375.

[78] Duhem (*Études sur Léonard de Vinci*, III, 478–79) would see in the *Liber calculationum* the influence of Oresme and would characterize the work of Suiseth as "l'œuvre d'une science sénile et qui commence à radoter." Wieleitner ("Zur Geschichte der unendlichen Reihen im christlichen Mittelalter," pp. 166–67) refers to Duhem in saying that Calculator rediscovered some of Oresme's examples. Thorndike has clearly shown, however, that Oresme's work was subsequent to that of Calculator.

[79] Stamm ("Tractatus de continuo," p. 23), apparently referring to an edition which appeared in 1520 and which he mistakenly holds to be the first, asserts that Suiseth illustrated his representation with geometrical figures. Such illustrations may well have been later interpolations, inasmuch as none such appear in the first edition (1477). Furthermore, French copyists sometimes drew diagrams on the edge of manuscripts of such Oxford works (Duhem, *Études sur Léonard de Vinci*, III, 449).

[80] "Quia formarum latitudines multipliciter variantur que multiplicitas difficiliter discernitur nisi ad figuras geometricas consideratio referatur." Oresme, *Tractatus de latitudinibus formarum*, fol. 201ʳ. I have made use of a photostat of a manuscript of this work in the Bayerische Staatsbibliothek, München, cod. lat. 26889, fol. 201ʳ.—fol. 206ʳ. See also Funkhouser, "Historical Development of the Graphical Representation of Statistical Data," p. 275; cf. also Stamm, *op. cit.*, p. 24.

[81] See Wieleitner, "Der 'Tractatus'," for a comprehensive description of this work. For comments on the work of Oresme and a study of this period as a whole, see Duhem, *Études sur Léonard de Vinci*, III, 346–405.

ascribed to Oresme and cited as his chief work on the subject. Although the views expressed in it are probably to be attributed to him, the treatise appears to be only a poor imitation, probably by a student,[82] of a larger work by Oresme entitled *Tractatus de figuratione potentiarum et mensurarum*.[83] This latter work, one of the fullest on the latitude of forms, was written probably before 1361.[84] It opens with the representation of variation by means of geometry, rather than with a dialectical exposition in terms of number, such as Calculator had given. Following the Greek tradition, which regarded number as discrete and geometrical magnitude as continuous, Oresme was led naturally to associate continuous change with a geometrical diagram. An intension, or rate of change by which a form acquires a quality, he imagined as represented by a straight line drawn perpendicular to a second line, the points of which represent the divisions of the time or space interval involved.[85] For example, a horizontal line or longitude may represent the time or duration of a given velocity, and the vertical height or latitude the intensity of the velocity.[86] It will be seen that the terms "longitude" and "latitude" are here used to represent, in a general sense, what we should now designate as the abscissa and the ordinate. The work of Oresme does not, of course, represent the earliest use of a coördinate system, for ancient Greek geography had employed this freely; nor can his graphical representation be regarded as equivalent to our analytic geometry,[87] for it lacks the fundamental

[82] Duhem, *Études sur Léonard de Vinci*, III, 399–400.

[83] This work is known also by other titles, such as *De uniformitate et difformitate intentionum*, and *De configuratione qualitatum*. See Wieleitner, "Ueber den Funktionsbegriff," for an extensive description of the portions of this work which are of significance in mathematics. For an analysis of those sections of the *De configuratione qualitatum* which deal with magic, see Thorndike, *History of Magic*, Vol. III, Chap. XXVI.

[84] Wieleitner, "Ueber den Funktionsbegriff," p. 198; Duhem, *Études sur Léonard de Vinci*, III, 375.

[85] "Omnis res mensurabilis extra numeros ymaginatur ad modum quantitates continue. Ideo opportet pro ejus mensuratione ymaginare puncta, superficies et lineas aut istorum proprietatis, in quibus, ut voluit Aristoteles, mensura seu proportio perprius reperitur . . . . Omnis igitur intensio successive acquisibilis ymaginanda est per rectam perpendiculariter erectam super aliquod puctum aut aliquot puncta extensibilis spacii vel subjecti." Wieleitner, "Ueber den Funktionsbegriff," p. 200.

[86] "Tempus itaque sive duratio erit ipsius velocitatis longitudo et ejusdem velocitatis intensio est sua latitudo." *Ibid.*, pp. 225–26. See also the diagrams in *Tractatus de latitudinibus formarum*, fols. 202^r and 205^r.

[87] Duhem has freely referred to Oresme as the inventor of analytic geometry. See, for example, his article on Oresme in the *Catholic Encyclopedia*; cf. also his *Études sur Léonard de Vinci*, III, 386.

notion that any geometric curve can be associated, through a co-
ordinate system, with an algebraic equation, and conversely.[88]

However, Oresme's work marks a notable advance in mathematical
analysis in that it associated the study of variation with the representation by coördinates. Although Aristotle had denied the existence
of an instantaneous velocity, the notion had continued to be invoked
implicitly, upon occasion, by Greek geometers and by Scholastic
philosophers. Oresme, however, was apparently the first to take the
significant step of representing an instantaneous rate of change by a
straight line.[89] He could not, of course, give a satisfactory definition
of instantaneous velocity, but he strove to clarify this idea by remarking that the greater this velocity is, the greater would be the distance
covered if the motion were to continue uniformly at this rate.[90]
Maclaurin, in attempting to clarify the Newtonian idea of a fluxion,
expressed himself very similarly almost four hundred years later, but
a rigorous and clear definition could be given only after the concept
of the derivative had been developed. Oresme, furthermore, was
confused by the perplexing problem of the indivisible and the continuum. In spite of his clear assertion that an instantaneous velocity
is to be represented by a straight line, he accepted the dictum of
Aristotle that every velocity persists throughout a time.[91] This view
implies a form of mathematical atomism which, although underlying
much of the thought leading to the calculus of Newton and Leibniz,
has been rejected by modern mathematics.

There is a widespread belief that the science of dynamics, which in
the seventeenth century played such a significant rôle in the formation
of the calculus, was almost entirely the product of the genius of
Galileo, who "had to create . . . for us"[92] the "entirely new notion
. . . of acceleration."[93] That such a view is a gross misconception
will be clear to anyone who makes even a cursory examination of the
fourteenth-century doctrine of the latitude of forms. Oresme, for

[88] Cf. Coolidge, "The Origin of Analytic Geometry," p. 233.
[89] "Sed punctualis velocitas instantanea est ymaginanda per lineam rectam." Wieleitner, "Ueber den Funktionsbegriff," p. 226.
[90] "Verbi gratia; gradus velocitatis descencus est major, quo subjectum mobile magis
descendit vel descenderet si continuaretur simpliciter." *Ibid.*, p. 224.
[91] "Omnis velocitas tempore durat." *Ibid.*, p. 225.
[92] Mach, *The Science of Mechanics*, p. 133. See also Hogben, *Science for the Citizen*,
p. 241.
[93] Mach, *op. cit.*, p. 145.

example, had a clear conception not only of acceleration in general but also of uniform acceleration in particular. This is evident from his statement that if the *velocitatio* (acceleration) is *uniformis*, then the *velocitas* (velocity) is *uniformiter difformis;* but if the *velocitatio* is *difformis*, then the *velocitas* is *difformiter difformis*. Oresme went further and applied his idea of uniform rate of change and of graphical representation to the proposition that the distance traversed by a body starting from rest and moving with uniform acceleration is the same as that which the body would traverse if it were to move for the same interval of time with a uniform velocity which is one-half the final velocity. Later, in the seventeenth century, this proposition played a central rôle in the development both of infinitesimal methods and of Galilean dynamics. It had been stated earlier and in a more

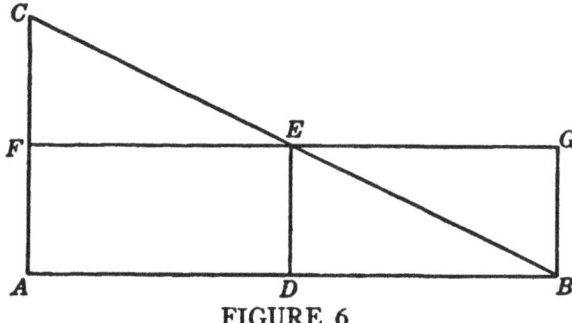

FIGURE 6

general form by Calculator, but Oresme and Galileo gave to it a geometrical demonstration which, although not rigorous in the modern sense, was the best which could be furnished before the integral calculus had been established.[94]

The proof of the theorem is based upon the fact that the motion under uniform velocity, inasmuch as the latitude is the same at all times, is represented by a rectangle such as *ABGF* (fig. 6) and that the uniformly accelerated motion, in which the ratio of the change in latitude to the change in longitude is constant,[95] corresponds to

[94] In the mistaken belief that the work of Oresme preceded that of Calculator, Duhem (*Études sur Léonard de Vinci*, III, 388–98) has called this proposition the law of Oresme.

[95] See Wieleitner, "Ueber den Funktionsbegriff," pp. 209–10; cf. also *Tractatus de latitudinibus formarum*, fol. 204ᵛ. Duhem, upon the basis of these observations, has unwarrantedly asserted that Oresme "gives the equation of the right line, and thus forestalls Descartes in the invention of analytical geometry." See his article, "Oresme," in the *Catholic Encyclopedia*.

the right triangle *ABC*. Oresme did not explicitly state the fact—
this is, of course, demonstrated in the integral calculus—that the areas
ABGF and *ABC* represent in each case the distance covered; but
this seems to have been his interpretation,[96] inasmuch as from the
congruence of the triangles *CFE* and *EBG* he concluded the equality
of the distances.[97] This is perhaps the first time that the area under
a curve was regarded as representing a physical quantity, but such
interpretations were to become before long commonplaces in the
application of the calculus to scientific problems. Oresme did not explain why the area under a velocity-time curve represented the
distance covered. It is probable, however, that he thought of the
area as made up of a large number of vertical lines or indivisibles,
each of which represented a velocity which continued for a very short
time. Such an atomistic interpretation is in harmony with the views
he expressed on instantaneous velocities and with the Scholastic
interest in the infinitesimal. Several centuries later this interpretation
was enunciated more boldly and vividly by Galileo, at a time when
atomistic conceptions enjoyed an even greater popularity in both
science and mathematics.

Arguments similar to those given by Oresme and Calculator appeared also in the work of other men of the time, particularly at Oxford
and Paris.[98] William of Hentisbery, a famous logician at Oxford who
perhaps outdid Calculator in propounding sophisms on the subject of
motion, clearly stated the law for uniformly difform variation.[99]
Marsilius of Inghen, at Paris, expounded this same law on the basis
of Oresme's geometrical representation.[100] The traditions developed
at Oxford and Paris were continued also in the fifteenth century in
Italy, where Blasius of Parma (or Biagio Pelicani), reputed the most
versatile philosopher and mathematician of his time, explained the
law similarly, in his *Questiones super tractatum de latitudinibus formarum*.[101] This same principle for the uniformly difform acquisition
of qualities was known at Paris in the sixteenth century to Alvarus

[96] Duhem, *Études sur Léonard de Vinci*, III, 394.
[97] Wieleitner, "Ueber den Functionsbegriff," p. 230.
[98] See Duhem, *Études sur Léonard de Vinci*, III, 405–81.
[99] Wieleitner ("Der 'Tractatus'," pp. 130–32, n.) says that Hentisbury's arguments are
accompanied by diagrams, but Duhem (*Études sur Léonard de Vinci*, III, 449) says there
is no geometrical demonstration given with his writings.
[100] Duhem, *Études sur Léonard de Vinci*, III, 399–405.
[101] See Amodeo, "Appunti su Biagio Pelicani da Parma."

Thomas,[102] John Major, Dominic Soto, and others.[103] The principles of uniformly accelerated motion thus seem to have been common knowledge to the Scholastics from the fourteenth century to the sixteenth, and it is very probable that Galileo was familiar with their work and made use of it in his development of dynamics.[104] At all events, when the famous *Discorsi* of Galileo appeared in 1638, it contained a diagram and a type of argument resembling strikingly that previously given by Oresme.[105]

The efforts toward the mathematical representation of variation which resulted, in the fourteenth century, in the flurry of treatises on the latitude of forms, were tied up in many ways with other questions related to the calculus. While he was discussing, in connection with *latitudo difformiter difformis*, a form represented graphically by a semicircle, the author of the *Tractatus de latitudinibus formarum* remarked that the rate of change of an intensity is least at the point corresponding to the maximum intensity.[106] Great things have been read into this casual remark, for historians of mathematics have ascribed to its author the sentiments that the increment in the ordinate of the curve is zero at its maximum point, and that the differential coefficient vanishes at this point![107] Ascribing such ideas to the author of the *Tractatus* is obviously unwarranted, inasmuch as they presuppose the clear conceptions of limit and differential quotient which were not developed until many centuries later.[108] Further-

[102] Wieleitner, "Zur Geschichte der unendlichen Reihen im christlichen Mittelalter," p. 154.

[103] Duhem, *Études sur Léonard de Vinci*, III, 531 ff.

[104] Zeuthen, *Geschichte der Mathematik im XVI. und XVII. Jahrhundert*, pp. 243–44.

[105] Mach (*The Science of Mechanics*, p. 131), probably unaware of the earlier work, has ascribed the diagram and ideas to Galileo.

[106] "In qualibet talis figura sua intensio terminatur ad summum gradum tarditatis et sua remissio incipit a summo gradu tarditatis ut in medio puncto aliqualis ubi terminatur intensio et incipit remissio, patet in figuris .c.d. et .d.c." *Tractatus de latitudinibus formarum*, fol. 205ʳ. The figures referred to (fol. 204ᵛ) are much like semicircles.

[107] Curtze ("Ueber die Handschrift R. 4°, 2," p. 97) says that Oresme noticed in general that "der Zuwachs . . . . der Ordinate einer Kurve in der unmittelbaren Nähe eines Maximums oder minimums gleich Null ist," and Moritz Cantor makes an even stronger statement when he says (*Vorlesungen*, II, 120) "Oresme's Augen offenbarte sich die Wahrheit des Satzes, den man 300 Jahre später in die Worte kleidete, an den Höhen- und Tiefpunkten einer Curve sei der Differentialquotient der Ordinate nach der Abscisse Null."

[108] Wieleitner ("Der 'Tractatus,'" p. 142) says "Ich halte aber die Bemerkung doch nur für rein anschauungsmässig"; and Timtchenko ("Sur un point du 'Tractatus de latitudinibus formarum' de Nicholas Oresme") agrees, saying "qu'il était encore assez loin du théorème exprimé par la formule $\frac{dy}{dx} = 0$ (pour le sommet de la figure)."

more, the writer had apparently no idea of generalizing the statement by extending the conclusion to other cases. His remarks do show clearly, however, how fruitful the idea of the latitude of forms was to be when it entered the geometry and algebra of later centuries, to become eventually the basis of the calculus. The Scholastic philosophers were striving to express their ideas in words and geometrical diagrams, and were not so successful as we who realize, and can make use of, the economy of thought which mathematical notation affords.

Further consideration by Oresme of a *latitudo difformiter difformis* led him, as it had also Suiseth somewhat earlier, to another topic essential in the development of the calculus—that of infinite series. In the same manner as Calculator, he considered a body moving with uniform velocity for half a certain period of time, with double this velocity for the next quarter of the time, with three times this velocity for the next eighth, and so ad infinitum. He found in this case that the total distance covered would be four times that covered in the first half of the time. However, whereas Calculator had had recourse to devious verbal argumentation in his justification of this result, the method of Oresme was here (as in his earlier work) geometrical and consisted in the comparison of the areas corresponding to the distances involved in the graphical representation of the motion.[109]

In this manner also he handled another similar case in which the time was divided into parts $\frac{2}{4}$, $\frac{3}{16}$, $\frac{3}{64}$, . . . the velocity increasing in arithmetic proportion as before. The total distance he found this time to be $\frac{16}{9}$ that covered in the first subinterval. Oresme then went on to still more complicated cases. For example, he assumed that during the first half of the time interval the body moved with uniform velocity, then for the next quarter of the time with motion uniformly accelerated until the velocity was double the original, then for the next eighth of the time uniformly with this final velocity, then for the next sixteenth with uniform acceleration until the velocity was again doubled, and so on. Oresme found that in this case the total distance (or *qualitas*) is to that of the first half of the time as seven is to two.[110] This is equivalent to summing geometrically to infinity the series $\frac{1}{2} + \frac{3}{8} + \frac{1}{4} + \frac{3}{16} + \frac{1}{8} + \frac{3}{32} + \frac{1}{16} + \frac{3}{64} + \ldots$.

[109] See Wieleitner, "Ueber den Funktionsbegriff," pp. 231–35.
[110] Wieleitner, *Ibid.*, p. 235.

Calculator and Oresme were not the only Scholastics who were concerned with such infinite series. In the anonymous tract written before 1390, *A est unum calidum*,[111] some of their results on series were rediscovered. More than two centuries later, in 1509, similar work appeared in the *Liber de triplici motu* of Alvarus Thomas, at Paris. This book was intended to serve as an introduction to the *Liber calculationum* of Suiseth and in it are handled several of the series given earlier by Calculator and Oresme. However, the author went on to similar but more complicated cases, in which he found (in the manner of Calculator and Oresme) the sums to infinity of the series

$$1 + \frac{7}{4}\cdot\frac{1}{2} + \frac{11}{8}\cdot\frac{1}{2^2} + \frac{19}{16}\cdot\frac{1}{2^3} + \ldots \text{ and } 1 + \frac{4}{3}\cdot\frac{1}{2} + \frac{7}{6}\cdot\frac{1}{2^2} + \frac{13}{12}\cdot\frac{1}{2^3}$$

+ . . . , these sums being $\frac{5}{2}$ and $\frac{20}{9}$ respectively.[112] Alvarus Thomas remarked that innumerable other such series can be found. In some of these the sums involve logarithms, so that he could not arrive at the sum exactly, but gave it approximately between certain integers. Thus he said that the sum of $1 + \frac{2}{1}\cdot\frac{1}{2} + \frac{3}{2}\cdot\frac{1}{2^2} + \frac{4}{3}\cdot\frac{1}{2^3} + \ldots$ (which results in $2 + \log 2$) lies between 2 and 4.[113]

It must be remembered that these infinite series were not handled by the Scholastic philosophers as they are now in the calculus, for they were given rhetorically, rather than by means of symbols, and were bound up with the concept of the latitude of forms. Furthermore, the results were found either through verbal arguments or geometrically from the graphical representation of the form, rather than by means of arithmetic considerations based upon the limit concept. Had their work been more closely associated with the geometrical procedures of Archimedes and less bound up with the philosophy of Aristotle, it might have been more fruitful. It could then have anticipated Stevin and the geometers of the seventeenth century in the elimination, from the classical method of exhaustion, of the argument by a reductio ad absurdum, and in the substitution of an analysis suggestive of the limit idea.

The fourteenth-century work at Paris and Oxford on the latitude

[111] Wieleitner, "Zur Geschichte der unendlichen Reihen," p. 167; cf. also Duhem, *Études sur Léonard de Vinci*, III, 474–77.

[112] Wieleitner, "Zur Geschichte der unendlichen Reihen"; Duhem, *Études sur Léonard de Vinci*, III, 531 ff.

[113] Wieleitner, "Zur Geschichte der unendlichen Reihen," pp. 161–64; see also Duhem, *Études sur Léonard de Vinci*, III, 540–41.

of forms and related topics was not forgotten during the subsequent
decline of Scholasticism, for the doctrines spread throughout the
Italian universities.[114] In particular, they were taught at Pavia,
Bologna, and Padua by Blasius of Parma.[115] He, in turn, was referred
to a century later by Luca Pacioli and Leonardo da Vinci; and Nicholas of Cusa was very likely influenced by him.[116] The work of Calculator and Oresme was much admired by the men of the fifteenth and
sixteenth centuries, several printed editions of both the *Liber calculationum* and the *De latitudinibus formarum*, as well as commentaries on each, appearing in the interval from 1477 to 1520.[117] In the
De subtilitate of Cardan, Calculator is classed among the great, along
with Archimedes, Aristotle, and Euclid.[118] Galileo, we have said, gave
in his dynamics a geometrical demonstration almost identical with
that of Oresme, and in another place referred to Calculator and
Hentisbery on activity in resisting media.[119] Descartes likewise, in
his attempt to determine the laws of falling bodies, employed arguments closely resembling those of Oresme.[120] As late as the end of the
seventeenth century the reputation of Calculator was such that
Leibniz on several occasions referred to him as almost the first to
apply mathematics to physics and as one who introduced mathematics into philosophy.[121]

In spite of the reputation maintained by the great proponents of
the doctrine of the latitude of forms, however, the type of work they
represented was not destined to be the basis of the decisive influence
in the development of the methods of the calculus. The guiding

[114] Duhem, *Études sur Léonard de Vinci*, III, 481–510.
[115] See Thorndike, *History of Magic*, IV, 65–79.
[116] See Amodeo, "Appunti su Biagio Pelicani da Parma," III, 540–53.
[117] The *Liber calculationum* was published at Padua in 1477 and 1498, and again at
Venice in 1520; *De latitudinibus formarum* appeared at Padua in 1482 and 1486, at Venice
in 1505, and at Vienne in 1515.
[118] Cardan, *Opera*, III, 607–8.
[119] "Secunda dubitatio: quomodo se habent primae qualitates in activitate et resistentia.
De hac re lege Calculatorem in tractatu De reactione, Hentisberum in sophismate." See
Opere, I, 172.
[120] *Œuvres*, X, 58–61.
[121] "Quis refert, primum prope eorum, qui mathesin ad physicam applicarunt, fuisse
Johan. [*sic*] Suisset calculatorem ideo, scholasticis appellatum": Letter to Theophilus
Spizelius, April 7/17, 1670. *Opera omnia* (Dutens), V, 347. "Vellem etiam edi scripta
Suisseti, vulgo dicti *calculatoris*, qui Mathesin in Philosophicam Scholasticam introduxit,
sed ejus scripta in Cottonianis non reperio." Letter to Thomas Smith, 1696. *Opera omnia*
(Dutens), V, 567. See also *Opera Omnia* (Dutens), V, 421.

principles were to be supplied by the geometry of Archimedes, although these were to be modified by kinematic notions derived from the quasi-Peripatetic disputations of the Scholastic philosophers on the subject of variation. As early as the beginning of the fifteenth century, the mensurational science and mathematics of Archimedes were becoming more influential, so that men like Blasius of Parma in some instances preferred his explanations to those of Aristotle. Blasius of Parma is said to have written on infinitesimals[122] as well as upon other mathematical topics, but no such works are known to be extant.

We do, however, have the work of Cardinal Nicholas of Cusa, which illustrates the influence both of the Scholastic speculations (perhaps through Blasius) and the work of Archimedes, as well as another tendency which grew up, particularly on German soil, in the fifteenth century. With the decline of the extremely rational and rigorous thought of Scholasticism (the overpreciseness of which was anathema to the growing Humanism of the time),[123] there was a trend toward Platonic and Pythagorean mysticism.[124] This may well have been in some measure responsible for the prevalence of mysticism[125] at the time. On the other hand, the development of science in Italy, culminating in the dynamics of Galileo, owed much to the mathematical philosophy of Plato and Pythagoras.[126] From the point of view of the rise of the calculus, this pervasive Platonism exerted for several centuries a not altogether unfortunate influence upon mathematics, for it allowed to geometry what Aristotle's philosophy and Greek mathematical rigor had denied—the free use of the concepts of infinity and the infinitesimal which Platonic and Scholastic philosophers had fostered. Nicholas of Cusa was not a trained mathematician, but he was certainly well acquainted with Euclid's *Elements* and had read Archimedes' works. Although he was primarily a theologian, mathematics nevertheless constituted for him, as for Plato,

[122] Hoppe ("Geschichte der Infinitesimalrechnung," p. 179) says "Er ist also in dieser Beziehung ein Vorläufer von Cavalieri, von dem ihn 148 Jahre trennen." Blasius died in 1416, but probably Hoppe has in mind the 1486 edition of Blasius' *Questiones*, which is separated from Cavalieri's *Geometria indivisibilibus* of 1635 by about 148 years.

[122] Thorndike, *History of Magic*, III, 370.

[124] Lasswitz, *Geschichte der Atomistik*, I, 264–65; see also Burtt, *The Metaphysical Foundations of Modern Physical Science*, pp. 18–42.

[125] Cf. Thorndike, *Science and Thought in the Fifteenth Century*, p. 22.

[126] Strong, *Procedures and Metaphysics*, pp. 3–4; see also Cohn, *Geschichte des Unendlichkeitsproblems*, p. 95.

the basis of his whole system of philosophy.[127] For him mathematics was not restricted, as for Aristotle, to the science of quantity, but was the necessary form for an interpretation of the universe.[128] Aristotle had insisted on the operational character of mathematics and had rejected the metaphysical significance of number, but Cusa revived the Platonic arithmology.[129] He again associated the entities of mathematics with ontological reality and restored to mathematics the cosmological status which Pythagoras had bestowed upon it. Furthermore, the validity of its propositions was regarded by him as established by the intellect, with the result that the subject was not bound by the results of empirical investigation.

This view of mathematics as prior to, or at least independent of, the evidence of the senses encouraged speculation and allowed the indivisible and the infinite to enter, so long as no inconsistency in thought resulted. Such an attitude enriched the subject and eventuated in the methods of the calculus, but this advantage was gained at the expense of the rigor which characterized ancient geometry. The logical perfection in Euclid found no counterpart in modern mathematics until the nineteenth century. From the point of view of the development of the calculus, this was not an unfortunate situation, for the infinite and the infinitely small were allowed almost free reign, even though it was about four hundred years before a logically satisfactory basis for these notions was found and the limit concept was definitely and rigorously established in mathematics.

In spite of Aristotle, the infinitely large and the infinitely small had been employed somewhat surreptitiously by Archimedes in his geometry, and the Scholastics had freely discussed these notions in their dialectical philosophy; but with Nicholas of Cusa these ideas became, together with an unnecessarily large admixture of Pythagorean and theological mysticism,[130] a recognized part of the subject matter of mathematics.

The definitions given by Nicholas of Cusa for the infinitely large (that which cannot be made greater), and for the infinitely small

[127] Schanz, *Der Cardinal Nicolaus von Cusa als Mathematiker*, pp. 2, 7.
[128] Löb, *Die Bedeutung der Mathematik für die Erkenntnislehre des Nikolaus von Kues*, pp. 36–37; cf. also Max Simon, *Cusanus als Mathematiker*.
[129] Cf. L. R. Heath, *The Concept of Time*, p. 83.
[130] Cohn, *Geschichte des Unendlichkeitsproblems*, p. 127.

(that which cannot be made smaller)[131] are unsatisfactory, and his mathematical demonstrations are not always unimpeachable. His work is significant, however, in that it made use of the infinite and the infinitesimal, not merely as potentialities, as had Aristotle, but as actualities which are the upper and lower bounds of operations upon finite magnitudes. Just as the triangle and the circle were to Nicholas of Cusa the polygons with the smallest and the greatest number of sides, so also zero and infinity were to him the lower and upper bounds of the series of natural numbers.[132] His view was colored, moreover, by his favorite philosophical doctrine—derived, perhaps, from the Scholastic quasi-theological disputations on the infinite—that a finite intelligence can approach truth only asymptotically. Consequently, the infinite was the source and means, and at the same time the unattainable goal, of all knowledge.[133] Cusa was led by this attitude to view the infinite as a *terminus ad quem* to be approached only by going through the finite[134]—an idea which represents a striving toward the limit concept[135] which was to be developed during the next five hundred years.

Upon the basis of this view, Nicholas of Cusa proposed a characteristic quadrature of the circle. If, contrary to Greek ideas of congruence, the circle belongs among the polygons as the one with an infinite number of sides and with apothem equal to the radius, its area may be found by the same means as that employed for any other polygon: by dividing it up into a number (in this case an infinite number) of triangles,[136] and computing the area as half the product of the apothem and the perimeter. Nicholas of Cusa added to this explanation of the measurement of the circle an Archimedean proof, using inscribed and circumscribed polygons, and a proof by the reductio ad absurdum; but when his method was used later by Stevin, Kepler, and others, the Archimedean proof was abandoned, an elementary equivalent of the limit concept being considered sufficient.

As Plato had opposed the Democritean atomic doctrine and yet attempted to link the continuum with indivisibles, so also Cusa objected to Epicurean atomism, but at the same time regarded the line

[131] Lorenz, *Das Unendliche bei Nicolaus von Cues*, p. 36.   [132] Lorenz, *op. cit.*, p. 2.
[133] Cohn, *op. cit.*, p. 87.   [134] Lasswitz, *Geschichte der Atomistik*, I, 282.
[135] Cf. Simon, "Zur Geschichte und Philosophie der Differentialrechnung," p. 116.
[136] Schanz, *Nicolaus von Cusa*, pp. 14–15.

as the unfolding of a point. In a similar manner, also, he held that while continuous motion is thinkable, in actuality it is impossible,[127] inasmuch as motion is to be regarded as composed of serially ordered states of rest.[138] These views call to mind the modern mathematical continuum and the so-called static theory of the variable, but to Nicholas of Cusa must not be ascribed any of the precision of statement which now characterizes these matters. He did not make clear in what manner the transition from the continuous to the discrete is made. He asserted that although in thought the division of continuous magnitudes, such as space and time, may be continued indefinitely, nevertheless in actuality this process of subdivision is limited by the smallness of the parts obtained—that is, of the atoms and instants.[139]

The looseness of expression found toward the middle of the fifteenth century in the formulation by Nicholas of Cusa of his views as to the nature of the infinitesimal and the infinitely large was paralleled about two centuries later by a similar lack of clarity in the multifarious methods of indivisibles to which they to some extent gave rise. These latter procedures, in turn, led to the differential and the integral calculus. It is not to be concluded, however, that the views of Cusa are to be construed as indicating any renaissance in mathematics, nor as heralding the rise of the new analysis. The cardinal, absorbed as he was in matters concerning church and state, did not contribute any work of lasting importance in mathematics. Regiomontanus was perfectly justified in his criticism[140] of the repeated attempts which were made by Cusa to square the circle. Nevertheless, as Plato's thought may have been instrumental in the use of infinitesimals made by Archimedes in his investigations preliminary to the application of the rigorous method of exhaustion, so also the speculations of Nicholas of Cusa may well have induced mathematicians of a later age to employ the notion of the infinite in conjunction with the Archimedean demonstrations.

Leonardo da Vinci, who was strongly influenced, through Nicholas

---

[127] L. R. Heath, *The Concept of Time*, p. 82.
[138] Cohn, *Geschichte des Unendlichkeitsproblems*, p. 94.
[139] L. R. Heath, *The Concept of Time*, p. 82; Cohn, *op. cit.*, p. 94; Lasswitz, *Geschichte der Atomistik*, II, 276 ff.
[140] See Kaestner, *Geschichte der Mathematik*, I, 572–76.

of Cusa, by Scholastic thought,[141] and who was acquainted also with the work of Archimedes, is said to have employed infinitesimal considerations in finding the center of gravity of a tetrahedron by thinking of it as made up of an infinite number of planes. We cannot, however, be sure of his point of view.[142] Cusa's ideas were more clearly expressed in the work of Michael Stifel, who held that a circle may correctly be described as a polygon with an infinity of sides, and that before the mathematical circle there are all the polygons with a finite number of sides, just as preceding an infinite number there are all the given numbers.[143] Somewhat later François Viète also spoke of a circle as a polygon with an infinite number of sides,[144] thus showing that such conceptions were widely held in the sixteenth century. The fullest expression of Nicholas of Cusa's mathematical thoughts on the infinite and the infinitesimal, however, are found in the work of Johann Kepler, who was strongly influenced by the cardinal's ideas—speaking of him as *divinus mihi Cusanus*—and who was likewise deeply imbued with Platonic and Pythagorean mysticism. It was probably the imaginative use by Cusa of the concept of infinity which led Kepler to his principle of continuity—which included normal and limiting forms of a figure under one definition, and in accordance with which the conic sections were regarded as constituting a single family of curves.[145] Such conceptions often led to paradoxical results, inasmuch as the notion of infinity had not yet received a sound mathematical basis; but Kepler's bold views suggested new paths which were be to very fruitful later. Kepler lived a century and a half after the time of Cusa. Moreover, unlike his predecessor, although his early training was in theology, he was primarily a mathematician. He was thus able, to the fullest extent, to take advantage

[141] Duhem, *Études sur Léonard de Vinci*, II, 99 ff.

[142] Libri, *Histoire des sciences mathématiques en Italie*, III, 41; Duhem, *Études sur Léonard de Vinci*, I, 35–36. The former asserts that Leonardo used indivisibles; the latter questions this. Duhem is wrong when he says in this connection (cf. *Les Origines de la statique*, II, 74) that Archimedes had restricted himself to plane figures in his work on the center of gravity, since in the *Method* the centers of gravity of segments of a sphere and of a paraboloid are determined.

[143] Cf. Gerhardt, *Geschichte der Mathematik in Deutschland*, pp. 60–74.

[144] Moritz Cantor, *Vorlesungen*, II, 539–40.

[145] This is indicated by his statement, ". . . dico, lineam rectam esse hyperbolarum obtussissimam. Et Cusanus infinitum circulum dixit esse lineam rectam. . . . Plurima talia sunt, quae analogia sic vult efferi, non aliter." Kepler, *Opera omnia*, II, 595.

of the accessibility of Archimedes' works resulting from their translation and publication during the middle of the sixteenth century.

It has been remarked that the medieval period added little to the classical Greek works in geometry or to the theory of algebra. Its contributions were chiefly in the form of speculations, largely from the philosophical point of view, on the infinite, the infinitesimal, and continuity, as well as of new points of view with reference to the study of motion and variability. Such disquisitions were to play a not insignificant part in the development of the methods and concepts of the calculus, for they were to lead the early founders of the subject to associate with the static geometry of the Greeks the graphical representation of variables and the idea of functionality.

Both Newton and Leibniz, as well as many of their predecessors, sought for the basis of the calculus in the generation of magnitudes— a point of view which may be regarded as the most notable contribution of Scholastic philosophy to the development of the subject. However, the sound mathematical basis for the seventeenth-century elaboration of the infinitesimal procedures into a new analysis was supplied by the mensurational treatises of Archimedes which had been composed much earlier. These had become scattered and some had been lost, although a number of them were familiar to the Arabs, and although Archimedean manuscripts were known in the Scholastic period.[146] No significant additions to his results were made until after the appearance of printed editions of his treatises. In 1543 the Italian mathematician, Nicolo Tartaglia, published at Venice portions of Archimedes' work, including that on centers of gravity, the quadrature of the parabola, the measurement of the circle, and the first book on floating bodies. The *editio princeps* of Venatorius appeared at Basel in 1544, and in 1558 the important translation by Federigo Commandino appeared at Venice. Commandino himself composed a treatise in the Archimedean manner—*Liber de centro gravitatis solidorum*—showing that the study of the methods of the great Syracusan mathematician had advanced to the point where new contributions could be made. Shortly after this, however, numerous attempts were made by Kepler and others to replace the almost tediously rigorous

---

[146] See Thorndike and Kibre, *Catalogue of Incipits*, for citations of some dozen manuscripts of the twelfth, thirteenth, fourteenth, and fifteenth centuries.

arguments of Archimedes by new methods which should be equivalent to the older ones but at the same time possess a simplicity and an ease of application to new problems which was lacking in the method of exhaustion.

It is precisely the search for such substitutes which was to lead within the next century, with the help of the Scholastic speculations, to the methods of the calculus. For the sake of convenience and unity, we shall make these anticipations the subject of the following chapter.

IV. A Century of Anticipation

THE DEVELOPMENT of the concepts of the calculus may be considered to have begun with the Pythagorean effort to compare—through the superposition of geometrical magnitudes—lengths, areas, and volumes, in the hope of thus associating with each configuration a number. Thwarted in this by the problem of the incommensurability of such magnitudes, it was left for later Greek geometers to circumvent such direct comparison through the method of exhaustion of Eudoxus. Through this method the need for the infinite and the infinitesimal had been obviated, although such notions had been considered by Archimedes as suggestive heuristic devices, to be used in the investigation of problems concerning areas and volumes which were preliminary to the intuitively clear and logically rigorous proofs given in the classical geometrical method of exhaustion.

During the later Middle Ages the conceptions of the infinite and the infinitesimal, and the related ideas of variation and the continuum, were more freely discussed and utilized. Because, however, the interest in such speculations remained less closely associated with geometry than with metaphysics or science, or even with attempts to explain the possibility of natural magic,[1] the primary influence leading to the calculus was not supplied by the doctrines of the Scholastic philosophers but by the greater enthusiasm for the methods of Archimedes which was manifest in the late sixteenth century and which continued throughout the whole of the century following. Whereas a multiplicity of printed editions of the works of Jordanus Nemorarius, Bradwardine, Calculator, Oresme, Hentisbery, and other medieval scholars had appeared in the late fifteenth and early sixteenth centuries, there arose toward the middle of the latter century a strong opposition, illustrated by the attitude of Ramus, to Aristotelianism and Scholastic methodology.[2] It was during the height of this reaction that the works of Archimedes appeared in numerous editions and, admiring Archimedes,

[1] Thorndike, *History of Magic*, III, 371.
[2] Johnson and Larkey, "Robert Recorde's Mathematical Teaching and the Anti-Aristotelian Movement."

the men of the time refused to recognize the work of the Middle Ages.[3]

In the mathematics of this period, nevertheless, the intrusion of influences from the Scholastic age is easily discernible in the attempts to reconcile with the thought of Archimedes the newly discovered infinitesimal methods which were fostered. There is noticeable in this development, as well, a tendency foreign alike to Archimedes and to the Scholastic period—a deeper interest in the Arabic algebra which had been developing in Italy, in which the concept of infinity did not figure. The algebra of Luca Pacioli's *Summa de arithmetica* was not greatly advanced over that to be found in Leonardo of Pisa almost three hundred years earlier,[4] but in the sixteenth century the subject was assiduously studied again. Before 1545 the cubic had been solved by Tartaglia and Cardan, and the quartic by Ferrari; and thereafter a freer use of irrational, negative, and imaginary numbers was made by Cardan, Bombelli, Stifel, and others. The Greeks had not regarded irrational ratios as numbers in the strict sense of the word, and the attitude in the medieval period had been similar. Bradwardine asserted that an irrational proportion is not to be represented by any number;[5] and Oresme, in discussing the popular question of whether the celestial motions were commensurable or incommensurable, concluded that geometry favored the latter but arithmetic the former.[6]

The Hindus and the Arabs, on the other hand, had not clearly distinguished between rational and irrational numbers. On adopting the Hindu-Arabic algebra, the sixteenth-century mathematicians had continued to employ the irrational ratio. They now recognized this as a number, but they stigmatized it, following Leonardo of Pisa, as a *numerus surdus*, and they continued to interpret it geometrically as a ratio of lines.[7] Negative quantities, admitted by the Hindus but not by the Greeks or the Arabs, were accepted in the sixteenth century as *numeri falsi* or *ficti*, but in the following century these were recog-

[3] Duhem, *Les Origines de la statique*, I, 212.   [4] Cajori, *History of Mathematics*, p. 128
[5] Hoppe, "Zur Geschichte der Infinitesimalrechnung," p. 158.
[6] Thorndike, *History of Magic*, III, 406.
[7] See Pringsheim, "Nombres irrationnels et notion de limite," pp. 137–40; Fink, *A Brief History of Mathematics*, pp. 100–1. However, Kepler, as late as 1615 spoke of the irrational as "ineffable." *Opera omnia*, IV, 565.

nized as numbers in the strict sense of the word.[8] Imaginary numbers also were regularly employed after the sixteenth century, although they continued to occupy an anomalous position in mathematics until the time of Gauss.

Such generalizations of number, although not at the time based upon satisfactory definitions, were influential later in leading to the limit concept and to the arithmetization of mathematics. More important than this, in the development of the algorithm of the calculus, was the systematic introduction, during the later sixteenth century, of symbols for the quantities involved in algebraic relations.

As early as the thirteenth century, letters had been used as symbols for quantities by Jordanus Nemorarius in his science and mathematics. Their establishment as symbols of the abstract quantities entering into algebra, however, was largely the work of the great French mathematician François Viète,[9] who used consonants to represent known quantities and vowels for those unknown. He distinguished arithmetic, or *logistica numerosa*, from algebra, or *logistica speciosa*, thus making the latter a calculation with letters rather than with numbers alone.

This literal symbolism was absolutely essential to the rapid progress of analytic geometry and the calculus in the following centuries,[10] for it permitted the concepts of variability and functionality to enter into algebraic thought. The improved notation led also to methods which were so much more facile in application than the cumbrous geometrical procedures of Archimedes, of which they were modifications, that these methods were eventually recognized as forming a new analysis —the calculus. The period during which this transformation took place may be considered as the century preceding the work of Newton and Leibniz.[11]

Numerous translations of the works of Archimedes had been made during the middle of the sixteenth century, and soon after this mathe-

[8] See Fine, *Number System*, p. 113; Paul Tannery, *Notions historiques*, pp. 333–34; Fehr, "Les Extensions de la notion de nombre dans leur développement logique et historique."

[9] Cajori, *History of Mathematics*, p. 139.

[10] Cf. Karpinski, "Is There Progress in Mathematical Discovery?" p. 47.

[11] For a comprehensive account of the methods developed during this period, see Zeuthen, *Geschichte der Mathematik im XVI. und XVII. Jahrhundert*. Zeuthen's account is an outline of the development of the subject in this interval, rather than of fundamental concepts, and so includes a large amount of mathematical detail.

maticians had reached the point at which original contributions could be made to the classical work of the Greeks. This is evidenced by the fact that Commandino in 1565 published a work of his own on centers of gravity. In this he proved, among other things, that the center of gravity of a segment of a paraboloid of revolution is situated on the axis two-thirds of the distance from the vertex to the base.[12] This was a proposition which Archimedes had, incidentally, demonstrated by infinitesimals in a treatise, the *Method*, which was apparently not known at that time. Commandino's proof followed the orthodox style of the method of exhaustion. Such extensions, however, had perhaps less influence on the development of the calculus than did certain striking innovations in method introduced by succeeding mathematicians. Perhaps the first of such significant modifications is that advanced by Simon Stevin of Bruges in 1586, very nearly a century before the first printed work on the calculus by Leibniz, in 1684, and that of Newton in 1687.

Stevin was essentially an engineer and a practical-minded scientist. For this reason he had perhaps less regard for the philosophy of science and the exigencies of mathematical rigor than for technological tradition and methodology.[13] As a result, Stevin did not merely imitate, as had Commandino, Archimedes' use of the method of exhaustion: he accepted the direct portion of his characteristic proof as sufficient to establish the validity of any proposition that required it, without adding in every case the formal reductio ad absurdum required by Greek rigor. Furthermore, he frequently omitted, as we do in the integral calculus, one of the approximating figures which Archimedes had used, being satisfied with either the inscribed or the circumscribed figure only. Stevin demonstrated as follows (in his work on statics, in 1586) that the center of gravity of a triangle lies on its median. Inscribe in the triangle ABC a number of parallelograms of equal height, as illustrated (fig. 7). The center of gravity of the inscribed figure will lie on the median, by the principle that bilaterally symmetrical figures are in equilibrium (a principle used by Archimedes in proving the law of the lever, and also by Stevin in his well-known demonstration of the law of the inclined plane). However, we may inscribe in the tri-

<hr/>

[12] Commandino, *Liber de centro gravitatis solidorum*, fol. 40ᵛ–41ᵛ.
[13] Strong, *Procedures and Metaphysics*, pp. 91–113.

angle an infinite number of such parallelograms, for all of which the center of gravity will lie on AD. Moreover, the greater the number of parallelograms thus inscribed, the smaller will be the difference between the inscribed figure and the triangle ABC. If, now, the "weights" of the triangles ABD and ACD are not equal, they will have a certain fixed difference. But there can be no such difference, inasmuch as each of these triangles can be made to differ by less than this from the sums of the parallelograms inscribed within them, which are equal. Therefore the "weights" of ABD and ACD are equal, and hence the center of gravity of the triangle ABC lies on the median AD.[14]

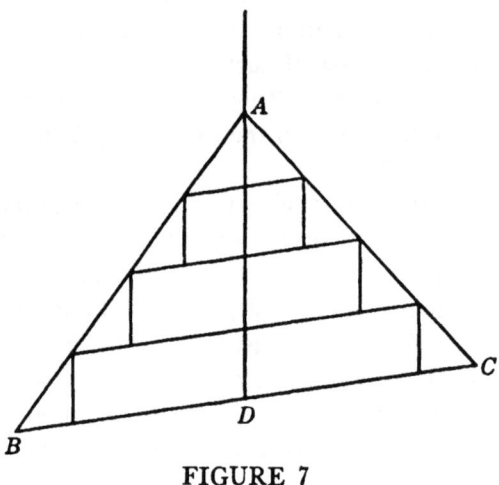

FIGURE 7

Exactly analogous demonstrations were given by Stevin of propositions on the centers of gravity of plane curvilinear figures, including the parabolic segment. These proofs given by Stevin indicate the direction in which the method of limits was developed as a positive concept. The Greek method of exhaustion had not boldly concluded, as did Stevin, that inasmuch as the difference could, by continued subdivision, be shown to be less than any given quantity, there could as a result be no difference. The Greeks felt constrained in every case to add the full reductio ad absurdum proof to show the equality. Stevin did not, of course, speak of the triangle as the limit of the sum of the

[14] Stevin, *Hypomnemata mathematica*, IV, 57–58; cf. also Bosmans, "Le Calcul infinitésimal chez Simon Stevin."

inscribed parallelograms; but it would require only slight changes in his method—largely in the nature of further arithmetization and the use of greater precision in terminology—to recognize in it our modern method of limits.

That Stevin regarded his approach to these problems as a significant modification of the method of Archimedes may be indicated in his proof of the proposition on the center of gravity of a conoidal segment —a proposition "the demonstration of which the ingenious and subtle mathematician Fredericus Commandinus gives in proposition 29 *de solidorum centrobaricis*, and which is arranged as follows in accordance

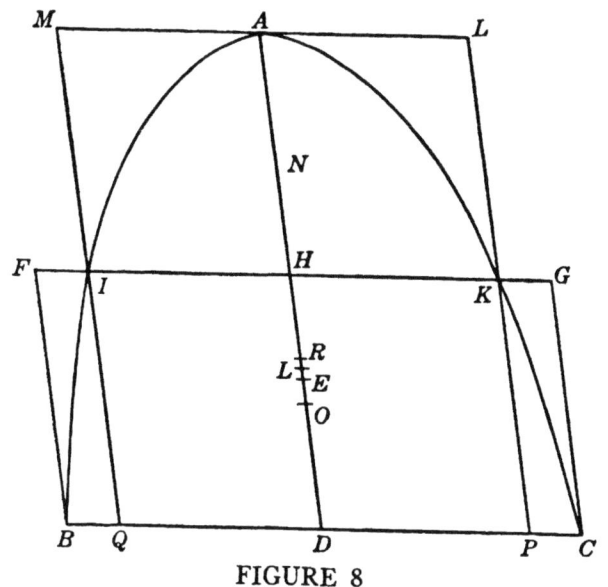

FIGURE 8

with our custom and method."[15] Circumscribe about the segment ABC two cylindrical segments $FGBC$ and $MLIK$ as illustrated (fig. 8). Now the centers of gravity of these cylindrical segments are, by the principle of the equilibrium of bilaterally symmetrical figures, at the mid-points, N and O, of their axes, AH and HD, respectively; and the center of gravity of the entire circumscribed figure is at R such that $NR = 2RO$. Letting E be the point such that $AE = 2ED$, it can be shown that $ER = \frac{1}{12}AD$. If, now, one similarly circumscribes about

[15] Stevin, *Hypomnemata mathematica*, II, 75–76.

the segment ABC four such cylindrical segments of equal height, the center of gravity of this circumscribed figure is found to lie above E at a point L such that $EL = \frac{1}{24}AD$. Successively doubling the number of these cylinders, the center of gravity of the circumscribed figure remains always above E and will differ from E by $\frac{1}{48}AD$, $\frac{1}{96}AD$, and so on. Thus the center of gravity descends, approaching E more and more closely. Modern mathematics would now conclude that E is the limit of the center of gravity of the circumscribed figure and therefore is the center of gravity of the conoidal segment. Stevin, however, reached this conclusion only after cautiously observing that similarly the center of gravity of the analogous inscribed figure ascended toward E in the same manner.

The demonstration given by Stevin, in the above proposition in terms of the sequence $\frac{1}{12}$, $\frac{1}{24}$, $\frac{1}{48}$, $\frac{1}{96}$, . . ., is comparable to that employed by Archimedes in his quadrature of parabola, in which the series $1 + \frac{1}{4} + \frac{1}{16} + \frac{1}{64} + \ldots$ figured. Both the Greek mathematician and the Flemish engineer, in their use of such sequences and series, stopped short of the limit concept. Neither thought of such a sequence or series as carried out to an infinite number of terms in the modern sense. Archimedes explicitly stopped with the nth term $\frac{1}{4^{n-1}}$ and added the term, $\frac{1}{3} \cdot \frac{1}{4^{n-1}}$, representing the remainder of the series; Stevin used the word infinite in the Peripatetic sense of potentiality only—the sequence could be continued as far as desired, and the error consequently made as small as one pleased. For this reason Archimedes had been constrained to complete his work through the demonstration by a reductio ad absurdum; Stevin, although he adduced no such formal argument, had recourse also to supplementary demonstrations, such as the inclusion, in the above proposition on the conoid, of a second sequence approaching the point from the other side. This hesitation on the part of Stevin to accept as sufficient the notion of the limit of an infinite sequence is apparent also in a proposition in his hydrostatics, which represented perhaps his nearest approach to the method of limits. Here he supplemented the "mathematical demonstration" of propositions, carried out as above, by a "demonstration by numbers," suggested, perhaps, by the then recent Italian work in algebra and mensuration, and encouraged by the neglect of geometry in favor of arithmetic found in the Netherlands during the sixteenth century.[16]

In supplementing the proof[17] that the average pressure on a vertical square wall of a vessel full of water corresponds to the pressure at its mid-point, he gave an "example with demonstration." He subdivided the wall into 4 horizontal strips and noted that the force on each is greater than 0, $\frac{1}{16}$, $\frac{2}{16}$, and $\frac{3}{16}$ units, and less than $\frac{1}{16}$, $\frac{2}{16}$, $\frac{3}{16}$, and $\frac{4}{16}$ units respectively; so that the total force is greater than $\frac{6}{16}$ and less than $\frac{10}{16}$. If the wall is subdivided into 10 horizontal strips, the force is found similarly to be greater than $\frac{45}{100}$ and smaller than $\frac{55}{100}$ units; on using 1,000 strips, it is determined as more than $\frac{499.500}{1.000.000}$ units, and less than $\frac{500.500}{1.000.000}$ units.[18] By increasing the number of strips, he then remarked, one may approach as closely as desired to the ratio one-half, thus proving that the force corresponds to that which would be observed if the wall were placed horizontally at a depth of half a unit.[19]

This "demonstration by numbers" would correspond exactly to that given in the calculus if Stevin had limited himself to one of his two sequences and had thought of the results given by his successive subdivisions of the wall as forming literally an infinite sequence with the limit $\frac{1}{2}$. Stevin, however, shared not only the Greek view with regard to infinity but also, although to a lesser extent than did most of his mathematical contemporaries, the classical apotheosis of geometry, so characteristic of ancient mathematics, which was to prevail in the seventeenth century. Even he considered his arithmetic proof, outlined above, as merely a mechanical illustration to be distinguished from the general mathematical demonstration.[20]

Perhaps the one tendency that did more than any other to conceal from mathematicians for almost two centuries the logical basis of the

---

[16] Cf. Struik, "Mathematics in the Netherlands during the First Half of the XVIth Century."

[17] Stevin, *Hypomnemata mathematica*, II, 121 ff.

[18] It will be recalled that Stevin had introduced his use of decimal fractions, in *De Thiende*, which had appeared in 1585, the year before the publication of his work in statics. See Sarton, "The First Explanation of Decimal Fractions and Measures (1585)."

[19] Stevin, *Hypomnemata mathematica*, II, 125–26; cf. also Bosmans, "Le Calcul chez Stevin," pp. 108–9.

[20] "Mathematicae & mechanicae demonstrationis a doctis annotatur differentia, neque injuria. Nam illa omnibus generalis est, & rationem cur ita sit penitus demonstrat, haec vero in subjecto duntaxat paradigmate numeris declarat." Stevin, *Hypomnemata mathematica*, II, 154.

calculus was the result of the attempt to make geometrical, rather than arithmetic, conceptions fundamental. This will be more true of Stevin's successors than it was of him. It must be borne clearly in mind, however, that although the logical basis of the calculus is arithmetic, the new analysis resulted largely from suggestions drawn from geometry.

The procedures substituted by Stevin for the method of exhaustion constituted a marked step toward the limit concept. The extent of his influence on contemporary thought, however, is difficult to determine. The work in statics which contains his anticipations of the calculus appeared in Flemish in 1586. It was included also in the Flemish, French, and Latin editions of his mathematical works published in 1605–8, and in a later French translation of his works by Girard in 1634.[21] With the exception of the last-mentioned edition, however, most of these were not easily accessible to mathematicians,[22] and, by the time of the French translation of 1634, there had already appeared in Italy and Germany a number of alternative methods which were destined to become much more widely known than those of Stevin. Nevertheless, the influence of the Flemish scientist is evident in the thought of a number of later mathematicians of the Low Countries.[23] Before we turn to these men, however, it may be well to indicate briefly the nature of the modifications of Archimedes' work which appeared in Italy and Germany shortly after Stevin had published his methods.

Luca Valerio likewise attempted, in his *De centro gravitatis solidorum* of 1604, a methodization of Archimedes' procedure which should obviate the need for the reductio ad absurdum and yet retain the necessary rigor of demonstration. His change was not so sweeping as that of Stevin, and his view is not so closely related to the modern. He merely tried to substitute for the method of Archimedes a few general theorems which could be cited instead of carrying through the details of proof in each and every case. No attempt had been made in Greek geometry to establish such propositions which might simply be quoted in particular cases in lieu of the double reductio ad absurdum demonstration.[24]

[21] See the article by Sarton, "Simon Stevin of Bruges (1548–1620)"; and that by Bosmans, "Simon Stevin," for biographies of Stevin, analyses of his work, and extensive bibliographical references.

[22] Bosmans, "Simon Stevin," p. 889.

[23] Bosmans, "Sur quelques exemples de la méthode des limites chez Simon Stevin."

[24] *The Works of Archimedes*, Introduction, p. cxliii.

Valerio professed that it was the appearance of Commandino's work on centers of gravity which had encouraged him,[25] as it had Stevin,[26] to attempt such a modification of the method of Archimedes. The significant generalization is found in the proposition that, given any figure in which the distance between points on opposite sides of a diameter vanishes, parallelograms can be inscribed and circumscribed in such a way that the excess of the circumscribed figure over that inscribed is less than any given area.[27] This is proved by observing that the excess is in each case equal to the area of the parallelogram BF (fig. 9) and that for suitably chosen approximating figures this "is less than a given area."[28] Then Valerio assumed without proof that if the difference be-

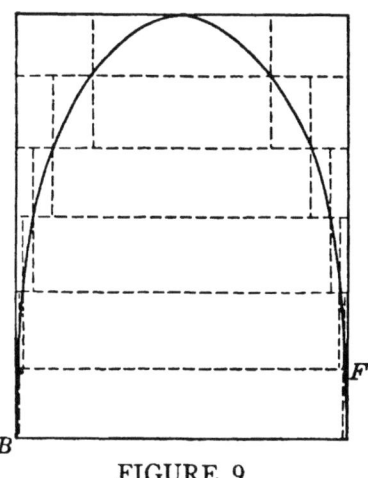

FIGURE 9

tween the inscribed and the circumscribed figures is smaller than any given area, this will be true also of the difference between the curve and either of these figures. This geometrical reasoning is strikingly similar to that presented in many present-day elementary textbooks on the calculus. Valerio did not, however, regard the area of the curve as necessarily defined by the limit of the area of either the inscribed or the circumscribed figure, as the number of such parallelograms becomes infinite. This is a sophisticated arithmetical conception which

[25] Valerio, *De centro gravitatis solidorum libri tres*, p. 1.

[26] It is doubtful whether Valerio knew of the work of Stevin. See Bosmans, "Les Démonstrations par l'analyse infinitésimale chez Luc Valerio," p. 211.

[27] Valerio, *De centro gravitatis*, p. 13.

[28] "Sed parallelogrammum BF est minus superficie proposita." *Ibid.*, p. 14.

was not established until two centuries later. Nevertheless, Valerio was in a sense anticipating the limit concept, in geometrical form, to the extent of indicating the necessary condition for the existence of such a limit—viz., that the difference in these areas can be made less than any specified area.

Having generalized, by the above demonstration, the method of inscribing areas, Valerio then proceeded to state general propositions which should replace, at least in so far as ratios were concerned, the argument by a reductio ad absurdum as found in the method of exhaustion. The intention of these may be stated in the following form: If four magnitudes, A, B, C, D, are given, and if two others, G and H, can be found which are at the same time greater or smaller than A and B by a magnitude less than any given magnitude, and if the ratio $\dfrac{G}{H}$ is at the same time greater or smaller than the ratios $\dfrac{A}{B}$ and $\dfrac{C}{D}$, then $\dfrac{A}{B}$ $= \dfrac{C}{D}$.[29] This proposition is significant not only as a methodization of Archimedes' procedure, but also as a vague striving toward the idea which is now expressed concisely by saying that the limit of a ratio of two variables is equal to the ratio of the limits of these variables. This latter concept is dependent, however, on considerations of the infinite —on speculations in which Valerio did not indulge, but which interested many of his contemporaries and successors, particularly those who combined mathematical investigations with theological interests.

Johann Kepler had been educated with the intention of entering the Lutheran ministry, but had been forced to turn later to the teaching of mathematics as a means of earning his livelihood. This may account in part for the fact that, although he owed as much to Archimedes as had Stevin, the nature of his work is so markedly different from that of the engineer of Bruges. There is in the thought of Kepler a deep strain of mysticism which was lacking in the attitude of Stevin, of whom Kepler may have known.[30] However, this speculative tendency on the part of

[29] *De centro gravitatis*, p. 69; see also Wallner, "Grenzbegriffes," p. 251.

[30] Bosmans ("Les Démonstrations sur l'analyse infinitésimale chez Luc Valerio," p. 211) holds that Kepler was evidently familiar with Stevin's work; and Hoppe ("Zur Geschichte der Infinitesimalrechnung," p. 160) that he may have known of it; Wieleitner ("Das Fortleben") believes that such an assumption is not justified.

Kepler is perhaps with more justice to be ascribed to the Platonic-Pythagorean influence, which had been strong in Europe in the previous century,[31] and which was so evident in the thought of Cusa, to whom Kepler owed at least part of his inspiration.[32] The belief that the universe was an ordered mathematical harmony, so strongly shown in Kepler's *Mysterium cosmographicum*, was combined with Platonic and Scholastic speculations on the nature of the infinite, giving him a modification of Archimedes' mensurational work which was to be a powerful influence in shaping the development of the calculus.

The ancient Greek philosophers, in their search for unity in this universe of perplexing multiplicity, had failed for two reasons to bridge the gap between the curvilinear and the rectilinear: first, they banned the infinite from geometry; and second, they hesitated, following the discovery of the irrational, to pursue further the Pythagorean association of numerical considerations with geometrical configurations. The pious enthusiasm of Kepler, however, saw in this impasse but one more evidence of the handiwork of the Creator, who had established all things in harmony. God wished quantity to exist so that the comparison between a curve and a straight line might be made. This fact was made clear to him by the "divine Cusanus" and by others who had regarded the forms of the curve and the straight line as complementary, daring to compare the curve to God and the line to his creatures. "For this reason those who have tried to relate the Creator to his handiwork and God to man and divine judgments to human represent an occupation which is by no means more useful than that of those who seek to compare the circle with the square."[33] Under such inspiration, guided as well by the speculations of Cusa and Giordano Bruno on the infinite in cosmology,[34] and fortified by the knowledge "that nature teaches geometry by instinct alone, even without ratiocination,"[35] Kepler was led to develop his modification of the procedures of Archimedes.

The task of writing a complete treatise on volumetric determinations seems to have been suggested to Kepler by the prosaic problem of determining the best proportions for a wine cask. The result was

[31] Burtt, *Metaphysical Foundations*, pp. 44–52; Strong, *Procedures and Metaphysics*, pp. 164 ff.
[32] Cf. Kepler, *Opera omnia*, I, 122; II, 490, 509, 595.
[33] *Ibid.*, I, 122.   [34] *Ibid.*, II, 509.   [35] *Ibid.*, IV, 612.

the *Nova stereometria*, which appeared in 1615.[36] This contains three parts, of which the first is on Archimedean stereometry, together with a supplement containing some ninety-two solids not treated by Archimedes. The second part is on the measurement of Austrian wine barrels, and the third on applications of the whole.

Kepler opened his work on curvilinear mensuration with the simple problem of determining the area of the circle. In this he abandoned the classical Archimedean procedures. He did not substitute for these the limiting consideration proposed by Stevin and Valerio, but had recourse instead to the less rigorous but more suggestive approach of Nicholas of Cusa. Like Stifel and Viète, he regarded the circle as a regular polygon with an infinite number of sides, and its area he therefore looked upon as made up of infinitesimal triangles of which the sides of the polygon were the bases and the center of the circle the vertex. The totality of these was then given by half the product of the perimeter and the apothem (or radius).[37]

Kepler did not limit himself to the simple proposition above, but with skill and imagination applied this same method to a wide variety of problems. By looking upon the sphere as composed of an infinite number of infinitesimal cones whose vertices were the center of the sphere and whose bases made up the surface, he showed that the volume is one-third the product of the radius and the surface area.[38] The cone and cylinder he regarded variously: as made up of an infinite number of infinitely thin circular laminae (as had Democritus two thousand years earlier), as composed of infinitesimal wedge-shaped segments radiating from the axis, or as the sum of other types of vertical or oblique sections.[39] The volumes of these he computed by the application of such views. In a similar manner he rotated a circle about a line and calculated, by infinitesimal methods, the volume of the anchor ring, or tore, thus generated.[40] This determination was equivalent to an application, for a special case, of the classical theorem of Pappus, later called Guldin's rule. Kepler then extended his work to solids not treated by the ancients. The areas of the segments cut from a circle by a chord he rotated about this chord, obtaining solids

[36] Followed, about a year later, by a popular German edition. The Latin appears, with notes, in Volume IV, *Opera omnia;* the German in Volume V.
[37] *Opera omnia,* IV, 557–58.
[38] *Ibid.,* IV, 563.   [39] *Ibid.,* IV, 564, 568, 576 ff.   [40] *Ibid.,* IV, 575–76, 582–83.

which he designated characteristically as apple or citron-shaped, according as the generating segment was greater or less than a semi-circle.[41] The volumes of these and other solids he likewise calculated by his infinitesimal methods. To Willebrord Snell, the editor of Stevin's works, Kepler proposed, as a challenge, the determination of the solids obtained similarly by the rotation of segments of conic sections.[42] This problem was significant in the later work of the Italian mathematician Cavalieri.

Some of Kepler's summations are remarkable anticipations of results found later in the integral calculus.[43] In his well-known *Astronomia nova* of 1609, for example, there is a computation[44] resembling that which is expressed in modern notation as $\int_0^\theta \sin \theta \, d\theta = 1 - \cos \theta$. Other calculations in this work correspond to approximations to elliptic integrals,[45] in one of which $\pi(a + b)$ is given as the approximate length of the ellipse with the semi-axes a and b.[46] Kepler, however, was far from clear on the point of the basic conceptions involved, with the result that his work is not free from errors.[47] He generally spoke of surfaces and volumes as made up of infinitesimal elements of the same dimension, but occasionally he lapsed into the language of indivisibles, which his successor Cavalieri was to find so congenial. In one place he spoke of the cone as though composed of circles,[48] and in the work by which he arrived in his astronomy at his famous second law he regarded the sector of an ellipse as the sum of its radius vectors.[49]

Kepler appears not to have distinguished clearly between proofs by means of the method of exhaustion, by ideas of limits, by infinitesimal elements, or by indivisibles. The conceptions which he held in his dem-

---

[41] Wolf (*History of Science, Technology, and Philosophy in the Sixteenth and Seventeenth Centuries*, pp. 204–5) has correctly pointed out that the word *citrium*, which Kepler used in this connection, is properly translated as "gourd"; but he has failed to add that Kepler himself, in his German edition, translated it as "citron." *Opera omnia*, V, 526.

[42] *Ibid.*, IV, 601, 656.

[43] See Struik, "Kepler as a Mathematician," in *Johann Kepler, 1571–1630*. This volume also contains an excellent bibliography of Kepler's works, by F. E. Brasch.

[44] See *Opera omnia*, III, 390; cf. also Gunther, "Über ein merkwürdige Beziehung zwischen Pappus und Kepler"; Eneström, "Über die angebliche Integration einer trigonometrischen Funktion bei Kepler."

[45] Cf. Struik, "Kepler as a Mathematician," p. 48; Zeuthen, *Geschichte der Mathematik im XVI. und XVII. Jahrhundert*, pp. 254–55.

[46] *Opera omnia*, III, 401.   [47] Moritz Cantor, *Vorlesungen*, II, 753.

[48] "Nam conus est hic veluti circulus corporatus." *Opera omnia*, IV, 568.

[49] *Opera omnia*, III, 402–3.

onstrations are a far cry from the notions held by the ancient geometers. The Greek thinkers saw no way of bridging the gap between the rectilinear and the curvilinear which would at the same time satisfy their strict demands of mathematical rigor and appeal to the clear evidence of sensory experience. Fortified by the scholastic disputations on the categorical infinite and by Platonic mathematical speculations, Kepler followed Nicholas of Cusa in resorting to a vague "bridge of continuity" which finds no essential difference between a polygon and a circle, between an ellipse and a parabola, between the finite and the infinite, between an infinitesimal area and a line.[50] This striving for an expression of the idea of continuity constantly reappears throughout the period of some fifty years preceding the formulation of the methods of the calculus. Leibniz himself, like Kepler, frequently fell back upon his so-called law of continuity when called upon to justify the differential calculus; and Newton concealed his use of the notion of continuity under a concept which was empirically more satisfying, though equally undefined—that of instantaneous velocity, or fluxion.

Kepler's *Doliometria*, or *Stereometria doliorum*, exerted such a strong influence in the infinitesimal considerations which followed its appearance, and which culminated a half century later in the work of Newton, that it has been called, with perhaps pardonable exaggeration, the source of the inspiration for all later cubatures.[51] Before passing on to these anticipations of the integral calculus, there should be cited a contribution which Kepler made to the thought leading to the differential calculus. The subject of the measurement of wine casks had led Kepler to the problem of determining the best proportions for these.[52] This brought him to the consideration of a number of problems on maxima and minima. In the *Doliometria* he showed, among other things, that of all right parallelepipeds inscribed in a sphere and having square bases, the cube is the largest;[53] and that of all right circular cylinders having the same diagonal, that one is greatest which has the diameter and altitude in the ratio of $\sqrt{2}$ to 1.[54]

These results were obtained by making up tables in which were

[50] Cf. Taylor, "The Geometry of Kepler and Newton."

[51] Moritz Cantor, *Vorlesungen*, II, 750.

[52] Kepler discovered, incidentally, that the Austrian barrels approximated very closely the desired proportions.

[53] *Opera omnia*, IV, 607–9.   [54] *Ibid.*, IV, 610–12.

listed the volumes for given sets of values of the dimensions, and from these selecting the best proportions. An inspection of such tables showed him an interesting fact. He remarked that as the maximum volume was approached, the change in the volume for a given change in the dimensions became smaller. Oresme, several centuries earlier, had made a similar observation, but had expressed it differently. Oresme had noticed that for a form which was represented graphically by a semicircle, the rate of change was least at the maximum point. This thought appeared again, in the seventeenth century, in the methods of the French mathematician Fermat. Whether Fermat was influenced in this direction by Kepler or Oresme is problematical; but a comparison of the distinctly different points of view which the latter men represent will aid later in understanding Fermat's approach. Kepler had made his remark upon the basis of numerical considerations. He was, moreover, more particularly concerned with static considerations as found in Greek geometry and in methods of indivisibles. He consequently expressed himself in terms of increments and decrements near the maximum point. On the other hand, the medieval problem of the latitude of forms and the graphical representation of continuous variability had led Oresme to state the conclusion in terms of the rate of change. The latter view has been made fundamental in mathematics through the concept of the derivative; but it is Kepler's mode of expression which appeared in the work of Fermat. Although the Scholastic views on variation played a significant rôle in the anticipations of the calculus, the static approach of Kepler predominated. Increments and decrements, rather than rates of change, were the fundamental elements in the work leading to that of Leibniz, and played a larger part in the calculus of Newton than is usually recognized. The differential became the primary notion and it was not effectively displaced as such until Cauchy, in the nineteenth century, made the derivative the basic concept.

Twenty years after the publication of the *Stereometria doliorum* of Kepler there appeared in Italy a work which rivaled it in popularity. So famous did the *Geometria indivisibilibus* of Bonaventura Cavalieri become that it has been maintained, with some justice, that the new analysis took its rise from the appearance, in 1635, of this book.[55] To

---

[55] Leibniz, *The Early Mathematical Manuscripts*, trans. by Child, p. 196.

what extent this is indebted to the earlier work of Kepler is difficult to determine. Cavalieri emphatically denied any inspiration from Kepler's method, other than "the names of a few solids and the admiration which frequently sets philosophers to reflecting."[56] It is not improbable, however, that the influence of Kepler upon Cavalieri may have resulted indirectly from the correspondence of both of these men with Galileo.[57] A work upon indivisibles which Galileo planned to write never appeared, but his views upon the subject are clearly brought out in his classic treatise, the *Two New Sciences*, which was published three years after the *Geometria* of Cavalieri. Galileo's opinions resemble strongly those expressed by his pupil, Cavalieri, and may well have been the source of the latter's inspiration.[58] It may be well, therefore, to examine at this point the views of Galileo, the teacher.

The forces molding the thought of Galileo and Cavalieri did not differ greatly from those which had shaped the ideas of Kepler. These men had mastered the Greek geometric methods, but they all betray the effects of Scholastic speculations and the Platonic view of mathematics which had exerted such a strong influence since the time of Nicholas of Cusa. Both Galileo and Cavalieri were probably acquainted with the modifications of the method of Archimedes by Valerio. Galileo referred to Valerio several times, in the *Two New Sciences*, as the great geometer and as the new Archimedes of his age.[59] Galileo gave in this work also an Archimedean demonstration of the quadrature of the parabola[60] and included in an appendix some work on centers of gravity in the manner of Commandino and Valerio.[61] Nevertheless both he and Cavalieri appear to be more significantly indebted, in attitude and method, to the later medieval speculations on motion, indivisibles, the infinite, and the continuum.[62]

The influence of Scholastic thought is clearly evidenced in the case of Galileo by his early writings. In these he considered, among other things, the Peripatetic doctrines of matter and form and of causes and qualities, and the Scholastic questions of intension and remission and of action and reaction.[63] On the latter he referred specifically to works

[56] *Exercitationes geometricae sex*, pp. 237–38.

[57] Paul Tannery, *Notions historiques*, p. 341; Moritz Cantor, *Vorlesungen*, II, 774–75.

[58] Paul Tannery, *ibid.*, p. 341.   [59] *Le opere di Galileo Galilei*, VIII, 76, 184, 313.

[60] *Opere*, VIII, 181 ff.   [61] *Opere*, VIII, 313; I, 187–208.

[62] Wallner, "Die Wandlungen des Indivisibiliensbegriffs von Cavalieri bis Wallis."

[63] *Opere*, I, "Iuvenalia," 111 ff., 119 ff., 126 ff.

by Calculator and Hentisbery.[64] In dynamics he made fundamental the doctrine of impetus, which had been suggested by the Scholastic philosopher Buridan in the fourteenth century and which was certainly familiar to Nicholas of Cusa, Leonardo da Vinci, and others of the fifteenth and sixteenth centuries.[65] Whether Cusa's influence on Galileo was significant (as it was on Kepler and probably also on Giordano Bruno) is difficult to determine;[66] but that the Scholastic discussions on motion were known to him will be very strongly indicated by an examination of one of the propositions in his *Two New Sciences.*

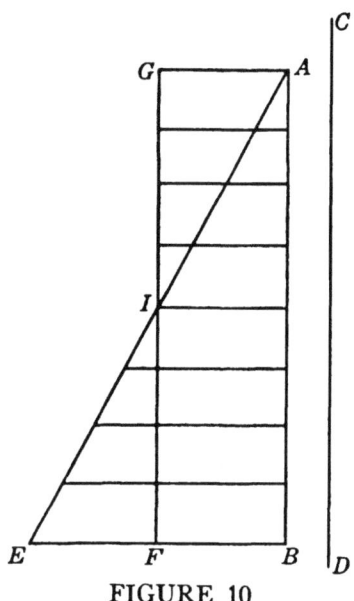

FIGURE 10

It will be recalled that Calculator and Hentisbery had demonstrated dialectically that the average velocity of a body moving with uniform acceleration is given by its velocity at the mid-point of the time interval. Oresme had given a geometrical demonstration of this proposition, in which he indicated that the area under the line representing the velocity was the measure of the distance. Galileo's demonstration of this proposition resembles strikingly that of Oresme. Let AB (fig. 10) represent the time in which the space CD is traversed by a body which

[64] *Ibid.*, I, 172.   [65] Duhem, *Études sur Léonard de Vinci*, Vol. III, *passim.*
[66] Goldbeck, "Galileis Atomistik und ihre Quellen."

starts from rest and is uniformly accelerated. Let the final speed be represented by *EB*. Then the lines drawn parallel to *EB* will represent the speeds of the body. It appears, then, that they may be interpreted also as the moments, or infinitesimal increments, in the distance covered by the moving body. Then the movements of the uniformly accelerated motion may be represented by the parallels of triangle *AEB*, whereas the parallels of the rectangle *ABFG* represent the corresponding moments of a body moving uniformly. But the sum of all the parallels contained in the quadrilateral *ABFG* is equal to the sum of those contained in the triangle *AEB*. Hence it is clear that the distances covered by the two bodies are equal, inasmuch as the triangle and the rectangle are equal in area if *I* is the midpoint of *FG*.[67]

Not only did Galileo reproduce with striking fidelity the diagram and argument of Oresme as outlined above, but he extended a remark made in this connection by Calculator and Hentisbery. Galileo could have read in the works of these men, or in commentaries upon them, that in uniformly accelerated motion the space covered in the second half of the time is three times that covered in the first half.[68] This observation Galileo extended to show that if we subdivide the time interval further, the distances covered, in each of these, will be in the ratio 1, 3, 5, 7, . . . [69] This is, of course, equivalent to the result expressed by the formula $s = \frac{1}{2}gt^2$, and is implied by the earlier work of the Scholastics.

The difference between Galileo's demonstration and that of Oresme is largely one of completeness. Oresme had been satisfied to say merely that inasmuch as the triangles *EIF* and *AIG* are equal, the distances must be the same, thus implying the infinitesimal considerations necessary to demonstrate that the areas represent distances. The geometrical demonstrations of Oresme and Galileo are based upon the supposition that the area under a velocity-time curve represents the distance covered.[70] Since neither one possessed the limit concept, each resorted, explicitly or implicitly, to infinitesimal considerations. Galileo expressed this view when he said that the moments, or small increments in the distance, were represented by the lines of the triangle and the

[67] *Opere*, VIII, 208–9.

[68] Duhem, *Études sur Léonard de Vinci*, III, 480, 513.   [69] *Opere*, VIII, 210 ff.

[70] Paul Tannery, "Notions historiques," pp. 338–39, mistakenly attributes this idea to Galileo, rather than Oresme.

rectangle, and that these latter geometrical figures were in actuality made up of these lines. Galileo did not make clear how the transition from the lines as velocities to the same lines as moments is to be made. Oresme likewise had begged the question when he had represented instantaneous velocities by lines and yet had maintained that all velocities act through a time. Galileo and Oresme patently employed the uncritical mathematical atomism which has appeared among mathematicians of all ages—in Democritus, Plato, Nicholas of Cusa, Kepler, and many others. It has been suggested that Galileo was influenced in his use of indivisibles by his strong Pythagorean and Platonic approach to science,[71] or by the revival in his day of interest in the atomism of Heron of Alexandria, in which Galileo failed to distinguish clearly between physical indivisibles and mathematical infinitesimals.[72] He may equally well have been led to his views by the Scholastic discussions of infinitesimals.

At any rate, in the dialogue of the first day, in the *Two New Sciences*, Galileo entered into an extended discussion of the subjects which had been so popular in the medieval period: the infinite, the infinitesimal, and the nature of the continuum. Galileo clearly admitted the possibility of the Scholastic categorematic infinity, but because of the numerous paradoxes to which this appeared to lead, Galileo concluded that "infinity and indivisibility are in their very nature incomprehensible to us."[73] Nevertheless, he made at least one trenchant observation upon this subject. It will be recalled that Calculator had remarked that there can be no ratio between an infinite magnitude and a finite one. Galileo asserted more generally "that the attributes 'larger,' 'smaller,' and 'equal' have no place either in comparing infinite quantities with each other or in comparing infinite with finite quantities."[74] In justifying this conclusion, Galileo indicated a significant shift of emphasis, for instead of considering the infinite from the point of view of magnitude, as had Aristotle and many medieval scholars, he focused attention, as had Plato, upon the infinite as multiplicity or aggregation. In this connection he indicated that the

[71] Wiener, "The Tradition behind Galileo's Methodology"; cf. also Brunschvicg, *Les Étapes de la philosophie mathématique*, p. 70.

[72] See Schmidt, "Heron von Alexandria im 17. Jahrhundert"; also other articles on Heron in the same volume.

[73] *Opere*, VIII, 76 ff.   [74] *Opere*, VIII, 82 ff.

infinite class of all positive integers could be put into a one-to-one correspondence with a part of this class—for example, with all the perfect squares.[75] This characteristic of infinite sets was rediscovered in the nineteenth century by Bolzano and later in that century was made fundamental in the establishment of the calculus upon a rigorously developed theory of infinite assemblages.[76]

In spite of the trenchant observations which Galileo made on the subject of the infinite, he felt strongly the inability of intuition to grasp this notion. He went so far as to conjecture that there might be a third possible type of aggregation between the finite and the infinite. This seems to have been suggested to him by the difficulties of making precise our vague ideas on the continuum. He maintained, contrary to Bradwardine, that continuous magnitudes are made up of indivisibles. However, inasmuch as the number of parts is infinite, the aggregation of these is not one resembling a very fine powder but rather a sort of merging of parts into unity, as in the case of fluids.[77] This analogy is a beautiful illustration of the effort which men made to picture in some way the transition from the finite to the infinite.

Galileo sought also to clarify to some extent the paradoxes on motion. This he did by regarding rest as an infinite slowness, thus resorting again to the vague feeling for the continuous to which Cusa and Kepler had sought to give expression. In enlarging upon this idea, Galileo applied to a falling body the argument which Zeno had given in the *dichotomy* and then answered this by an appeal to intuition in reversing the descent. In ascent, the body passes through an infinite number of grades of slowness, finally to come to rest. The beginning of motion is precisely the same, except that the order is reversed.[78] This argument constitutes a recognition—found also in Aristotle—of the similarity of the difficulties in the *dichotomy* and the *Achilles*. It represents, as well, an attempt to clarify the sense in which an infinite series may be said to have a sum. Later Newton in his calculus, when he spoke of his instantaneous velocities as *prime* or *ultimate* ratios, made an analogous appeal to the fact that the beginning and the end of motion are to be similarly conceived. Galileo appears not to have realized, as Newton did somewhat vaguely, that only in terms of the

[75] *Ibid.*, VIII, 78.
[76] Kasner, "Galileo and the Modern Concept of Infinity," pp. 499–501.
[77] *Opere*, VIII, 76 ff.   [78] *Opere*, VIII, 199–200.

limit concept can precise meaning be given to either the sum of an infinite series or to a first or last ratio.

Whatever the influences which shaped the thought of Galileo, those felt by his friend and student, Cavalieri, are not likely to have been greatly different. Although familiar with the views of Valerio and perhaps also with those of Stevin,[79] Cavalieri did not develop the limit idea which these men had adumbrated. Instead he had resort to the less subtle notion of the indivisible which had been adopted by Galileo.[80] However, whereas Galileo had employed this in physical explanation, Cavalieri made it the basis of a geometrical method of demonstration which achieved remarkable popularity. This method had been developed by Cavalieri as early as 1626, for in that year he wrote to Galileo, saying that he was going to publish a book on the subject.[81] This work appeared in 1635, as the *Geometria indivisibilibus continuorum nova quadam ratione promota*,[82] and the method it presented was further developed a dozen years later in the *Exercitationes geometricae sex*.

Cavalieri at no point in his books explained precisely what he understood by the word *indivisible*, which he employed to characterize the infinitesimal elements used in his method. He spoke of these in much the same manner as had Galileo in referring to the parallel lines representing velocities or moments as making up the triangle and the quadrilateral. Cavalieri conceived of a surface as made up of an indefinite number of equidistant parallel lines and of a solid as composed of parallel equidistant planes,[83] these elements being designated the indivisibles of the surface and of the volume respectively. Although he recognized that the number of these must be indefinitely great, he did not follow his master Galileo in speculations as to the nature of the infinite. The attitude of Cavalieri toward infinity was one of agnosticism.[84] He did not share the Aristotelian view of infinity as indicating a potentiality only—a conception which, in conjunction with the work of Stevin and Valerio, pointed toward the method of limits. On the other hand, Cavalieri did not join Nicholas of Cusa and Kepler in re-

[79] Bosmans, "Les Démonstrations par l'analyse infinitésimale," p. 211.
[80] Marie, *Histoire des sciences mathématiques et physiques*, III, 134.
[81] Moritz Cantor, *Vorlesungen*, II, 759.
[82] I have used the later edition, Bononiae, 1653.
[83] *Exercitationes geometricae sex*, p. 3.
[84] Cf. Brunschvicg, *Les Étapes de la philosophie mathématique*, p. 166.

garding the infinite as possessing a metaphysical significance. It was employed by him solely as an auxiliary notion, comparable to the "sophistic" quantities of Cardan. Inasmuch as it did not appear in the conclusion, its nature need not be made clear. That the infinite did not enter explicitly into the arguments of Cavalieri was due to the fact that at every stage he centered attention upon the correspondence between the indivisibles of two configurations, rather than upon the totality of indivisibles within a single area or volume. The proposition still known in solid-geometry textbooks as *Cavalieri's Theorem* is characteristic of his approach: If two solids have equal altitudes, and if sections made by planes parallel to the bases and at equal distances from them are always in a given ratio, then the volumes of the solids are also in this ratio.[85]

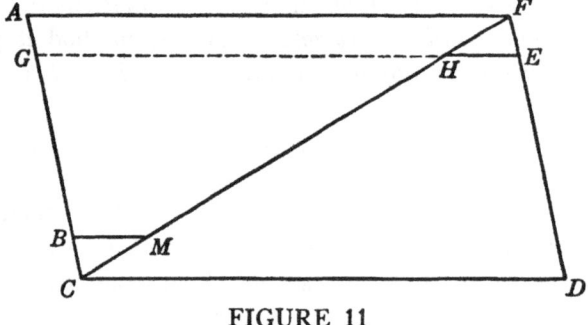

FIGURE 11

Typical of the propositions in the method of indivisibles, and of far-reaching significance in later developments, were a number of theorems on the lines of a parallelogram and those of its constituent triangles. One of these propositions consisted in an extended demonstration that if a parallelogram *AD* (fig. 11) is divided by the diagonal *CF* into two triangles, *ACF* and *DCF*, then the parallelogram is double either triangle.[86] This is proved by showing that if one lay off *EF* = *CB* and draw *HE* and *BM* parallel to *CD*, then the lines *HE* and *BM* are equal. Therefore all the lines of triangle *ACF* taken together are equal to all those of *CFD*; consequently triangles *ACF* and *CDF* are equal, and

[85] *Geometria indivisibilibus*, pp. 113–15; *Exercitationes geometricae sex*, pp. 4–5. Cf. also Evans, "Cavalieri's Theorem in His Own Words."
[86] *Geometria indivisibilibus*, Proposition XIX, pp. 146–47; *Exercitationes geometricae sex*, Proposition XIX, pp. 35–36.

the sum of the lines of the parallelogram AD is double the sum of the lines of either triangle.

From here Cavalieri went on to prove by a similar, but considerably more involved, argument that the sum of the squares of the lines in the parallelogram is three times the sum of the squares of the lines in each of the constituent triangles.[87] Using this latter proposition, he then easily demonstrated, among other things, that the volume of a cone is $\frac{1}{3}$ that of the circumscribed cylinder, and that the area of a parabolic segment is $\frac{2}{3}$ the area of the circumscribed rectangle.[88] These results were of course known to Archimedes, but a problem which Kepler had proposed some years before now led Cavalieri to a use of indivisibles bolder than that which is found in the *Geometria*, and to a new result of significance.[89]

Kepler had, in his *Stereometria*, challenged geometers to find the volume of the solid obtained by rotating a segment of a parabola about its chord.[90] Cavalieri determined this volume by basing the problem on the discovery that the sum of the fourth powers of the lines of a parallelogram is 5 times the sum of the fourth powers of the lines of one of the constituent triangles. Then he recalled that in his *Geometria indivisibilibus* he had found the ratios of the lines to be 2 to 1 and the ratio of the squares of the lines to be 3 to 1. In order not to leave a gap in his results on the ratios of the powers of the lines of a parallelogram and of the triangle, Cavalieri sought the ratio for the sums of the cubes of the lines, and found this to be as 4 to 1. He then concluded by analogy that for the fifth powers it would be 6 to 1, for sixth powers 7 to 1, and so on, the sum of the nth powers of the lines of the parallelogram being to the sum of the nth powers of the lines of the triangle as $n + 1$ to 1.[91]

The method of Cavalieri here employed was based upon several lemmas which are equivalent to special cases of the binomial theorem.[92] For example, to prove that the sum of the cubes of the lines of a paral-

---

[87] *Geometria indivisibilibus*, Proposition XXIV, pp. 159–60; *Exercitationes geometricae sex*, Proposition XXIV, pp. 50–51.

[88] *Geometria indivisibilibus*, pp. 185, 285–86; *Exercitationes geometricae sex*, pp. 78 ff.

[89] For a summary of this work, see Bosmans, "Un Chapitre de l'œuvre de Cavalieri."

[90] *Opera omnia*, IV, 601.

[91] *Exercitationes geometricae sex*, pp. 290–91. For a statement of the manner in which he was led to this result, see pp. 243–44.

[92] *Exercitationes geometricae sex*, p. 267.

lelogram is four times the sum of the cubes of the lines of one of the constituent triangles, he began with $(a + b)^3 = a^3 + 3a^2b + 3ab^2 + b^3$. Then he proceeded somewhat as follows: setting $AF = c$, $GH = a$, $HE = b$, in figure 11, we have $\Sigma c^3 = \Sigma a^3 + 3\Sigma a^2b + 3\Sigma ab^2 + \Sigma b^3$, where the sums are taken over the lines of the parallelogram and triangles. By symmetry this can be written $\Sigma c^3 = 2\Sigma a^3 + 6\Sigma a^2b$. Now $\Sigma c^3 = c\Sigma c^2 = c\Sigma (a + b)^2 = c\Sigma a^2 + 2c\Sigma ab + c\Sigma b^2$. But in an earlier proposition on lines Cavalieri had shown that $\Sigma a^2 = \Sigma b^2 = \frac{1}{3}\Sigma c^2$. This gives us $\Sigma c^3 = \frac{2}{3}c\Sigma c^2 + 2c\Sigma ab$

$$= \tfrac{2}{3}\Sigma c^3 + 2(a + b)\Sigma ab = \tfrac{2}{3}\Sigma c^3 + 2\Sigma a^2b + 2\Sigma b^2a$$
$$= \tfrac{2}{3}\Sigma c^3 + 4\Sigma a^2b$$

or $\Sigma a^2b = \frac{1}{12}\Sigma c^3$. Substituting this in the equation above, we obtain $\Sigma c^3 = 2\Sigma a^3 + \frac{1}{2}\Sigma c^3$, or $\Sigma c^3 = 4\Sigma a^3$, and the proposition is proved.[93]

Cavalieri realized that this method can be generalized for all values of n,[94] but he gave complete demonstrations only up to and including $n = 4$. For higher values he gave only some "cossic" indications, which had been given him by Beaugrand.[95] These may well have been derived from the contemporaneous work of Fermat,[96] whose methods we shall consider later.

The results of Cavalieri outlined above are in a general sense equivalent to what we would now express by the notation $\int_0^a x^n dx = \dfrac{a^{n+1}}{n+1}$. We have seen that Archimedes, through the use of series in connection with his stereometric work, had recognized the truth of this proposition for $n = 1$ and $n = 2$. He may have known of it for $n = 3$ also, and the Arabs proved it for $n = 4$ as well.[97] Cavalieri's work, although based upon somewhat different views, was a generalization of these results of Archimedes and the Arabs. It will be seen later that in this respect Cavalieri had been anticipated by several contemporary mathematicians. Nevertheless, his statement, which appeared in 1639,[98] represents the first publication of this theorem, which played a signifi-

[93] *Ibid.*, pp. 273–74.   [94] "Et sic in infinitum." *Ibid.*, p. 268.

[95] *Ibid.*, pp. 286–89.   [96] See Fermat, *Œuvres*, Supplement, p. 144.

[97] See Paul Tannery, "Sur le sommation des cubes entiers dans l'antiquité," and Ibn al-Haitham (Suter), "Die Abhandlung über die Ausmessung des Paraboloids."

[98] *Centuria di varii problemi*, pp. 523–26. Simon ("Zur Geschichte und Philosophie der Differentialrechnung," p. 118) has said—very probably through some error—that Cavalieri found this result in 1615.

cant rôle in the development of infinitesimal methods during the period from 1636 to 1655. Within this interval, the mathematicians Torricelli, Roberval, Pascal, Fermat, and Wallis arrived, all more or less independently and by varying methods, at this fundamental result and extended it, as well, to include negative, rational fractional, and even irrational values of n. It was perhaps the first theorem in infinitesimal analysis to point toward the possibility of a more general algebraic rule of procedure, such as that which, formulated a generation later by Newton and Leibniz, became basic in the integral calculus. Cavalieri himself had no vision of such a new analysis; neither he nor Galileo appears to have been seriously interested in algebra, either as a manner of expression or as a form of demonstration. The proposition remained for Cavalieri a geometrical theorem concerning the ratio subsisting between a sum of powers of the lines of a parallelogram and that of one of the constituent triangles. Furthermore, he did not express any clear conception—such as is basic in our idea of the definite integral—of his ratios as limits of sums of infinitesimal parallelograms. Cavalieri never did make clear his interpretation of the indivisible and in this respect laid his method open to attack.

Cavalieri's use of indivisibles in his *Geometria* had been criticized by the Jesuit, Paul Guldin, who asserted not only that the method had been taken from Kepler, but also that it was incorrect, inasmuch as it led to paradoxes and fallacies. Cavalieri defended himself from the first charge by pointing out that his method differed from that of Kepler in that it made use only of indivisibles, whereas Kepler had thought of a solid as made up of very tiny solids.[99] In answering the charge that the method was invalid, Cavalieri maintained that although the indivisibles may correctly be considered as having no thickness, nevertheless, if one wished, he could substitute for them small elements of area and volume in the manner of Archimedes. Guldin had said that since the number of indivisibles was infinite, these could not be compared with one another. Furthermore, he had

[99] "Ex minutissimis corporibus." *Exercitationes geometricae sex*, p. 181; see also Kepler, *Opera omnia*, IV, 656–57, for notes, including extracts from Cavalieri on this point. Cavalieri's assertion in this connection should be sufficient to refute the statement often made (see e. g., Wolf, *A History of Science, Technology, and Philosophy in the Sixteenth and Seventeenth Centuries*, p. 206) that he was probably well aware of the fact that his indivisibles must be of the same dimension as that of the figure which they constitute.

pointed out a number of fallacies to which the method of indivisibles appeared to lead. In answering these arguments, Cavalieri said that difficulties in the method are avoided by observing that the two infinities of elements to be compared are of the same kind. If, for example, the altitudes of two figures are unequal, their horizontal sections are not to be compared, because the corresponding indivisibles of one are not the same distance apart as are those in the other. Whereas in one figure there may be 100 indivisibles between two sections, between the corresponding sections of the other there may be 200.[100]

This explanation Cavalieri followed with an ingenuous comparison of the indivisibles of a surface with the threads of a piece of cloth, and those of a solid with the pages of a book.[101] Although in geometrical solids and surfaces the indivisibles are infinite in number and lacking in all thickness, nevertheless they may be compared in the same manner as in case of the cloth and the book, if one observes the precaution mentioned above. Cavalieri did not explain how an aggregate of elements without thickness could make up an area or volume, although in a number of places he linked his method of indivisibles with ideas of motion. This association had been suggested somewhat vaguely by Plato and the Scholastic philosophers, and Galileo had followed them in associating dynamics with geometrical representation. Napier, in 1614, had likewise employed the idea of the fluxion of a quantity to picture by means of lines the relation between logarithms and numbers.[102] Cavalieri followed this trend in holding that surfaces and volumes could be regarded as generated by the flowing of indivisibles.[103] He did not, however, develop this suggestive idea into a geometrical method. This was done by his successor Torricelli, with the result that it eventuated later in Newton's method of fluxions. Cavalieri's indivisible itself was also to find a counterpart in the thought of both Newton and Leibniz—in the former's conception of moments and in the latter's notion of differentials. Cavalieri's vague suggestions were thus to play a large part in the development of the calculus.

The *Geometria indivisibilibus* of Cavalieri achieved popularity almost immediately, and became, except for the works of Archimedes,

[100] *Exercitationes geometricae sex*, pp. 238–39; cf. also p. 17.
[101] *Ibid.*, pp. 239–40; cf. also pp. 3–4.
[102] Zeuthen, *Geschichte der Mathematik im XVI. und XVII. Jahrhundert*, pp. 134–35.
[103] *Exercitationes geometricae sex*, pp. 6–7; *Geometria indivisibilibus*, p. 104.

the most quoted source for mathematicians dealing with infinitesimal considerations in geometry. The significance of the work in the history of mathematics has been justly recognized, but such recognition has on occasion led to attempts to impute to Cavalieri views which he was far from possessing. It has been asserted that in his work one discerns clearly "the fundamental notion of the differential calculus," inasmuch as "the indivisible is nothing but the differential,"[104] and that one discovers in it "definite integrals in the sense of Cauchy and Riemann."[105] An examination of his ideas and methods will show that such judgments are not sound. Cavalieri was far from possessing the views which are expressed in the terms "differential" and "integral." He himself appears to have regarded his method only as a pragmatic geometrical device for avoiding the method of exhaustion; the logical basis of this procedure did not interest him. Rigor, he said, was the affair of philosophy rather than geometry.[106] The limit idea, toward which Stevin and Valerio were working, was more completely concealed in Cavalieri's method than in Kepler's. Furthermore, there is in the *Geometria indivisibilibus* a complete lack of emphasis on the algebraic and arithmetical elements which were to lead, first to the rules of procedure of the calculus and later to the satisfactory definitions of the differential and the integral. Cavalieri regarded area and volume as intuitively clear geometrical concepts, and invariably determined the ratio of these, rather than a numerical value associated with a single area or volume. This preoccupation with ratios was to be one of the chief causes of the confusion in the basic ideas of the calculus during the following two centuries.

The fact that Cavalieri paid so little heed to the demands of mathematical rigor made geometers chary of accepting the method of indivisibles as valid in demonstrations, although they employed it readily in preliminary investigations. This element of hesitation is seen to particularly good advantage in the work of Evangelista Torricelli, the friend of Cavalieri and the pupil of Galileo.

Torricelli fully realized the advantages and disadvantages of the method of indivisibles; and he suspected that the ancients possessed

---

[104] Milhaud, "Note sur les origines du calcul infinitésimal," pp. 37–38.

[105] Bortolotti, "La scoperta e le successive generalizzazioni di un teorema fondamentale di calcolo integrale," p. 210, n.

[106] *Exercitationes geometricae sex*, p. 241.

some such method for discovering difficult theorems, the proofs of which they cast in another form either "to hide the secret of their method or to avoid giving occasion for contradiction to jealous detractors."[107] The work of Archimedes indicates how correct Torricelli was in his assumption of the existence of this method, but the motive behind the proofs by exhaustion lay rather in an effort to satisfy the Greek demands for intuitive clarity and logical rigor. Torricelli himself was not quite satisfied with demonstrations by the method of indivisibles, for he usually supplemented these by proofs in the manner of Archimedes, or of Valerio, to whom he refers (echoing Galileo) as the "Archimedes of our century."[108]

The *De dimensione parabolae* of Torricelli is an interesting exercise, in which the author offers twenty-one demonstrations of the quadrature of the parabola. In ten of these the proposition is established by the method of the ancients,[109] and in the other eleven by the new geometry of indivisibles.[110] Included among the former[111] is the well-known demonstration by the method of exhaustion, given by Archimedes in his *Quadrature of the Parabola*. There is, as well, a proof[112] closely resembling the basic proposition of Luca Valerio on inscribed and circumscribed figures; viz., that it is possible to inscribe within the parabolic segment a figure, made up of parallelograms of equal height, which shall differ from the segment by less than any given magnitude. Torricelli did not, of course, consider the area of the parabola as defined by the limit of the inscribed figure, but he approached more closely to this idea than had Valerio. The latter had been satisfied to state that the circumscribed and inscribed figures differed by less than a given magnitude and to imply that the figure itself would therefore differ from either of these by less than this magnitude; Torricelli clearly stated this implication, from which the limit concept would easily follow on arithmetizing the quantities involved. However, Torricelli followed Cavalieri in restricting himself to geometrical considerations, and in consequence he was attracted more toward indivisibles than toward limits.

[107] *Opere di Evangelista Torricelli*, I (Part 1), 140.   [108] *Opere*, I (Part 1), 95.

[109] "More antiquorum absoluta." *Opere*, I (Part 1), 102.

[110] "Per novam indivisibilium geometriam pluribus modis absoluta." *Opere*, I (Part 1), 139.

[111] Proposition V. *Opere*, I (Part 1), 120–21.   [112] Lemma XV. *Opere*, I (Part 1), 128–29.

Among the eleven demonstrations by the geometry of indivisibles, he included one[113] which is—oddly enough—almost identical with the mechanical quadrature given by Archimedes in the *Method*, a work not known in the seventeenth century. This coincidence shows how closely Cavalieri's geometry of indivisibles resembled the mathematical atomism upon which Archimedes' method was probably based.

Torricelli far outdid his master Cavalieri in the flexibility and perspicuity of his use of the method of indivisibles in making new discoveries. One of the novel results which pleased him greatly was the

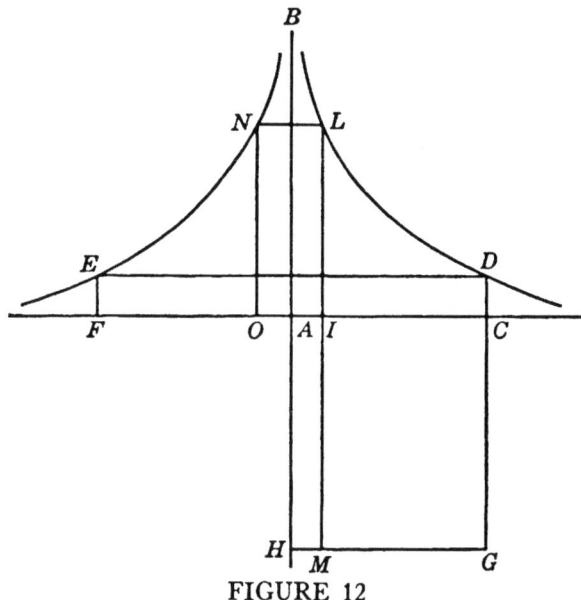

FIGURE 12

determination, in 1641,[114] that the volume of an infinitely long solid, obtained by revolving about its asymptote a portion of the equilateral hyperbola, was finite.[115] Torricelli believed that he was the first to discover that a figure with infinite dimensions could have a finite magnitude; but in this respect he had been anticipated, probably by Fermat and Roberval, and certainly in the fourteenth century by Oresme. It will be recalled that Oresme, in applying geometric repre-

[113] Proposition XX. *Opere*, I (Part 1), 160–61.
[114] Bortolotti, "La memoria 'De infinitis hyperbolis' di Torricelli," p. 49.
[115] "De solido hyperbolico acuto." *Opere*, I (Part 1), 173–221.

sentations to the considerations of Calculator on the latitude of forms and infinite series, had shown that a figure with an infinite altitude (velocity) could nevertheless have a finite area (distance). Torricelli's proof is interesting in that it makes use of the idea of cylindrical indivisibles, whereas those of Cavalieri had invariably been plane. Let the hyperbola be rotated about BA, and let ED be a fixed horizontal line (fig. 12). Let $ACGH$ be a right circular cylinder with AC as altitude and AH as the diameter of its base. Then Torricelli showed that for any position of the line NL parallel to ED, the cylinder with altitude NO and diameter OI has a lateral surface area equal to the cross-sectional area IM of the cylinder $ACGH$. But the cylindrical surfaces $NLIO$ make up the volume of the infinitely long solid of revolution $FEBDC;$ and similarly the areas of the circles of diameter IM constitute the volume of the cylinder $ACGH$. Therefore the two volumes are equal.[116]

The demonstration given by Torricelli shows strikingly the facility offered by the comparison of indivisibles, and the extent to which it resembles the procedure employed in the integral calculus, in which, of course, the cylindrical and the circular elements are given a thickness, and the limit of the sum, as this thickness approaches zero, is determined. Torricelli said that he himself was satisfied that the truth of this theorem was sufficiently clear; but for the benefit of those not so kindly disposed toward indivisibles,[117] he added, as usual, a demonstration by the method of the ancients.[118]

The work of Torricelli is so eminently catholic in its application of the ideas and methods suggested by his predecessors and contemporaries that his name on numerous occasions became the center of disputes concerning priority. Probably no century presents more charges of plagiarism than does the seventeenth. This is accounted for largely by the fact that the method of indivisibles and related procedures were used extensively and effectively by mathematicians of many nationalities, all working on similar problems which were leading toward the calculus. Inasmuch as there was no logically established justification for these heuristic infinitesimal methods, the interrelations of the di-

---

[116] "Propterea omnes simul superficies cylindricae, hoc est ipsum solidum acutum EBD, una cum cylindro basis FEDC, aequale erit omnibus circulis simul, hoc est cylindro ACGH." *Opere*, I (Part 1), 194.

[117] *Ibid.*   [118] *Ibid.*, pp. 214–21.

verse points of view were but vaguely realized, and were frequently denied. Added to this was the further difficulty that many contributors to the theory of the subject—notably Roberval, Fermat, and Newton —either did not publish their results, or did so only very tardily. Failure of mathematicians to date their works further added to the confusion experienced in attempting to attribute specific contributions to individual men. No attempt will be made in this essay to consider at length such charges of plagiarism, many of which were not sufficiently substantiated. Nevertheless, an effort will be made to make clear, so far as is possible in this difficult period, the part each man played in the development of the ideas of the calculus.

It will be recalled that Cavalieri had enunciated in geometrical terminology what may be considered the first general theorem of the calculus: that is, the equivalent of $\int_0^a x^n dx = \dfrac{a^{n+1}}{n+1}$, for all positive integral values of n. The generalization and proof of this theorem for all rational values of n (except $n = -1$) has been commonly attributed to Fermat.[119] However, inasmuch as Fermat did not publish his result during his lifetime, it is difficult to determine the relationship between his work and the analogous and approximately concurrent results of Torricelli and Roberval. It would appear[120] that the enunciation, at least, of the generalization is to be attributed to Fermat. The date of this is doubtful. It may have been as early as 1635,[121] or as late as 1643.[122] Whether the demonstrations of the rule given by Roberval for positive integers and by Fermat for the general rational case anticipated that given by Torricelli in 1646, in his unorganized treatise, *De infinitis hyperbolis*, is not clear.[123] However, the form of proof employed

---

[119] See, for example, Zeuthen, *Geschichte der Mathematik im XVI. und XVII. Jahrhundert*, p. 265.

[120] Bortolotti, "La scoperta," p. 215.

[121] See Walker, *A Study of the Traité des Indivisibles of Roberval*, pp. 142–64.

[122] Bortolotti, "La Scoperta," p. 215.

[123] Zeuthen asserts that it is to Fermat that we owe the study of the higher parabolas. He says that Fermat was probably in possession of the rule for positive integers in 1636 and that it is almost certain that he had a general demonstration for all cases in 1644. "Notes sur l'histoire des mathématiques," IV. "Sur les quadratures avant le calcul intégral, et en particulier sur celles de Fermat," (1895), pp. 43 ff. Surico, on the other hand, maintains the priority of Torricelli in this connection, placing the discovery in 1641 and the generalization by 1646. He concludes that Fermat's quadratures were posterior to 1654, perhaps about 1656. "L'integrazione di $y = x^n$ per n negativo, razionale, diverso da -1 di Evangelista Torricelli."

is in each case so peculiar to the ideas of its author that, in considering the development of the concepts of the calculus, we may, without fear of misdirection, consider them as independent of each other.

The demonstration offered by Torricelli is in the manner of Archimedes—purely geometrical and employing the method of exhaustion. Given any hyperbola DC (fig. 13), he proved that "the quadrilineum $EDCF$ is to the frustum $DCBG$ as the power (*dignitas*) of BA is to the power of AE."[124] The proof of this he carried out by the use of inscribed and circumscribed figures and the application of several lemmas, including the fundamental proposition of Valerio, in which it is shown that these figures can be made to differ by less than any given area. Torricelli remarked, incidentally, that the same procedure and conclusion will also be found to apply, with slight alterations, in the case of parabolas.

This result is equivalent to the analytic statement that if the curve is $x^m y^n = k$, then the ratio of the areas $EDCF$ and $DCBG$ is $\frac{n}{m}$. The determination of this ratio is in a general sense equivalent to the evaluation of what would now be written as $\int_a^b x^{-\frac{m}{n}} dx$. In such a representation of the problem in terms of modern symbolism,[125] however, there is a strong temptation to read into the author's work the concepts which are called to mind by the newer notation. One must not attribute to Torricelli any such algebraic notion as that implied by the modern integral sign. Analytic considerations at no point enter into his thought, and there is no indication that he had any desire to establish an algorithmic rule of procedure applicable to other cases. The result remained a simple geometrical proposition on the ratio of areas, although the particular quadrature involved was later to be of fundamental significance in the calculus.

The danger in interpreting Torricelli's work in terms of modern notations and ideas is apparent in an evaluation of his work on tangents. Torricelli discovered that if DT is tangent at D to the hyperbola considered above (fig. 13), then TE is to EA as the ratio of the powers of

[124] *Opere*, I (Part 2), 256.
[125] Such as that given at some length by Bortolotti, "La memoria."

AB and AE;[126] that is, if the hyperbola is $x^m y^n = k$, the ratio of the subtangent to the abscissa is $\dfrac{n}{m}$. An interpretation of this proposition has been made,[127] in which it is implied that Torricelli regarded the tangent as determined by the secant through C and D as C approaches D, and that he was therefore not far from the idea of a differential quotient. However, there is nothing in Torricelli's language to justify such a conclusion. Torricelli's proof is based, not on the modern idea of the tangent as the limit of a variable secant, but upon the ancient

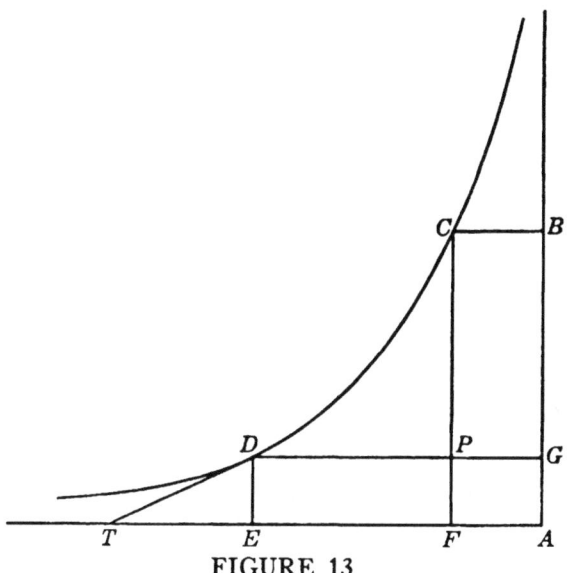

FIGURE 13

static definition: a line touching the curve at only one point. He showed that the assumption that the line TD, determined by the ratio stated above, is not tangent at D—that is, that it intersects the curve in another point—leads to a contradiction. There is in this proof, of course, nothing of the modern idea of limits.[128] However, in another connection Torricelli made use of a dynamic conception of tangents which was significantly suggestive in the development of Newton's fluxional calculus—that based on the parallelogram of virtual velocities.

[126] *Opere*, I (Part 2), 257.
[127] Bortolotti, "La memoria," pp. 143–44. [128] Cf. also *Opere*, I (Part 2), 304 ff.

The principle of the parallelogram of velocities may be considered to be implied[129] in Peripatetic science, but inasmuch as Aristotle did not develop the notion of instantaneous velocity, the doctrine long failed to be widely employed. Archimedes appears to have applied it in geometry, and much later it was again suggested by Leonardo da Vinci,[130] and used by Stevin,[131] who, because of a predominant interest in statics, thought of it in terms of virtual displacements rather than velocities. Upon the clarification by Galileo of trajectories in terms of the notion of inertia and of the doctrine of the independence of superimposed effects, the idea of the composition of motions was destined to play a significant rôle in science (particularly in dynamics and optics) and mathematics during the seventeenth century.

Torricelli's determination, by means of the composition of motions, of the tangents to parabolas of any positive integral degree furnishes a striking illustration of his application of the methods of Galileo and Cavalieri. It had been recognized since the fourteenth century that the motion of a freely falling body is uniformly accelerated—that is, that the velocity increases in proportion to the time elapsed—and Galileo had incorporated this fact in his dynamics. It had been stated by Calculator, Oresme, and others that the distance covered was consequently one-half that which would be traversed by a body moving uniformly for the same length of time with half of the final velocity of the falling body. Galileo reasserted this and added that this implied that the distance covered by the falling body varied as the square of the time. Torricelli pursued this idea further and inquired what would be the state of affairs if the velocity were to vary as the square of the time. In this case the distance covered would be given by the sums of the squares of the lines in the triangle ABE (fig. 14), where these squares represent the velocities of the body for a given time interval AB. But Cavalieri had demonstrated that the sum of the squares of the lines of the triangle ABE is $\frac{1}{3}$ the sum of the squares in the parallelogram $ABEI$. Therefore the distance covered will be $\frac{1}{3}$ that which would have been covered by a second body moving for the time AB with a uniform velocity equal to the final velocity of the first body—

[129] See Duhem, *Les Origines de la statique*, II, 245.

[130] Duhem, *op. cit.*, II, 245–65, 347–48; cf. also Dühring, *Kritische Geschichte der allgemeinen Principien der Mechanik*, p. 15.

[131] Lasswitz, *Geschichte der Atomistik*, II, 12–13.

or, inversely, the final velocity will be given by three times the distance covered. The distance, moreover, will vary as the cube of the time.[132]

If, therefore, we imagine a projectile moving with a composite motion made up of a uniform horizontal velocity and of a vertical velocity which varies as the square of the time, the curve traversed will be a cubical parabola; if the vertical velocity varies as the cube of the time, the curve will be a quartic parabola; and so on. The tangents to

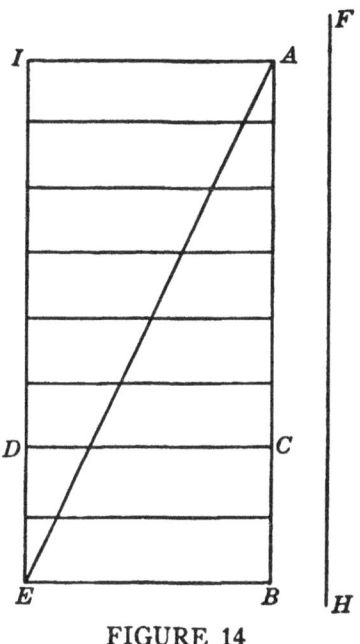

FIGURE 14

these curves many now be determined as follows: Let the curve *ABC* (fig. 15) be, for example, the cubic parabola. Then if *EB* is tangent to the curve, we will have *ED* = 3*AD*. This is clear from the fact that the moving point will possess at *B* a double impetus: one horizontal, and given by the distance *BD:* and the other vertical, and given (from the considerations above) by three times the vertical distance *AD*. Therefore the direction of the point *B*, by the composition of these two velocities, will be that of the line *BE*, which is consequently the tan-

[132] *Opere*, I (Part 2), 311.

gent.[122] Torricelli remarked that the same type of argument can likewise be applied to other parabolas, the ratio $\dfrac{ED}{AD}$ being the degree of the parabola.

Torricelli's method, employing as it does the idea of instantaneous direction and implying, therefore, the limit concept, represents a marked advance over the stultifying definition of the tangent given

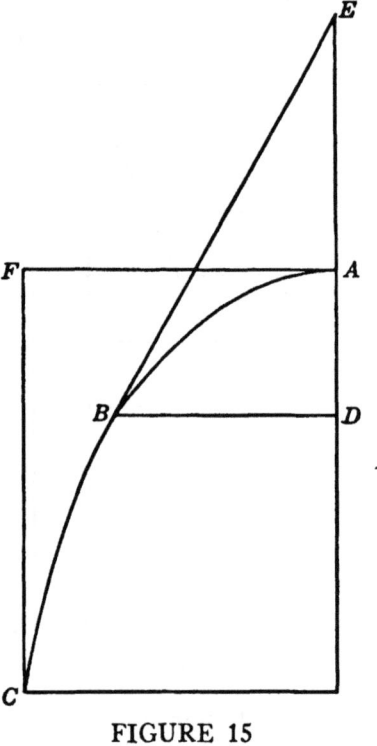

FIGURE 15

by the ancient geometers. It indicates, as well, a departure from the classical tradition, in the intrusion of the notion of instantaneous velocity into geometrical demonstrations. Dynamical considerations had occasionally forced themselves upon the attention of Greek geometers. Archimedes may well have used Torricelli's very method in determining the tangent to his spiral; but he did not, any more than his con-

[122] *Ibid.*, pp. 310–11.

temporaries, regard the idea of motion as sufficiently rigorous for application in his static and formal geometrical proofs, nor did he even develop a science of kinematics. The Scholastic philosophers of the fourteenth century, however, had given quantitative representations of dynamics, work which was elaborated by Galileo. Torricelli now employed these in pure geometry in the determination of tangents to parabolas.

He made use of them also in finding the tangents to a large class of curves suggested to him by the work of Archimedes. Torricelli considered the curves generated by a point which moves along a uniformly rotating line with a velocity which is not necessarily uniform, but is, instead, a function of its distance from the fixed point about which the line rotates.[134] If the velocity is such that in equal intervals of time the distances of the variable point from the fixed point are in continued proportion, Torricelli called the curve a "geometric spiral," to distinguish it from the "arithmetic spiral" of Archimedes.[135] The equation of the geometric spiral may be written in polar coördinates as $\rho = ae^{b\theta}$, but Torricelli did not treat these curves analytically. Instead, he employed considerations from synthetic geometry and mechanics to give him the tangents, as well as the lengths of the curves and the areas bounded by them.

In his use of kinematic representations in mathematics, he may have been anticipated by the French mathematicians Roberval[136] and Descartes (whom we shall consider shortly); but it was largely through Torricelli's work that the notion gained such popularity that it was accepted by Barrow in his geometry and by the latter's student, Newton, in the method of fluxions.

Among the results achieved by Torricelli through the application of the methods of exhaustion, of indivisibles, and of the composition of motions are to be found a number of remarkable anticipations of those found in the calculus. They include, as well as numerous theorems on quadratures and tangents, some of the earliest results on rectification. Torricelli appears also to have recognized and made use of the fact that the problem of tangents was the inverse of that of quadratures.[137]

[134] "De infinitis spiralibus," *Opere*, I (Part 2), 349–99.   [135] *Ibid.*, p. 361.
[136] See Jacoli, "Evangelista Torricelli ed il metodo delle tangenti detto *Metodo del Roberval*."
[137] Bortolotti, "La memoria," pp. 150–52.

He did not, however, attempt to establish upon these methods any general rules of procedure which might be applied in all cases. He did not regard them as constituting a new type of analysis and consequently did not seek for a universal algorithm. The composition of velocities, for example, could be employed only in the case of curves for which the generating motions were known beforehand. Only with the analytic methods of Fermat, Descartes, and Barrow, or with the calculus of Newton and Leibniz did it become possible to determine in general, from the equation of a curve, the motions by which the curve could be regarded as traced, or the instantaneous direction.

Although Torricelli's work marked a significant step toward the calculus, the basic concepts employed in it were still far from the modern point of view. Torricelli had no more idea of defining the notion of instantaneous velocity in terms of limits than had his master, Galileo—or, for that matter, than his successor, Newton. Similarly, although Torricelli was well aware of the fact that the method of indivisibles gave results consonant with those obtained by the methods of the ancients, he appears to have been far from realizing that the two are to be associated through the limit concept. His view of the indivisible resembled strongly the vague mathematical atomism of Democritus. Torricelli agreed with the assertion, which he attributed to Galileo,[138] that a point is equal to a line; and in "confirming" Galileo's demonstration that the area under a velocity-time graph represents distance, Torricelli affirmed that in the case of unequal lines the number of points on each was the same, but that the points themselves were unequal.[139] The lack of a suitable basis for indivisibles was perhaps more serious than the omission of a definition of instantaneous velocity, for whereas intuition may serve to clarify the use of velocity as an undefined element, there is no such safe guide in the use of indivisibles. Torricelli realized the difficulties involved; but, although he wrote a work on paradoxes in the use of infinitesimals,[140] he was unable to resolve the logical perplexities.

The date 1647 is significant in the development of the calculus for a number of reasons: in the first place Cavalieri and Torricelli died in this year, both at early ages; secondly, Cavalieri's more ambitious

[138] *Opere*, I (Part 2), 321.
[139] *Ibid.*, p. 259.   [140] *De indivisibilium doctrina perperam usurpata. Ibid.*, pp. 415–23.

work on indivisibles—the *Exercitationes geometricae sex*—appeared; thirdly, the ponderous *Opus geometricum* of the greatest of circle-squarers, Gregory of St. Vincent, was published at Antwerp.

Let us now turn from the development of infinitesimals in Italy, to consider briefly a concurrent trend in the Low Countries, of which the point of view was somewhat different. Gregory of St. Vincent and Cavalieri probably worked independently of each other[141] and were in possession of their methods at practically the same time—about 1625 and 1626 respectively.[142] Both men were directly connected, in their mathematical inspiration, with the tradition of Archimedes, and each was directly indebted, as well, to others—Cavalieri to Galileo (from whom he probably adopted his indivisibles), Valerio, and perhaps Kepler; Gregory to Stevin and Valerio, from whom he borrowed the idea of giving to propositions involving the method of exhaustion a direct and yet rigorous demonstration in place of the Greek reductio ad absurdum.[143] Gregory, however, added an element not found in their works, for he connected the question with the Scholastic discussions on the nature of the continuum and the result of infinite division. Archimedes, Stevin, and Valerio had subdivided only until the error was less than a certain amount, but Gregory interpreted this as meaning an actually infinite subdivision. Instead of the parallelograms of Stevin and Valerio, he used, in his *Opus geometricum* (or *Problemum austriacum*, as it is sometimes called), infinitely many infinitely thin rectangles;[144] and for the *n*-sided polygons used by Archimedes he substituted an inscribed polygon of an infinite number of sides,[145] as had Nicholas of Cusa.

Gregory applied his conceptions to problems of cubatures by a process which he called *ductus plani in planum*. This phrase referred to a means of constructing a geometrical solid from two given plane figures. Let the figures be, for example, a semicircle and a rectangle having the diameter of the circle as one side (fig. 16). Then, by "applying the rectangle to the semicircle," Gregory had in mind the following: Place the rectangle $ABA'B'$ perpendicular to the plane of the semicircle. Then for any point on AB erect in the two figures perpen-

[141] Moritz Cantor, *Vorlesungen*, II, 818.
[142] *Ibid.*, II, 759; see also Bosmans, "Grégoire de Saint-Vincent."
[143] Bosmans, *op. cit.*, pp. 250–56.   [144] *Opus geometricum*, p. 961.
[145] Wallner, "Über die Entstehung des Grenzbegriffes."

dicular lines XX' and XZ and upon these complete the rectangle $XZX'Z'$. The geometrical solid, of which this rectangle is a section for all positions of X on AB, is the one sought. In this case it is a portion of a cylinder, but Gregory applied the process to innumerable other figures[146] and found the volumes of the solids obtained.

It is obvious that in this procedure Gregory of St. Vincent was implicitly making use of infinitesimals. His view of the nature of these, although perhaps less naïve than that of Cavalieri, was not clear or rigorous.[147] The manner in which he constructed his solid figures would suggest that he was thinking in terms of indivisibles. Nevertheless, we

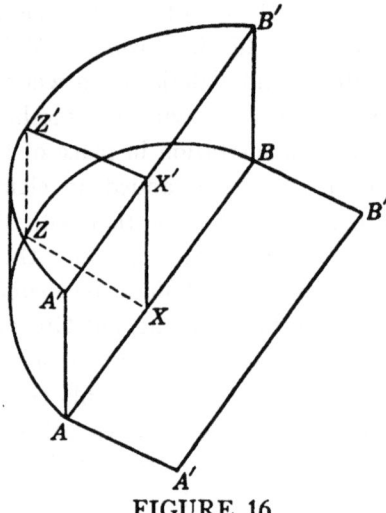

FIGURE 16

see in other connections that he did not regard them, as had Cavalieri' as being without thickness, but rather in the manner of Kepler, as literally making up the geometrical figure. After inscribing in two three-dimensional figures very thin parallelepipeds, he added that "these parallelepipeds can be so multiplied that they *exhaust* the body within which they are inscribed."[148]

This is perhaps the first use of the word "exhaust" in this sense and the earliest example in which a configuration was literally exhausted in this manner. In Greek demonstrations by the method of exhaustion

---

[146] *Opus geometricum,* Book VII.   [147] Moritz Cantor, *Vorlesungen,* II, 818.
[148] "Parallelepipeda illa ita posse multiplicari ut corpora ipsa, quibus inscribuntur, exhauriant." *Opus geometricum,* p. 739.

the figure was thought of simply as approximated, to within a given degree of accuracy, by the inscribed or circumscribed figure. The Greek method is therefore, in a sense, improperly designated as the method of exhaustion; but Gregory apparently used the word in its literal sense, allowing the subdivision to continue to infinity. He did not explicitly state how one is to visualize the exhaustion of the body by means of infinitesimal elements, but he was certainly nearer to the modern view than Democritus and the Scholastics, or even Cavalieri and Torricelli, had been. Instead of thinking of static indivisibles, he reasoned in terms of a varying subdivision, thus approximating the method of limits. This fact led Gregory in the direction which was ultimately to supply the rigorous basis for the calculus, for his infinite subdivision brought him to the notion of the limit of an infinite geometrical progression.[149]

The Greeks had not attempted to *define* a curve as the *terminus ad quem*, or limit, approached by the inscribed or circumscribed figure. Valerio had made the gap between the method of exhaustion and the limit idea narrower by stating directly that the approximating inscribed and circumscribed figures could be made to differ by less than any magnitude; and Stevin had aided significantly in this direction by his omission, on occasion, of one of the two approximating figures and by his use of arithmetic sequences. It was Gregory of St. Vincent, however, who gave perhaps the first explicit statement that an infinite series defines in itself a magnitude which may be called the limit of the series. "The terminus of a progression is the end of the series to which the progression does not attain, even if continued to infinity, but to which it can approach more closely than by any given interval."[150] As an illustration, Gregory gave a line segment AK, which was subdivided by points $B, C, \ldots$ with K as their limit and such that the segments AB, BC, $\ldots$ were in continued proportion. Gregory, however, lost somewhat the force of the limit concept by remarking that "the magnitude AK is equal to the magnitude of the whole progression . . . continued to infinity; or, which is the same thing, K is

[149] Wallner ("Über die Entstehung des Grenzbegriffes," pp. 251–52) says this was the first use of a truly *infinite* series: but in this he is mistaken, inasmuch as Calculator, Oresme, and other Scholastics had employed them long before, as we have seen, in connection with the latitude of forms.

[150] *Opus geometricum*, p. 55.

the terminus of the ratio AB to BC continued to infinity."[151] In other words, AK was not defined as the sum, *because* K was the limit: Gregory pointed out, rather, on the basis of geometrical intuition, that the remark that AK is the sum, is equivalent to the statement that K is the limit. Nevertheless, he stated more clearly than had anyone previously that an infinite series can be strictly considered as having a sum.

Gregory of St. Vincent recognized also that the paradox in the *Achilles* is to be explained in terms of the limit of an infinite series. Assuming, as had the Scholastics and Galileo, that "motion is a kind of quantity," he asserted that the speeds of Achilles and the tortoise must have a proportion and calculated by geometric progressions at what point the positions of the two will coincide.[152] He failed to recognize in this connection that the question of Zeno was not *when* or *where* Achilles would overtake the tortoise, but rather *how* he could do it. This appeal in the paradox to sensory experience, instead of reasoning, was to be without doubt the chief obstacle to the development of the calculus in terms of the limit of an infinite sequence. Although Gregory of St. Vincent did not express himself with the rigor and clarity characteristic of the nineteenth century, his work is to be kept in mind as constituting the first attempt explicitly to formulate in a positive sense—although still in geometrical terminology—the limit doctrine which had been implicitly assumed by both Stevin and Valerio, as also, probably, by Archimedes in his method of exhaustion.

Gregory of St. Vincent maintained that he had squared the circle.[153] Perhaps it is on account of this fact[154] that it has been said[155] that Gregory received only disdain on the part of his contemporaries, his memory being rehabilitated by Huygens and Leibniz.[156] On the other hand, there can be no doubt that his work exerted a strong influence on many of the mathematicians of his time. As a teacher in various Jesuit schools, he numbered among his disciples Paul Guldin, Andreas Tacquet,[157] Jean-Charles della Faille,[158] and others who used infinites-

---

[151] *Ibid.*, p. 97.   [152] *Ibid.*, pp. 101–3.   [153] *Ibid.*, pp. 1099 ff.

[154] Bosmans, "Grégoire de Saint-Vincent," pp. 254–55.

[155] Marie, *Histoire des sciences mathématiques*, III, 187.

[156] Cf. Leibniz, *Mathematische Schriften*, V, 331–32.

[157] See Kaestner, *Geschichte der Mathematik*, III, 266–84, 442–49, for summaries of Tacquet's works.

[158] See Bosmans, "Le Mathématicien anversois Jean-Charles della Faille de la Compagnie de Jésus."

imal considerations, particularly in the then-popular problem of determining centers of gravity.

Although Guldin wrote in the old Archimedean manner, his work has become well known for two reasons. In the first place, the so-called Guldin theorem—that the volume of a solid of revolution is given by the product of the rotating area and the distance through which the center of gravity moves in one revolution[159]—has been much discussed from the point of view of possible plagiarism from Pappus.[160] Secondly, Guldin became, as we have seen, the chief critic of the lack of rigor in the use of infinitesimals by Kepler and of indivisibles by Cavalieri. From the point of view of the development of ideas, however, the work of Tacquet was more significant than that of Guldin.

André (or Andreas) Tacquet resembles his contemporary, Torricelli, in the generality of his adoption from his predecessors of varied infinitesimal methods. In his *Cylindricorum et annularium libri IV* he gave, for example, four demonstrations of the proposition that the volume of a sphere is equal to that of a cylindrical wedge whose base is half a great circle of the sphere, and whose altitude is equal to the circumference of the sphere. This theorem had been given by a number of mathematicians since Kepler, as well as by Archimedes in the *Method*, probably not then extant. Tacquet, however, after proving the theorem in two ways by the use of inscribed and circumscribed figures, gave two further demonstrations by indivisibles, based on the equality of triangles and circular sections. Torricelli had himself been satisfied with the rigor of proofs by means of indivisibles, although he supplied alternative demonstrations for the benefit of others. Tacquet, on the other hand, said that he did not consider that the method of Cavalieri was to be admitted as either legitimate or geometrical.[161] He maintained that the cylindrical wedge could not, in all strictness, be considered as made up of triangles; nor could the sphere be regarded

---

[159] Kepler had given a special case of this theorem in 1615, in his determination of the volume of the tore.

[160] For a bitter discussion on this point, see Smith and Miller, "Was Guldin a plagiarist?" The editor of the recent French edition of Pappus' works, Paul Ver Eecke, exonerates Guldin in no uncertain terms, saying that the theorem of Guldin cannot be deduced from the form in which it is given by Pappus, and indeed that a link by inspiration between the two is open to serious doubt. See Ver Eecke, "Le Théorème dit de Guldin consideré au point de vue historique."

[161] *Cylindricorum et annularium*, pp. 23–24; or *Opera*, 3d pagination, p. 13.

as composed of circles. Neither was the wedge generated by the flux of a triangle, nor the sphere by the motion of a circle. A geometrical magnitude, he asserted, is made up only of *homogenea*, that is, parts of like dimension—a solid of small solids, an area of small areas, and a line of small lines—and not of *heterogenea*, or parts of a lower dimension, as Cavalieri had maintained. He therefore felt that a proposed magnitude was to be exhausted (a word which he undoubtedly acquired from Gregory of St. Vincent) by inscribing *homogenea* within them "as in the manner of the ancients."[162]

Tacquet's criticism of the method of indivisibles and his insistence on the use, instead, of *homogenea* was, of course, quite justified. Had he tried to reconcile the two points of view by employing the word "exhaust" in its literal sense in terms of the doctrine of limits, his work might have served to clarify the method of indivisibles which men continued to employ, not because they understood its significance, but because it invariably gave them correct results. It is the more strange that Tacquet did not do this, inasmuch as he developed the thoughts of Gregory of St. Vincent on limits. For example, in his *Arithmeticae theoria et praxis* of 1656, he explained the *Achilles* in terms of geometrical progressions;[163] and in another passage he remarked that in a progression continued ad infinitum, in which the terms decrease in a given proportion, the smallest term vanishes,[164] thus applying the criterion of a limit. This arithmetical work of Tacquet appeared almost simultaneously with the *Arithmetica infinitorum* of John Wallis, in which we shall find the notion of limit more vigorously proposed. Before reviewing the ideas of Wallis, however, it will be necessary to consider at some length the significant anticipations of methods of the calculus made by a group of illustrious contemporary French mathematicians—Roberval, Pascal, and Fermat—upon at least one of whom—Pascal—Tacquet exerted an influence.

Giles Persone de Roberval held the chair of Ramus at the Collège Royal, a position that depended upon supremacy in an examination held every three years, the questions for which were propounded by the incumbent at the time. To this fact Roberval attributed the secrecy, with respect to his methods and results, which eventuated in his loss

---

[162] *Ibid.*   [163] *Arithmeticae theoria et praxis*, pp. 502–3.
[164] "Minimus terminus evanescat." *Ibid.*, p. 475; cf. also Bosmans, "André Tacquet et son traité d'arithmétique théorique et pratique."

to other men of credit for priority. He said that he had considered attentively the "divine Archimedes," and that from this study he worked out for himself "the sublime and never to be sufficiently praised doctrine of the infinite."[165] Roberval seems to have worked out his method of indivisibles between the years 1628 and 1634,[166] that is, only a few years after Gregory of St. Vincent and Cavalieri had developed theirs and before the work of either had been published. The almost simultaneous appearance of these procedures indicates how widespread was the tendency toward infinitesimal considerations during the early seventeenth century.

Roberval admitted no inspiration for his work other than that of Archimedes. It is quite probable, nevertheless, that he was to some extent influenced by Kepler,[167] and portions of his work greatly resemble the ideas of Stevin and Valerio; but we can point to no clear indications of indebtedness.[168] It would be interesting to know to just what extent the work of Stevin was known in France during the first half of the seventeenth century. The similarity of some of the ideas expressed by Pascal and Roberval to those found in the writings of Stevin half a century earlier is striking; but the former scientists made no acknowledgment of indebtedness to the engineer of Bruges, whose works had appeared several times in Flemish, Latin, and French.[169]

Roberval was without doubt familiar with the work of Cavalieri, which he defended from carping critics;[170] but his view of indivisibles appears to have been far less naïve than that of the Italian. He said quite clearly that in his method he did not regard a surface as really composed of lines or a solid as made up of surfaces, but as in actuality built up of small pieces of surfaces and solids respectively, these "infinite things" being regarded "just as if they were indivisibles."[171] In his *Traité des indivisibles*, Roberval asserted that throughout the discussion it was to be understood that the phrase "the infinite number of

[165] Walker, *A Study of the Traité des Indivisibles of Roberval*, pp. 15–16.
[166] *Ibid.*, pp. 142–64.
[167] *Ibid.*, p. 81.
[168] Duhem (*Les Origines de la statique*, I, 290–326) asserts that Roberval knew the work of Stevin and Valerio.
[169] It has been suggested (see *The Physical Treatises of Pascal*, p. 4, n.) that Stevin was the victim, because of his liberality of thought, of a conspiracy of silence on the part of Catholics in the Low Countries.
[170] "Divers ouvrages de M. Personier de Roberval," p. 444.
[171] Walker, *A Study of the Traité des Indivisibles of Roberval*, p. 16.

points" is used for the infinity of little lines which make up the whole line, and that "the infinite number of lines" represents the infinity of little surfaces which make up the whole surface, and so on.[172]

Roberval attributed like views to Cavalieri, saying that the latter did not really mean that a surface was made up of lines; but in this he was, apparently, overgenerous. Influenced principally by the classic works of Archimedes, Roberval did not recognize that in Cavalieri's work, as in that of Galileo and Torricelli, the atomic and Scholastic traditions had operated to modify the author's thought and result in the method of indivisibles. Unlike Archimedes, however, Roberval substituted the conception of infinity for the method of exhaustion, somewhat in the manner of Gregory of St. Vincent, but without explicitly formulating the limit concept. He did, however, supply the essential element found in our conception of the definite integral, in that, after dividing a figure into small sections, he allowed these continually to decrease in magnitude, the work being carried out largely arithmetically and the result being obtained by summing an infinite series. This method, which resembles to a considerable extent the procedures of Stevin, contrasts strongly with that of the fixed indivisibles of geometric character which is found in the work of Cavalieri.

It has been indicated that the equivalent of the theorem $\int_0^a x^n dx = \dfrac{a^{n+1}}{n+1}$ had been anticipated by Cavalieri for positive integral values of n, and by Torricelli for rational values of n (except, of course, for $n = -1$). At about the same time Roberval had arrived at this result, perhaps on Fermat's suggestion, through investigations which bring out nicely the somewhat different emphasis found in his work. Whereas Cavalieri and Torricelli had proceeded on the basis of the purely geometrical considerations involved in the method of exhaustion and in the method of indivisibles, the great French mathematical triumvirate of Roberval, Fermat, and Blaise Pascal[173] combined their interest in the geometry of Archimedes with an enthusiasm for the theory of numbers, and this colored their work. As a consequence, Roberval was led to make an association between numbers and geometrical magni-

---

[172] "Divers ouvrages," pp. 249–50.

[173] Their famous contemporaries and countrymen, Desargues and Descartes, in this respect display a somewhat different spirit.

tudes, which resembles strongly that of the Pythagoreans, particularly that of Nicomachus. A line segment, it has been remarked, Roberval regarded as made up of an infinite number of little lines, represented by points, which can be made to correspond to the positive integers. If, now, we consider successively the right isosceles triangles with sides made up of 4, 5, 6, . . . points or indivisibles respectively, the total number of such units in the triangles will be given as follows:

The triangle of 4 is $10 = \frac{1}{2}(4)^2 + \frac{1}{2}(4)$

The triangle of 5 is $15 = \frac{1}{2}(5)^2 + \frac{1}{2}(5)$

The triangle of 6 is $21 = \frac{1}{2}(6)^2 + \frac{1}{2}(6)$

The second term on the right in each line is half of the side, and represents the excess of the triangle over the half square. This continues to diminish, in proportion to the first term, as the number of points and lines is increased. Since the number of lines in a geometrical triangle or square is infinite, the excess or half of one line, "does not enter into consideration."[174] It is therefore clear that the triangle is half the square. This argument is evidently roughly equivalent to that indicated by the notation $\int_0^a x\,dx = \dfrac{a^2}{2}$.

Roberval continued this type of work by remarking similarly that if the lines should follow one another in the order of the square, the sum of all these little lines, or the points which represent them, would be to the last, taken an equal number of times, as the pyramid is to the prism, that is, as 1 to 3. For example, if we take a square pyramid of dots having four on an edge, we have $1^2 + 2^2 + 3^2 + 4^2 = 30 = \frac{1}{3}(4)^3 + \frac{1}{2}(4)^2 + \frac{1}{6}(4)$; if there are five on an edge, we have $1^2 + 2^2 + 3^2 + 4^2 + 5^2 = 55 = \frac{1}{3}(5)^3 + \frac{1}{2}(5)^2 + \frac{1}{6}(5)$; and so on. In these equations the first term on the right is one-third the cube, the second is one-half the square, and the last is one-sixth the number of points in the edge of the base of the pyramid. Inasmuch as the number of squares in a geometrical cube is infinite, the last two terms are as nothing, so

[174] "Divers ouvrages," pp. 247–48.

that the sum is $\frac{1}{3}$ the cube.[175] In the same way, the sum of the cubes is one-fourth of the fourth power; the sum of the fourth powers is one-fifth of the fifth power; the sum of the fifth powers is one-sixth of the sixth power, and so on.[176] In other words, Roberval has in this manner indicated the equivalent of the theorem, $\int_0^a x^n dx = \dfrac{a^{n+1}}{n+1}$, for positive integral values of n. He appears not to have given a demonstration for other values of n,[177] as did Torricelli and Fermat.

The arguments of Roberval resemble the "demonstrations by arithmetic" given by Stevin half a century earlier, and similar ones of Wallis a few years after Roberval's work.[178] All of these represent efforts to express the notion of limit, but Roberval in his work obscured

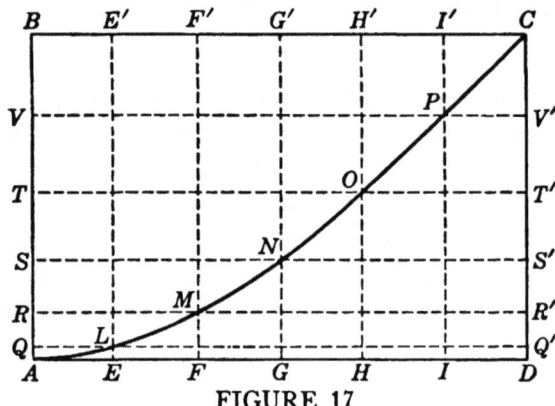

FIGURE 17

the limit idea somewhat by resorting to his notion of indivisibles. Instead of drawing his conclusion from the limits of the arithmetic sequences involved, he had recourse, as did most of his contemporaries, to geometrical intuition. Making use of the Pythagorean and Nicomachean association of numbers with geometrical points, he remarked that since "the side has no ratio to the cube, . . . adding or subtracting a single square has no effect."[179] Intuition of this type led,

---

[175] *Ibid.*, p. 248.   [176] *Ibid.*, pp. 248–49.   [177] Zeuthen, "Notes," 1895, p. 43.

[178] Walker (*A Study of the Traité des Indivisibles of Roberval*, p. 165) makes no mention of Stevin in this respect, incorrectly representing Roberval's work as the first of its type. The statement that the idea of an arithmetic limit appeared for the first time in the seventeenth century (*ibid.*, p. 35) is, of course, not accurate, inasmuch as Stevin's arithmetical work in limits appeared in 1586.

[179] "Divers ouvrages," p. 249.

through the neglect of infinitesimals of higher order, to the differential calculus of Leibniz, rather than to the method of limits which Newton suggested and which triumphed in the end.

Roberval successfully applied this quasi-arithmetical method of indivisibles to varied problems in quadratures. Typical of these is his quadrature of the parabola.[180] The procedure which he employed is quite different from any given by Torricelli in his twenty-one quadratures of the parabola. Roberval's method resembles, rather, Stevin's demonstration by numbers, reinforced by intuition of indivisibles. Let $AE = 1, AF = 2, AG = 3, \ldots$ (fig. 17). Then from the definition of the parabola we know that $\dfrac{EL}{FM} = \dfrac{AE^2}{AF^2}$, and similarly for the other points of division. Hence

$$\begin{aligned}
\frac{\text{area } ADC}{\text{area } ABCD} &= \frac{AE(EL + FM + GN + \ldots)}{AD \cdot DC} \\
&= \frac{AE(AE^2 + AF^2 + AG^2 + \ldots)}{AD \cdot AD^2} \\
&= \frac{1^2 + 2^2 + 3^2 + \ldots + AD^2}{AD \cdot AD^2} = \tfrac{1}{3}
\end{aligned}$$

Roberval by similar methods found the areas under other curves, such as parabolas of higher degrees, the hyperbola, the cycloid, and the sine curve, as well as various volumes and centers of gravity associated with these. He may in this connection have anticipated Torricelli in finding the volumes of infinitely long solids. He also used an ingenious transformation of one figure into another which came to be spoken of as the method of Robervallian lines and which resembled the *ductus plani in planum* of Gregory of St. Vincent. Such transformations played a large part in the geometry of the seventeenth century because of the lack of a facile method for handling curves whose equations involved radicals, but after the development of the calculus these lost their popularity, as well as their significance.

Roberval in his work showed a remarkable flexibility, using divers infinitesimal elements, such as triangles, parallelograms, parallelepipeds, cylinders, and concentric cylindrical shells. Throughout it all,

[180] *Ibid.*, pp. 256–59.

the idea of limits is implied, but is concealed under the terminology of Roberval's method of indivisibles. An adumbration of the method of limits is indicated also by the manner in which Roberval reconciled the demonstrations by indivisibles with those of the ancient geometers. First Roberval showed that the unknown quantity lies between inscribed and circumscribed figures which differ "by less than every known quantity proposed." Then he showed that the quantity in question bore to the circumscribed figure a ratio less, and to the inscribed figure a ratio more, than the proposed ratio. Finally Roberval proved the proposition by the application of a general lemma: "If there is a true ratio $R:S$ and two quantities A and B such that for a small (quantity) added to A, then this sum has to B a greater ratio then $R:S$ and for a small (quantity) subtracted from A, the remainder has to B a ratio less than $R:S$; then I say that $A:B$ as $R:S$."[181] This form of argument, resembling strikingly the corresponding propositions of Valerio (with whose work he may have been familiar), is equivalent to a statement that the limit of a quotient of variables is equal to the quotient of their respective limits.

In Roberval's propositions on indivisibles one recognizes numerous anticipations of the integral calculus, some of which are equivalent to the determination of definite integrals of algebraic and trignometric functions. Roberval was concerned, as well, with problems of the differential calculus—for he developed a method of tangents so much like that of Torricelli that charges of plagiarism arose.[182] He regarded every curve as the path of a moving point, and accepted as an axiom that the direction of motion is also that of the tangent.[183] By looking upon the motion of the point as made up of two component movements, he found the tangent by determining the resultant of these. Thus he found the tangent to the parabola by making use of the fact that, since this curve is the locus of points equidistant from the focus and the directrix, it may be regarded as generated by a point moving with a compound motion made up of a uniform motion of translation away from the directrix and an equal uniform radial motion away from the focus. By the parallelogram (in this case a rhombus) of velocities it is

[181] Walker, *A Study of the Traité des Indivisibles of Roberval*, pp. 38–39.
[182] "Divers ouvrages," pp. 436–78; Walker, *op. cit.*, pp. 142–64. Moritz Cantor (*Vorlesungen*, II, 808–14) concludes that the charges are not substantiated.
[183] "Divers ouvrages," p. 24.

therefore determined that the resultant velocity—and consequently the tangent to the parabola at any point—will be in the direction of the bisector of the angle between the focal radius at the point and the perpendicular from the point to the directrix. This direction being known, the tangent can be drawn.[184] The motions involved here are different from those used for the same curve by Torricelli, but the underlying idea of the composition of movements is essentially the same. The method is, of course, subject to the difficulty that one must in some way discover the laws of motion before one can determine the tangent. Roberval appears, from his correspondence with Fermat in 1636, to have had another method of tangents which proceeded analytically and which he says was connected with the problem of quadratures.[185] This might have been significant in the history of the calculus, but it apparently was lost.

It is difficult to determine the extent of Roberval's influence on contemporary mathematicians inasmuch as his *Traité des indivisibles* was not published until 1693—that is, only after the calculus itself had been made known. It is very probable, however, that he had a strong influence upon Pascal the younger, whose father, Etienne Pascal, was a close friend of Roberval.

Blaise Pascal in a sense represents the highest development of the method of infinitesimals carried out under the traditions of classical geometry. He was not so much a creative genius as a mathematician, scientist, and philosopher, with a remarkable flair for clarifying ideas which had been somewhat vaguely set forth by others, and for supplying these with a reasonable basis.[186] This penchant of Pascal's is well illustrated in science by his lucid organization of the principles of hydrostatics; in mathematics one sees it in his exposition of the nature of infinitesimals, in which, to be sure, one perceives also a touch of his characteristic mystical turn.

Pascal was not a professional geometer, and as a result his geometrical work was accomplished in two periods which were separated by an interval of mathematical inactivity (from 1654 to 1658) during which he devoted his interests to theology. These two periods, moreover, are characterized by somewhat different views as to the nature

---

[184] *Ibid.*, pp. 24–26.   [185] See Moritz Cantor, *Vorlesungen*, II, 812.
[186] Bosmans, "Sur l'œuvre mathématique de Blaise Pascal."[1]

of infinitesimals. Pascal had two predominating interests in mathematics: geometry and the theory of numbers. Toward the end of the first of his two periods of mathematical work the latter was dominant, and at this time he applied the theory of infinitesimals to his work on the arithmetic triangle. Although this is usually called Pascal's triangle, the ordering of binomial coefficients had been known to Stifel long before.[187]

In this connection he enunciated, in the *Potestatum numericarum summa* of 1654, the theorem on the integral of x^n, which we have met with in the work of Cavalieri, Torricelli, and Roberval. Pascal's demonstration of this[188] is derived not from classic geometrical propositions alone, but from an examination of the figurate numbers represented in the arithmetic triangle—a form of proof suggestive of that of Roberval, which appears not to have been generally known at Paris at the time.[189] In the arithmetic triangle the numbers in the first row (or column) may be considered units or points making up a line. Those in the second row represent the sums of the numbers in the first row and may be

$$
\begin{array}{cccccc}
1 & 1 & 1 & 1 & 1 & . & . \\
1 & 2 & 3 & 4 & . & . \\
1 & 3 & 6 & . & . \\
1 & 4 & . & . \\
1 & . & . \\
\end{array}
$$

considered therefore as the sums of points or units—that is, as lines. The numbers in the third row, which are in turn the sums of those in the second row, may therefore be considered as the sums of lines, that is, as triangles. The numbers in the fourth row similarly represent pyramids. Geometrical intuition now fails, but one can continue by analogy.[190]

From such geometrical considerations and from the numerical relationships within the triangle, Pascal was led immediately to consider the sums of *powers* of the positive integers. Recalling the results by the geometrical procedures of the ancients for the sums of the

---

[187] See Bosmans, "Note historique sur le triangle arithmétique de Pascal."
[188] *Œuvres*, III, 346–67, 433–593.   [189] Zeuthen, "Notes," 1895, p. 43.
[190] Cf. Bosmans, "Sur l'interprétation géométrique, donnée par Pascal à l'espace à quatre dimensions."

squares and of the cubes, he recognized that these were not immediately applicable to powers of higher degree. Pascal, however, developed a general arithmetical method for determining the sum, not only in the case of terms which are integral powers (of the same degree) of the first N natural numbers, but also for powers (of the same degree) of any integers in arithmetic progression. Pascal expressed his result rhetorically, but this may be given in symbolic form by the equation:

$$^{n+1}C_1 d\Sigma^{(n)} + {}^{n+1}C_2 d^2\Sigma^{(n-1)} + \ldots + {}^{n+1}C_n d^n\Sigma^{(1)} = (a + Nd)^{n+1} - a^{n+1} - Nd^{n+1},$$

where a is the first term of the progression, d the common difference, N the number of terms, n the degree of the powers in question, $^{n+1}C_i$ the number in the $(i + 1)$st column and the $(n - i + 2)nd$ row in Pascal's triangle, and $\Sigma^{(j)}$ the sum of the jth powers of the terms of the progression.

As Pascal remarked, it will be obvious to any one who is at all familiar with the doctrine of indivisibles that this result can be applied to the determination of curvilinear areas. In order to find the area under the curve $y = x^n$, for example, the surface in question is to be regarded as the sum of ordinates which are the nth powers of abscissae chosen in arithmetic progression (with first term zero and with common difference equal to unity), of which in this case there will be an infinite number. Moreover, a single point adds nothing to the length of a line: nor does the addition of a line to a surface cause any difference in area, for the former is an indivisible with respect to the latter. Or, speaking arithmetically, roots do not figure in a ratio of squares, nor squares in a ratio of cubes, and so on.[191] The rule given above consequently becomes, on calling the greatest abscissa b and on neglecting as zero the terms of lower order, $(n + 1)\Sigma^{(n)} = b^{n+1}$. In a general sense this is, of course, the equivalent of the expression $\int_0^b x^n dx = \dfrac{b^{n+1}}{n+1}$.

Or, translating this into the terminology often used at the time, the sum of lines in a triangle is half the square of the longest; the sum of the squares of the lines is one-third the cube; the sum of the cubes is one-fourth the "square-square," and so on.

The essential point in Pascal's demonstration is the omission of

[191] *Œuvres*, III, 366–67.

terms of lower dimension. This type of argument has frequently been attributed to Cavalieri,[192] but there appears to be no basis for such a view. Cavalieri's method was based upon a strict correspondence of the indivisibles in two figures, and there were no unpaired or omitted elements. The method of dropping terms seems to have entered in the work of Roberval and Pascal through the association of the indivisible of geometry with arithmetic and the theory of numbers. The geometrical intuition of indivisibles of lower dimension was, in their work, carried over into arithmetic to justify the neglect of certain terms of lower degree. Pascal went so far as to compare the indivisible of geometry with the zero of arithmetic, much as Euler later regarded the differentials of the calculus as nothing but zeros.

This neglect of quantities, as found in Pascal, has been characterized[193] as the basic principle of the differential calculus. Such a designation is indeed misleading, for the subject is no longer explained in terms of the omission of fixed infinitesimals. Nevertheless, the work of Pascal exerted perhaps the strongest influence in shaping the views of Leibniz, who adopted into his calculus as fundamental the doctrine that "differences" of higher order could be neglected. Newton also occasionally lapsed into this type of argument in dropping out of the calculation "moments" which did not add significantly to the result. For almost two centuries mathematicians tried to justify such procedures, but in the end the basis of analysis was found, not in these, but rather in the method of limits toward which the geometrical method of exhaustion and arithmetical modifications of this by Stevin, Tacquet, Roberval, and others had pointed.

In answering the objections of those of his contemporaries who held that the omission of infinitely small quantities constituted a violation of common sense, Pascal had recourse to a favorite theme—that the heart intervenes to make this work clear. In this case what is necessary is the "esprit de finesse," or intuition, rather than the "esprit de géométrie," or logical thought, much as the action of grace, as well as physical experience, is above reason. In this respect the paradoxes of geometry are to be compared to the apparent absurdities of Christi-

[192] Ball, *History of Mathematics*, p. 249; Cajori, *History of Mathematics*, p. 161; Marie, *Histoire des sciences mathématiques*, IV, 72; Milhaud, "Note sur les origines du calcul infinitésimal," p. 35.

[193] Simon, "Zur Geschichte und Philosophie der Differentialrechnung," p. 120.

anity, the indivisible being to geometrical configurations as our justice to God's.[194]

The mysticism which Pascal often displayed in his attitude toward the infinitesimal does not appear in all of his work. Particularly in the later period of his mathematical activity—in which his interest centered about the cycloid, the curve which Montucla called "la pomme de discorde"[195] because it engendered so many quarrels with respect to priority—his view appears to be modified. In connection with problems such as those in his *Traité des sinus du quart de cercle* of 1659 in which he balanced elements as Archimedes had done in his mechanical method, he used the language of infinitesimals in speaking of the sum of all the ordinates; but he added that one need not fear to do this, inasmuch as what is really meant is the sum of arbitrarily small rectangles.[196] In his later numerical demonstrations also Pascal sought to avoid arguments based upon the neglect of infinitely small quantities. The Aristotelian view had denied the existence in the realm of number of the infinitely small, while admitting, as a potentiality, the infinitely great. Pascal, on the contrary, maintained in *De l'esprit géométrique* that in the sphere of number the infinitely great and small are complementary. Corresponding to every large number, such as 100,000, there existed a small one, the reciprocal $\frac{1}{100,000}$ so that the existence of the indefinitely large implied that of the indefinitely small. Number, he held, was as much subject to the two infinities—in greatness and in smallness—as were such other undefined primitive terms in geometry as time, motion, and space.[197] The contrast between discrete and continuous magnitudes was not so great as Aristole had felt, and was, in fact, vanishing with the spread of analytic methods in geometry.

The change in Pascal to a clear point of view with respect to infinitesimals may have been the result of friendship with Roberval, who had said that Cavalieri did not really think of indivisibles as lines. It may equally well have come from Pascal's reading of Tacquet's *Cylindricorum et annularium*,[198] in which the author denied the validity of concluding anything about the ratio of surfaces from the ratio of their indivisibles, or lines. Tacquet had been particularly emphatic in deny-

[194] See *Œuvres*, XII, 9, XIII, 141-55.
[195] Montucla, *Histoire des mathématiques*, II, 52.
[196] *Œuvres*, IX, 60–76.   [197] *Œuvres*, IX, 247, 253, 256, 268.
[198] Bosmans, "La Notion des indivisibles chez Blaise Pascal."

ing that a configuration could be thought of as composed of *heterogenea*, or elements of a lower dimensionality. Pascal was in general agreement with him on the question of homogeneity, but his view of the transition from the finite to the infinite was different. Tacquet was inclined toward the limit idea of Gregory of St. Vincent, although he preferred to avoid the difficulty by returning to the clarity afforded by the method of exhaustion.

Pascal, on the other hand, looked upon the infinitely large and the infinitely small as mysteries—something which nature has proposed to man, not to understand, but to admire.[199] Furthermore, Tacquet had made use of the limits of infinite series, as had also Stevin and Roberval. The work of Pascal, however, was carried out in connection with the older theory of numbers and classical geometry of which he considered his method an elaboration. The newer analytic procedures of Fermat and Descartes did not appeal to him, and for them he substituted a remarkable facility in the manipulation of geometric transformations, similar to those of Gregory of St. Vincent and Roberval. Through these he related the figurate sums of his number theory to problems in the synthetic geometry of continuous magnitudes and anticipated numerous results of the integral calculus, including the equivalent of integration by parts. His underestimation of the value of the algebraic and the analytic viewpoints may have been responsible not only for his inability to define the central and unifying concept of the integral calculus—that of a limit of a sum—but also for his failure to recognize the inverse nature of the problems of quadratures and tangents.

The idea and figure of what is now called the differential triangle had appeared on several occasions before the time of Pascal, and even as early as 1624. Snell, in his *Tiphys Batavus*, had thought of a small spherical surface bounded by a loxodrome, a circle of latitude, and a meridian of longitude as equivalent to a plane right triangle.[200] Numerous diagrams somewhat resembling the differential triangle are to be found in the geometrical works of the middle seventeenth century, such as those in the *De infinitis hyperbolis* of Torricelli and the *Traité des indivisibles* of Roberval, with which Pascal may have been familiar.

[199] *Œuvres*, IX, 268.
[200] Aubry, "Sur l'histoire du calcul infinitésimal entre les années 1620 et 1660," p. 84.

In all of these, however, the significance of the quotient of two sides of the triangle for the determination of tangents appears to have escaped emphasis. It was much the same with Pascal. In connection with a diagram (see fig. 18) from his *Traité des sinus du quart de cercle* of 1659, he remarked that *AD* is to *DI* as *EE* is to *RR* or *EK*, and that for small intervals the arc may be substituted for the tangent. Pascal made use of these lemmas to determine the sum of the sines (ordinates) of a portion of the curve, that is, the area under this portion. If Pascal had at this point only been more interested in arithmetic considerations and in the problem of tangents, he might have anticipated the important concept of the limit of a quotient and have discovered the

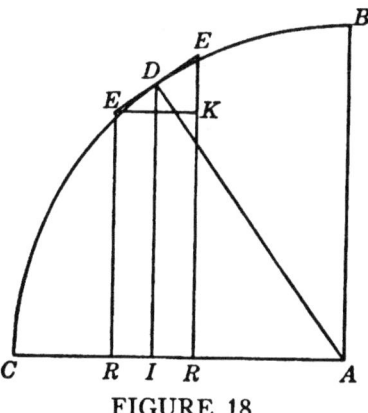

FIGURE 18

significance of this for the determination of both tangents and quadratures. Had he done this, he would have hit upon the crucial point in the calculus some seven years before Newton and about fourteen years before Leibniz. The latter, who later, as we shall see, made use of this very diagram to establish his infinitesimal calculus, said, in 1703, in a letter to James Bernoulli, that sometimes Pascal seemed to have had a bandage over his eyes.[201] This apparent lack of imagination was very likely the result of a predilection for the classical, such as later restrained the scientist Huygens also from making full use of the new procedures.

[201] See Leibniz, *The Early Mathematical Manuscripts*, pp. 15–16, and *Mathematische Schriften*, III, 72–73, n.

Pierre de Fermat, the friend of Pascal and perhaps the greatest French mathematician of the century, possessed a singular erudition and displayed an enthusiastic interest in Greek and Latin philology. This led him to study carefully such classic mathematical works as those of Archimedes, Apollonius, and Diophantus. The influence of the first of these three had been very strong for almost two centuries, but in the work of Viète and of Fermat we have evidence of the impression made by the other two also, as well as by the Arabic and Italian development of algebra. Viète realized the facility to be gained in the handling of geometric problems by their reduction to the solution of algebraic equations, a procedure which he therefore followed whenever possible.[202] Viète's equations betrayed their origin in geometry, in that he was always careful to have them all homogeneous; but his work was nevertheless, in a sense, an inversion of the Greek view, in accordance with which algebraic equations were reduced to geometric constructions for purposes of solution.

Fermat was familiar with the methods of Viète and developed them into an analytic geometry at about the time that Descartes was preparing his famous *Géométrie* of 1637. The work of Fermat and Descartes went much further than did either the algebraic solution of geometric problems by Viète or the graphical representation of variables by Oresme, for it associated with each curve an equation in which are implied all of the properties of the curve. This recognition, which Fermat expressed in calling the equation the "specific property" of the curve, constitutes the basic discovery of analytic geometry. Although in publication Fermat was anticipated by Descartes, he far outdid his rival in the application of the new point of view to the problems of infinitesimal analysis, which the books of Kepler and Cavalieri had popularized.

All of the anticipations of methods of the calculus which we have so far considered were related to geometry. Infinite series had sometimes been employed, but they were derived from the geometrical representation of the problem. Infinitesimal lines, surfaces, and solids had been used, but not infinitesimal numbers. Aristotle had denied the infinitely small in arithmetic, for the obvious reason that since number was a

---

[202] For an unusually extensive account of this work, see Marie, *Histoire des sciences mathématiques*, III, 27–65.

collection of unities, no number could be smaller than one. As a result of the algebra and the analytic geometry of the sixteenth and seventeenth centuries, this attitude had been modified, as has already been seen in the case of Pascal. The present-day view of the symbols entering into an equation is that they represent, in general, continuous variables; for Fermat and Descartes, however, they represented indeterminate constants [203] to which line segments could be associated, the tacit assumption being made that to every segment there corresponded some number. To such a view there was nothing incongruous in the idea of infinitesimal constants or numbers, since they would correspond to the geometrical infinitesimals which were being used so successfully. These numerical infinitesimals arose first through some interesting problems considered by Fermat.

Pappus had spoken of a "minima et singularis proportio," and this led Fermat to dwell on the fact, as he explained in a letter of 1643, that in a problem which in general has two solutions, the maximum or minimum value gives but a single solution.[204] Thus if a line of length a is divided by a point P into two parts, x and $a - x$, there are in general two positions of P which will make the area of the rectangle on x and $a - x$ a given quantity, A. For the maximum area, however, there is only one position, the mid-point.

From this fact Fermat was led to formulate his remarkably ingenious and fruitful method for determining maximum and minimum values. His method first appeared in an article in 1638; but Fermat said that the discovery went back some eight or ten years previous.[205] The argument in this is as follows: Given a line segment of length a, mark off from one end a distance x. The area on the segments x and $a - x$ will then be $A = x(a - x)$. If instead of the distance x one were to mark off the distance $x + E$, the area would be $A' = (x + E)(a - x - E)$. For the maximum area the two values will be the same, from Pappus' observation, and the points x and $x + E$ will coincide. Consequently,

[203] Wallner, "Entwickelungsgeschichtliche Momente bei Entstehung der Infinitesimalrechnung," p. 119.

[204] See Giovannozzi, "Pierre Fermat. Una lettera inedita."

[205] See Wieleitner, "Bermerkungen zu Fermats Methode der Aufsuchung von Extremenwerten." In a letter to Roberval, written in 1636, Fermat said that in 1629 he was in possession of his method of maxima and minima. See Paul Tannery, "Sur la date des principales découvertes de Fermat"; cf. also Henry, "Recherches sur les manuscrits de Pierre de Fermat."

setting the two values of A equal to each other and letting E vanish, the result is $x = \dfrac{a}{2}$.[206]

The procedure which Fermat here employed is almost precisely that now given in the differential calculus, except that the symbol $\triangle x$ (or occasionally h) is substituted for E. In his work there appeared, for perhaps the first time, the idea which has become basic in such problems—that of changing the variable slightly and then letting this change vanish. However, the reasoning by which Fermat supported his method is far less clear than that given at the present time. Modern analysis makes use of the concept of the limit, as the change $\triangle x$ *approaches* zero. Fermat, however, seems to have interpreted the operation as one in which E vanishes in the sense of actually *being* zero. For this reason, as Berkeley remarked in the following century, it is difficult to see by what right he took the positions x and $x + E$ to be different and yet in the end said that they coincide. Fermat's argument has frequently been interpreted in terms of the limit concept,[207] in which E is to be regarded as a variable quantity approaching zero. Fermat, however, does not appear to have thought of it in this way.[208] In fact the function concept and the idea of symbols as representing variables does not seem to enter into the work of any mathematician of the time.

In answering criticisms of his method, Fermat presented a statement of his reasoning which appears to link it with the remarks of Oresme and Kepler on the change at a maximum point.[209] He justified the equating of the two values of A by remarking that at a maximum point they are not really equal but they should be equal. He therefore formed the pseudo-equality[210] which became equality on letting E be zero. From this it is clear that he was thinking in terms of equations and the infinitely small, rather than of functions and the limit concept.

[206] Fermat, *Œuvres*, I, 133–34, 147–51; III, 121–22; Supplement, pp. 120–25. Cf. also Voss, "Calcul différentiel," p. 246.

[207] Duhamel, "Mémoire sur la méthode des maxima et minima de Fermat, et sur les méthodes des tangentes de Fermat et Descartes"; cf. also Mansion, "Méthode, dite de Fermat, pour la recherche des maxima et minima."'

[208] See Wallner, "Entwickelungsgeschichtliche Momente," pp. 122–23; cf. also Paul Tannery, *Notions historiques*, p. 344.

[209] Paul Tannery, however, thinks Fermat borrowed nothing from Kepler, whose works he probably had not read. See his Review of Vivanti, *Il concetto d'infinitesimo*, p. 232.

[210] "Adaequalitas." See *Œuvres*, I, 133–79, for his justification.

None the less, the method worked so beautifully that it found a ready acceptance among mathematicians. As a result, infinitesimals were uncritically introduced into analysis, to become firmly intrenched as the basis of the subject for about two centuries before giving way, as the fundamental concept of the calculus, to the rigorously defined notion of the derivative. Even now the subject is generally known as the "infinitesimal calculus," in spite of the fact that the infinitesimal, while of great pragmatic value in adding to the facility of manipulation of the subject in exercises, is logically secondary and even unnecessary.

Fermat was led by the success of his method to apply it, about 1636, to the determination of tangents to curves. This he did as follows: Let the curve be a parabola (fig. 19). Then from the "specific property" of

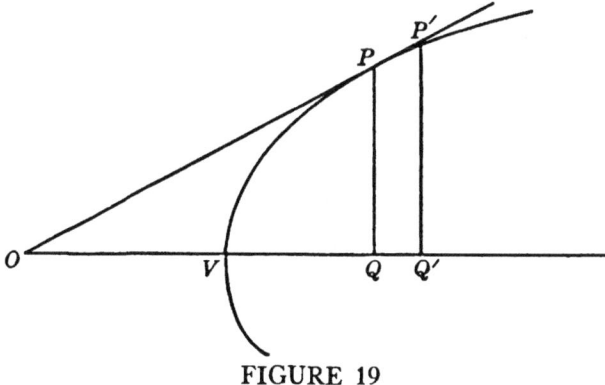

FIGURE 19

the curve it is clear that if we set $OQ = a$, $VQ = b$, and $QQ' = E$, we shall have $\dfrac{b}{b + E} > \dfrac{a^2}{(a + E)^2}$. Torricelli, in his work on the tangents to parabolas, had frequently set down such inequalities.[211] However, whereas Torricelli had made use of arguments by a reductio ad absurdum, Fermat's characteristic procedure resembles more closely the method of limiting values. Inasmuch as for small values of E the point P' may be regarded as practically on the curve as well as on the tangent line, the inequality becomes, as in the method for maximum values, a pseudo-equality. By allowing E to vanish, this pseudo-equality becomes a true equality and gives the desired result, $a = 2b$.[212]

[211] See *Opere*, I (Part 2), 304 ff., 315 ff.   [212] *Œuvres*, I, 134–36; III, 122–23.

The method of tangents Fermat believed to be an application of his method for maxima, but he was unable to explain what quantity he was maximizing. Descartes naturally supposed that it was the length of the line from the curve to a fixed point O on the axis of the parabola. However, on carrying out Fermat's method on this assumption, the result he obtained was, of course, different from that of Fermat. What Descartes had really found was the normal to the curve—that is, the minimum distance from a point on the axis to the curve. This would have furnished an excellent illustration of the method for determining a maximum or a minimum value, but inasmuch as Fermat had not given a rule for distinguishing maxima from minima, neither he nor Descartes recognized it as such. Descartes simply concluded that although the result Fermat had obtained was correct, the method was not generally applicable.

Perhaps as the result of the unnecessarily bitter criticism and supercilious attitude of Descartes, Fermat later modified his explanation for determining tangents.[213] Instead of interpreting the method in terms of maxima, he said that the point P' was indifferently taken as on the curve or on the tangent line. Then after forming the pseudo-equality, the quantity E was to vanish, to give the desired result. This procedure is strictly comparable to that now employed in the calculus, the theoretical justification of which is given in terms of limits; but the explanation of Fermat resembles rather the neglect of infinitesimals to be found in the work of Leibniz. It is also strikingly suggestive of the doctrine of perfect and imperfect equations, presented almost two hundred years later by Carnot in his attempted concordance of the conflicting views of the calculus then prevailing.

Fermat applied analogous considerations to the problem of determining the center of gravity of a segment of a paraboloid, again under the misapprehension that he was employing the method of maxima and minima. In this he let the center of gravity, O, of the segment be a units from the vertex. On decreasing the altitude, h, of the segment by E, the center of gravity is changed. Fermat, however, knew from a number of lemmas that the distances of the centers of gravity of the two segments are proportional to the altitudes and that the volumes of the segments are to each other as the squares of the altitudes. By

<hr>

[213] Duhamel, "Mémoire sur la méthode des maxima et minima de Fermat," pp. 310–16.

taking moments about O, he was able to make use of these facts to set up a pseudo-equality involving a, h, and E. In accordance with his general principle, he allowed E to vanish and obtained the result $a = \frac{2}{3}h$.[214]

The determination of the center of gravity of the paraboloidal segment did not constitute a new result. Archimedes had calculated this some nineteen hundred years earlier in the *Method*, and Commandino and Maurolycus had rediscovered it only a century before. Nevertheless, this exercise of Fermat is significant in the history of the calculus as the first determination of the center of gravity by means of methods equivalent to those of the differential calculus, instead of by means of a summation resembling those of the integral calculus. Fermat's friend Roberval was astonished that one should be able to obtain by means of this maximum and minimum method a result which had generally been derived from summation considerations. The integral calculus is, of course, implied in the lemmas which Fermat employed in this connection, and the method of maxima served somewhat indirectly to determine simply the value of the constant of proportionality entering into these. Nevertheless, this theorem might have led Fermat to a recognition of the significance of the inverse nature of summation and tangent problems. That this escaped him is the more strange, in that he developed remarkable procedures for the determination of quadratures as well as for tangents.

The equivalents of what we express as $\int_0^a x^n dx = \dfrac{a^{n+1}}{n+1}$ had appeared in various forms in the work of Cavalieri, Torricelli, Roberval, and Pascal. Fermat also gave demonstrations of this rule—in fact, he may well have anticipated all the others in this respect—one of which is strikingly different from those given earlier.[215] In his earlier investigations, about 1636, he appears to have made use of the inequalities

$$1^m + 2^m + 3^m + \ldots + n^m > \frac{n^{m+1}}{m+1} > 1^m + 2^m + 3^m + \ldots + (n-1)^m$$

to establish the result for positive integral values of n. This constituted a generalization of the inequalities of Archimedes which was known also

[214] *Œuvres*, I, 136–39; III, 124–26.
[215] For an excellent exposition of this method of quadratures, see Zeuthen, "Notes," 1895, pp. 37–80.

to Roberval. Fermat may also have given a proof, based upon the formation of figurate numbers, similar to that of Pascal.[216] However, before 1644 he had found the quadratures, cubatures, and centers of gravity of the "parabolas" of fractional degree $a^m y^n = b^n x^m$,[217] curves which he seems to have been the first to propose, but which were investigated also by Cavalieri, Torricelli, Roberval, and Pascal. Fermat, therefore, probably possessed as early as that date a general proof of the theorem for rational fractional values as well, although the definitive redaction of this was not made until 1657.[218]

In this connection, Fermat's procedure constitutes a generalization of one found in the *Opus geometricum* of Gregory of St. Vincent, although Fermat may not have known of this work, for he here mentions only Archimedes. Gregory had shown that if along the horizontal asymptote of a rectangular hyperbola points are marked off whose distances from the center are in continued proportion, and if at these points ordinates are erected to the curve, then the areas intercepted between these are equal.[219] Fermat modified this process in such a way that it could be applied to both the general fractional hyperbolas and parabolas. In finding the area under $y = x^{\frac{p}{q}}$,[220] for example, from 0 to x, he would take points on the axis with abscissas $x, ex, e^2x, \ldots$ where $e < 1$ (fig. 20). Then erecting the ordinates at these points, the areas of the rectangles constructed upon successive ordinates will form an infinite geometric progression. As Gregory of St. Vincent and Tacquet had earlier found the sums of such progressions, so here Fermat determined the sum of the rectangles as $x^{\frac{p+q}{q}}\left(\dfrac{1 - e}{1 - e^{\frac{p+q}{q}}} \right)$. In order to find the area under the curve, however, one must have not only an infinite number of such rectangles, but the area of each must be infinitely small. This can be brought about by setting $e = 1$. Before

[216] *Ibid.*, pp. 42–43.

[217] Mersenne, *Cogitata physico-mathematica;* see preface to *Tractatus mechanicus.* See also *Œuvres de Fermat*, I, 195–98.

[218] Zeuthen, "Notes," 1895, pp. 44 ff.

[219] *Opus geometricum*, Proposition CIX, p. 586.

[220] The notation of Fermat has been here slightly modified to make the meaning more clear. The equations of Fermat retained the homogeneous character found in Viète.

doing this, however, Fermat made the substitution $e = E^q$ in order to evaluate the indeterminate form. The sum then becomes

$$x^{\frac{p+q}{q}}\left(\frac{1 - E^q}{1 - E^{p+q}}\right) = x^{\frac{p+q}{q}} \cdot \frac{(1 - E)(1 + E + E^2 + \ldots + E^{q-1})}{(1 - E)(1 + E + E^2 + \ldots + E^{p+q-1})}.$$

When e "vanishes," E does likewise, and the sum is then $\dfrac{qx^{\frac{p+q}{q}}}{p + q}$, which is the area under the curve. By taking $e > 1$, Fermat applied the same method to the fractional hyperbolas also, finding the area under these from any abscissa to infinity.[221]

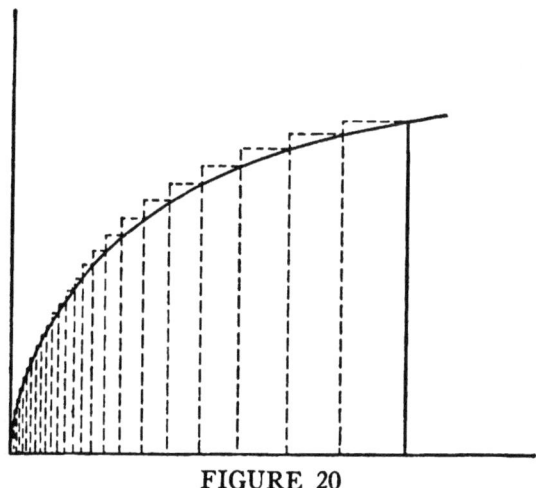

FIGURE 20

In these quadratures we see most of the essential aspects of the definite integral—the division of the area under the curve into small elements of area, the approximate numerical determination of the sum of these by means of rectangles and the analytic equation of the curve, and finally an attempt by Fermat to express the equivalent of what we would call the limit of this sum, as the number of elements is indefinitely increased and as the area of each becomes indefinitely small. One is almost tempted to say that Fermat recognized all the aspects except that of the integral itself; that is, he did not recognize the operation involved as significant in itself. The procedure was for him, as

[221] *Œuvres*, I, 255–88; III, 216–40.

it had been for all of his predecessors, simply that of finding a quadrature—of answering a specific geometrical question. Only with Newton and Leibniz were the processes involved in infinitesimal considerations recognized as constituting operations, independent of any geometrical or physical considerations, to which characteristic names were applied.[222]

That a curvilinear area, such as those under Fermat's parabolas and hyperbolas, could be equal to one bounded only by straight lines had been known in antiquity. It had long been held impossible, however, that a curved line could be exactly equal in length to a straight line,[223] and Fermat shared this view with a number of his contemporaries. Sluse and Pascal in this connection expressed admiration for the order of nature, which refused to allow a curve to equal a line.[224] Gregory of St. Vincent, Torricelli, Roberval, and Pascal, nevertheless, had by infinitesimal and kinematic means compared the arcs of spirals with those of parabolas. Then, shortly before 1660, there suddenly appeared a number of rectifications of curved lines by William Neil, Christopher Wren, Heinrich van Heuraet, John Wallis, and others.[225]

These were, in general, based upon approximations to the curve by means of polygons, followed by applications of infinitesimal or limit considerations. Upon hearing of them, Fermat himself carried out a rectification of the semicubical parabola. His procedure in this connection is typical of his general approach and indicates well the interrelation of the various aspects of his work. For any point P on the curve with abscissa $OQ = a$ and ordinate $PQ = b$, the subtangent, $TQ = c$, is known by his tangent method to be $c = \frac{2}{3}a$ (fig. 21). If then an ordinate $P'Q'$ to the tangent line is erected at a distance E from the ordinate PQ, the segment PP' is known in terms of a and E. For the curve $ky^2 = x^3$, this is $PP' = E\sqrt{\dfrac{9a}{4k} + 1}$. But the point

---

[222] Simon appears to be overenthusiastic in saying of this work ("Geschichte und Philosophie der Differentialrechnung," p. 119): "F. hatte auch bereits in ähnlicher Weise wie später Riemann den Begriff des bestimmten Integrals erfasst bei der Berechnung von $\int x^d dx$. Hier ist Grenzübergang, hier ist Bestimmung des Werthes $\frac{0}{0}$, hier is völliges Bewusstsein des continuitäts gesetzes."

[223] Kaestner, *Geschichte der Mathematik*, I, 498; III, 283.

[224] See Zeuthen, "Notes," 1895, pp. 73–76; and Pascal, *Œuvres*, VIII, 145, IX, 201.

[225] Moritz Cantor, *Vorlesungen*, II, 827 ff.

P', for small values of E, may be regarded as on the curve as well as on the tangent line, so that the length of the curve can be thought of as the sum of segments such as PP'. The sum of these segments, in turn, can be taken as the area under the parabola $y^2 = \dfrac{9x}{4k} + 1$. Inasmuch as the quadrature of this is known, the length of the curve is determined.[226]

It is surprising that Fermat, who used his method of maxima and minima for finding centers of gravity, who reduced a problem of rectification which involved tangents to a question of quadratures, who

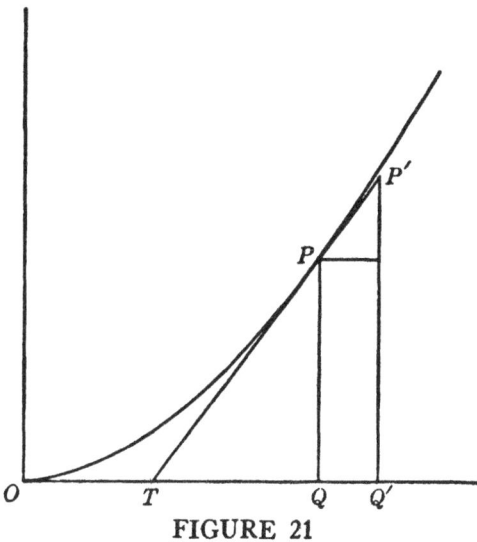

FIGURE 21

used infinitesimals geometrically and analytically in such a wide variety of problems, should have missed seeing, as Pascal had also, the fundamental connection between the two types of questions. Because these two men did not see this, in problems in which integrations by parts would now be employed they had recourse to clever geometrical transformations. Fermat, in his problems, made use of diagrams which are much the same as the one of Pascal's which Leibniz later found so suggestive of his differential triangle, and yet he did not perceive their deeper significance. Had Fermat only observed more

[226] For Fermat's rectification, see *Œuvres*, I, 211–54; III, 181–215.

closely the results for the tangents and the quadratures of his parabolas and hyperbolas, he might have discovered the fundamental theorem of the calculus and have become, what he has sometimes been unwarrantedly called, the "true inventor of the calculus."[227]

Fermat, of course, realized in a sense that the two types of problems had an inverse relationship. That he did not pursue this thought further may well have been due to the fact that he thought of his work simply as the solution of geometrical problems and not as representing a type of argument significant in itself. His methods of maxima and minima, of tangents, and of quadratures he regarded as constituting characteristic approaches to these questions, rather than a new type of analysis. Furthermore, they were apparently restricted in application. Fermat knew how to make use of them only in the case of rational expressions, whereas Newton and Leibniz, through their application of infinite series, recognized the universality of such procedures. Nevertheless, no mathematician, with the possible exception of Barrow, so nearly anticipated the invention of the calculus as did Fermat.

The influence of Fermat on his contemporaries and immediate successors[228] is difficult to determine. Probably his work was not so well known as that of Cavalieri, for the latter was widely read in his two famous books, whereas Fermat did not publish either his methods or his results. For this failure to publish his work, Fermat, it has been said,[229] lost the credit for the invention of the calculus; but such an assertion is incorrect. In the first place, it is clear that he cannot be thought of as its inventor. Secondly, his work was collected and printed posthumously as the *Opera varia*, in 1679, before either of the earliest published accounts of the calculus by Newton or Leibniz had appeared.

In spite of the fact that Fermat did not himself publish his methods,

[227] Lagrange, Laplace, and Fourier have so called him, but Poisson has correctly pointed out that Fermat does not deserve such a designation, inasmuch as he failed to recognize the problem of quadratures as the inverse of that of tangents. The relevant statements of these four men may all be found in Cajori, "Who Was the First Inventor of the Calculus?" Cf. also Marie, *Histoire des sciences mathématiques*, IV, 93 ff. Sloman has very unfairly said that "Fermat hardly deserves to be named at all" in this connection. See his *Claim of Leibnitz to the Invention of the Differential Calculus*, pp. 45–47.

[228] See Genty, *L'Influence de Fermat sur son siècle*. This book is more concerned, however, to show the priority and independence of Fermat's results than specifically to point out their effects on others.

[229] Dantzig, *Number, the Language of Science*, pp. 131–32.

these became known through his correspondence with such men as Roberval, Pascal, and Mersenne, as well as through the publication by others during his lifetime of portions of it. As a consequence, his work was in large part responsible for a number of transition methods which appeared just before the advent of the calculus. It is therefore scarcely correct to say, with Lagrange,[230] that "Fermat's contemporaries did not seize the spirit of this new type of calculus." Infinitesimal considerations such as those of Fermat constituted a large portion of the mathematical activity of the period. Nevertheless, there was indeed one great mathematician who remained somewhat cool toward the new views, even though in his early years he had made effective use of them. This was René Descartes, the severest critic of Fermat.

The very first mathematical production of Descartes was an attempt, in 1618, to deal with the laws of falling bodies by means of infinitesimals. In this he made an error, in that he assumed, as Galileo had also in 1604,[231] that the velocity was proportional to the distance rather than the time; but if the distance axis in his demonstration were changed to the time axis, his procedure would be essentially that which Oresme and other Scholastics had used.[232] Descartes was probably familiar with their works and may have derived this form of proof, as well as suggestions on analytic geometry, from reading Oresme.[233]

At any rate, Descartes was acquainted with ancient, medieval, and modern views on infinitesimals, and used them. In a second memoir of about the same time he wrote on fluid pressure. In this connection he may have known through Beekman of Stevin's work on infinitesimals. At all events, in considering the force drawing a body, he used such phrases as the "first instant of its movement," and the "first imaginable speed."[234] Some years later, in 1632, he answered correctly a number of

[230] See Brassine, *Précis des œuvres mathématiques de P. Fermat*, p. 4.
[231] Duhem, *Études sur Léonard de Vinci*, III, 564.
[232] See Descartes, *Œuvres*, X, 219; cf. also X, 59, 76–77.
[233] There is a great difference of opinion on this subject. Wallner ("Entwickelungsgeschichtliche Momente," p. 120) sees not the least influence of Oresme on Descartes; Stamm ("Tractatus de continuo," p. 24) says that the problem of the latitude of forms of Oresme was undoubtedly the most important influence on Descartes; Wieleitner ("Ueber den Funktionsbegriff," p. 242) says that Descartes undoubtedly knew of Oresme's work, but that the essential idea of the dependence of quantities found in analytic geometry is missing in Oresme; Duhem (*Études sur Léonard de Vinci*, III, 386) has claimed that Oresme created analytic geometry.
[234] Milhaud, *Descartes savant*, pp. 162–63.

questions, which Mersenne had sent him, on areas, volumes, and centers of gravity connected with the parabolas $y^n = px$—problems similar to those which Fermat had solved. Descartes did not tell what method he had used, but it was probably a skillful application of the methods of Archimedes, Kepler, and Cavalieri.[235] However, after the publication of his famous *Géométrie* in 1637, Descartes' interest in the subject began to wane,[236] for his mathematical work was only an episode in the development of his philosophy. Consequently he did not participate effectively in the development of the infinitesimal methods which preoccupied the minds of most mathematicians at this time.[237] The acrimonius quarrel with Fermat, however, sustained his interest in the problem of tangents and led him to considerations which, if pursued further, might have been more effective than the infinitesimal methods in leading to a clearer understanding of the basis of the calculus.

The ancients, with the possible exception of Archimedes, had not developed a general definition of a tangent to a curve nor any method of determining it. Descartes, however, realized more fully than a number of his contemporaries that this constituted not only "the most useful and general problem that I know but even that I have ever desired to know in geometry."[238] He thereupon elaborated his celebrated method of tangents in terms of the equality of roots. Descartes' method consisted in passing through two points of the curve a circle with its center on the x axis, and then making the points of intersection coincide. The center of the circle thus becomes the point on the x axis through which the normal to the curve passes, and the tangent is consequently known. This may be illustrated in somewhat simplified form as follows: Let the tangent at the point (a, a) on the parabola $y^2 = ax$ be required. The equation of the circle going though (a,a) and having its center on the axis is $x^2 + y^2 - 2hx + 2ah - 2a^2 = 0$, where h is undetermined. Substituting ax for y^2, the quantity h may then be so determined that the resulting equation has equal roots—that is, so that the intersections of the circle with the parabola

[235] *Ibid.*, pp. 164–68.
[236] *Ibid.*, p. 246; cf. also Marie, *Histoire des sciences mathématiques et physiques*, IV, 21.
[237] Milhaud (*Descartes savant*, pp. 162–63) denies that it was the insistence on clear ideas which made Descartes avoid the use of infinitesimals.
[238] Descartes, *Œuvres*, VI, 413.

coincide.[239] This value, $h = \frac{4}{3}a$, is the abscissa of the point on the axis through which the normal to the parabola passes. The tangent is then the line through (a, a) perpendicular to this normal.

It is to be remarked that Descartes' method is purely algebraic, no concepts of limits or infinitesimals being manifestly involved. However, any attempt to interpret geometrically the significance of the case in which the roots are equal, to explain what is meant by speaking of coincident points, or to define the tangent to a curve, would necessarily lead to these conceptions. If Descartes in his geometry had thought in terms of continuous variables rather than of a correspondence between symbols which represented lines in a geometrical diagram,[240] he might have been led to interpret his tangent method in terms of limits, and so have given a different direction to the anticipations of the calculus. His algebra, however, was still grounded in the geometry of lines, and the idea of a continuously varying quantity was not really established in analysis until the time of Euler.[241]

In criticising Fermat's method of tangents, Descartes attempted to correct the method by interpreting it in terms of equal roots and coincident points, a procedure which was practically equivalent to defining the tangent as the limit of a secant.[242] Descartes did not express himself in this manner, however, inasmuch as the concept of a limit was far from clear at this time. Fermat, who was thinking of infinitesimals, could not see that his method had anything in common with the algebraic (limit) method of Descartes and so precipitated a quarrel as to priority, one of the many which the seventeenth century produced as the result of the confusion of thought as to the basis of infinitesimal methods. Descartes preferred his method of tangents to that of Fermat because of the apparent freedom from the concept of infinitesimals, although its application was frequently much more tedious and was limited to algebraic curves.

In finding the tangent to a nonalgebraic or "mechanical" curve, such as the cycloid, Descartes in 1638 made use of the concept of an instantaneous center of rotation. This is, of course, also directly connected with the use of limits and infinitesimals, but is expressible

[239] *Ibid.*, VI, 413–24; cf. also Voss, "Calcul différentiel," p. 244.
[240] Descartes, *Œuvres*, VI, 369; cf. also VI, 411–12.
[241] Fine, *Number-System*, p. 121.
[242] See Duhamel, "Mémoire," pp. 298–308; Milhaud, *Descartes savant*, pp. 159–62.

without the use of such terminology, through the circumlocution afforded by the notion of instantaneous velocity. Supposedly, this notion is intuitively clear, but at the time it was not rigorously defined. It had been made acceptable, moreover, by the work of Galileo,[243] and Roberval and Torricelli were employing it in geometry at practically the same time as Descartes. Descartes' reasoning was as follows: If a polygon is rolled along a straight line, any vertex will describe a series of circular arcs, the centers of which are the points on the line which the vertices of the polygon touch: *i. e.*, in rolling the polygon along the line, we rotate the polygon about each of these points in turn. The cycloid, now, is the curve generated by a point on a circle: *i. e.*, by a vertex of a polygon with an infinite number of sides, as it is rolled along a line. The cycloid is therefore made up of an infinite number of circular arcs, and the tangent at any point, P is therefore perpendicular to the line joining P to the point Q in which the generating circle touches the base line. Inasmuch as Q can easily be determined, the tangent at P can be drawn.[244]

In Descartes' work one sees an avoidance of the idea of an infinitely small quantity in mathematics and the use, instead, of algebraic and mechanical conceptions. Whereas Fermat saw only the practical advantages of the infinitesimal methods, Descartes saw better the risks they entailed. Descartes' evasion of them was of course justified by the lack of a clear theoretical basis for infinitesimal reasoning, but it was opposed to the mathematical trend of the time. We have seen that in the years following his *Discours de la méthode*, there appeared in print an unprecedentedly large number of works devoted to infinitesimal methods. In most of them the work was largely based on synthetic geometry, although Roberval and Pascal showed an arithmetizing tendency in their quadratures. In France only Fermat made effective use, in the anticipations of the calculus, of the new analytic methods which he and Descartes were developing. In England, however, the mathematician and theologian John Wallis applied analytic geometry to the problem of quadratures with comparable success.

John Wallis had become familiar with analytical methods largely

[243] De Giuli ("Galileo e Descartes") asserts that Descartes owed to Galileo much of his philosophic method.
[244] Walker, *A Study of the Traité des Indivisibles of Roberval*, pp. 137–39.

through Harriot. In his work on conic sections, Wallis followed Viète, Descartes, Fermat, and Harriot in the application of literal algebra to the problems of geometry. Wallis, however, went far beyond these men in that he sought to free arithmetic completely from geometric representation, a goal which he thought would be easily reached.[245] He first showed how all the theorems of Euclid V could be derived arithmetically without difficulty, and then in algebra broke away from the idea, derived from geometry, that the terms of an equation must be homogeneous. Luckily, Wallis did not worry overmuch about mathematical rigor: we know now how difficult the arithmetizing of mathematics was to be.

Instead of observing the caution which classical ideas of rigor exacted, Wallis was influenced by prevailing thought to make free use of analogy and incomplete induction in his work, as well as of the concepts of infinity and infinitesimals, which had not yet been rigorously established. We have seen these tendencies developing in the work of Cavalieri and Fermat and, in continuing this tradition, Wallis came nearer to the limit concept than did any other of Newton's predecessors. It is clear that this notion is implicit in the work of most of his French and Italian contemporaries, but it was not expressed by them. Instead, the concept of the infinitesimal was employed. With Wallis' arithmetical point of view, however, one is brought again into more direct touch with the limit idea which the mathematicians of the Low Countries—Stevin, Gregory of St. Vincent, and Tacquet—had sought to formulate.

To what extent Wallis was influenced by the ideas of these other men is difficult to determine. He wrote the most complete treatise on statics since the time of Stevin,[246] and may well have been familiar with the arithmetical limit methods of the latter, either directly or through the somewhat similar work of Roberval.[247] Wallis admitted that upon the advice of Wren he read, about 1652, part of the *Opus geometricum* of Gregory of St. Vincent, but he added that in this he did not run across any propositions new to him.[248] On the other hand, Wallis and Tacquet were probably independent of each other in their

[245] Prag, "John Wallis." [246] Duhem, *Les Origines de la statique*, II, 211.
[247] Walker, *A Study of the Traité des Indivisibles of Roberval*, p. 77; cf. also p. 165.
[248] Wallis, *Opera mathematica*, Vol. II, *Arithmetica infinitorum*, Preface.

work, inasmuch as their books on this subject appeared almost simultaneously—in 1655 and 1656 respectively. Oddly enough, moreover, although Wallis displays the arithmetizing and limit tendencies of Tacquet, the chief inspiration for his work came rather from reading, in 1650, the geometrical method of indivisibles of Cavalieri as expounded by Torricelli. This he professed in the prefaces of two of his books: *De sectionibus conicis tractatus*, and *Arithmetica infinitorum sive nova methodus inquirendi in curvilineorum quadraturam.*[249] Cavalieri's work, however, had been almost purely geometrical, whereas Wallis proceeded largely arithmetically, and in the end abstracted from the geometry of indivisibles the arithmetic notion of a limit. "Following Oughtred, Descartes, and Harriot," he applied to his demonstrations the symbolism of arithmetic in order to give to them "at the same time the maximum brevity and perspicuity." The use of arithmetic calculation he held to be simpler and not less "legitimate or scientific" than that by lines.[250]

The manner in which Wallis made the transition from the geometry of lines to the arithmetic of numbers is brought out clearly in his proof that the area of a triangle is the product of the base by half the altitude.[251] He assumed at the outset, as had Cavalieri, that a plane figure may be regarded as made up of an infinite number of parallel lines— or rather, as he preferred, of an infinite number of parallelograms, the altitudes of which are equal, that of any one of them being $\frac{1}{\infty}$, or an infinitely small aliquot part of the altitude of the figure.[252] Here we have not only the first appearance of the symbol ∞ for infinity,[253] but also the earliest use of the Scholastic categorematic infinity in the field of arithmetic. Furthermore, the treatment by Wallis of the infinitely small is far more daring and decisive than that of Fermat. Whereas the latter had not expressly called his symbol E an infinitesimal, Wallis for his part said that $\frac{1}{\infty}$ represented an infinitely small quantity, or *non-quanta*. A parallelogram whose altitude is infinitely

[249] *Opera mathematica*, Vol. II.

[250] *Opera mathematica*, Vol. II, *De sectionibus conicis*, "Dedicatio," and also p. 3.

[251] *Ibid.*, pp. 4–9. A good summary of the work of Wallis along these lines is to be found in Sloman, *The Claim of Leibnitz to the Invention of the Differential Calculus*, pp. 8 ff.

[252] *Opera mathematica*, Vol. II, *De sectionibus conicis*, p. 4.

[253] "Esto enim ∞ nota numeri infiniti." *Ibid.*

small or zero is therefore "scarcely anything but a line," except that this line is supposed "extensible, or to have such a small thickness that by an infinite multiplication a certain altitude or width can be acquired."[254]

Returning to the proposition on the area of the triangle, Wallis supposed this to be divided into an infinite number of lines, or infinitesimal parallelograms, parallel to the base. The areas of these, taken from the vertex to the base, form an arithmetic progression beginning with zero. Moreover, there is a well-known rule that the sum of all the terms in such a progression is the product of the last term by half the number of terms. Since "there is no cause for discrimination between finite and infinite numbers," it can be applied to the areas in the triangle. If the altitude and base of the triangle are taken as A and B respectively, the area of the last parallelogram in the progression will then be $\frac{1}{\infty} A.B.$ The area of the whole triangle is therefore $\frac{1}{\infty} A.B. \cdot \frac{\infty}{2}$, or $\frac{1}{2}A.B.$[255] He then applied a similar type of argument to numerous quadratures and cubatures involving cylinders, cones, and conic sections.

Wallis realized that his procedure was highly unorthodox, but he said that it could be verified by "that very well-known apagogic method" of in-and-circumscribed figures. To give this Wallis felt would be superfluous, because "the frequent iteration would create nausea in the reader." Furthermore, he said that anyone versed in mathematics could supply such proof, since it occurred frequently among the ancients and the moderns.[256] Modern mathematics has found it necessary to modify greatly the view of the infinite which Wallis held and to banish entirely his infinitely small magnitudes. Nevertheless, the development of the calculus is the result of efforts, such as his, to substitute for the prolixity of the method of exhaustion a direct arithmetical analysis.

The procedure of Wallis in *De sectionibus conicis* was based largely upon crude manipulations of his symbol ∞. In the *Arithmetica infinitorum*, however, he pursued similar investigations from a somewhat different point of view—one resembling more closely the arithmetic

[254] *Ibid.*
[255] *Ibid.*, pp. 8–9; cf. also *Opera mathematica*, Vol. II, *Arithmetica infinitorum*, p. 2.
[256] *Opera mathematica*, Vol. II, *De sectionibus conicis*, p. 6.

methods of Stevin and Roberval and the limit concept. In this connection he demonstrated the equivalent of the theorem: $\int_0^a x^n dx = \dfrac{a^{n+1}}{n+1}$, apparently unaware that this proposition had appeared in numerous forms during the preceding twenty years. Wallis reached this conclusion by observing first the equalities

$$\frac{0+1}{1+1} = \tfrac{1}{2}; \frac{0+1+2}{2+2+2} = \tfrac{1}{2}; \frac{0+1+2+3}{3+3+3+3} = \tfrac{1}{2}; \dots$$

In these the ratio is $\tfrac{1}{2}$ for any finite number of terms, from which Wallis concluded that this will be the ratio likewise for an infinite number. Through this, Wallis arrived at an alternative form of demonstration for the theorem above on the area of a triangle.[267]

In proceeding further, Wallis noted that in the equalities

$$\frac{0+1}{1+1} = \tfrac{1}{3} + \tfrac{1}{6}; \frac{0+1+4}{4+4+4} = \tfrac{1}{3} + \tfrac{1}{12}; \frac{0+1+4+9}{9+9+9+9} = \tfrac{1}{3} + \tfrac{1}{18}; \dots,$$

the greater the number of terms, the more closely does the ratio approximate to $\tfrac{1}{3}$, so that "at length it differs from it by less than any assignable magnitude." If this is continued to infinity, the difference "will be about to vanish completely." Consequently, the ratio for an infinite number of terms is $\tfrac{1}{3}$.[268]

Wallis then proceded to observe, in a similar manner, that the analogous ratios for the third, fourth, fifth, and higher powers of the integers are respectively $\tfrac{1}{4}, \tfrac{1}{5}, \tfrac{1}{6}$, and so on. He then affirmed the validity of the rule for all powers, rational or irrational (except, of course, -1).[269] This extension of the work of his French and Italian predecessors was made upon the basis of what Wallis spoke of as interpolation and induction. By the former, he seems to have had in mind a principle of continuity,[260] or of permanence of form, by which the rule could be asserted to hold for values intermediate between those for which it was known to be valid. By the latter, Wallis meant not mathematical or complete induction but induction in the scientific sense,

[267] *Opera mathematica*, Vol. II, *Arithmetica infinitorum*, pp. 1–3; cf. also p. 157.

[268] *Ibid.*, pp. 15–16; cf. also p. 158.

[269] *Ibid.*, pp. 31–53. Wolf (*A History of Science, Philosophy, and Technology in the Sixteenth and Seventeenth Centuries*, p. 209) has incorrectly stated in this connection that the predecessors of Wallis confined themselves to positive integral powers. Furthermore, he neglects to state that Wallis extended the rule to irrational powers also.

[260] Nunn, "The Arithmetic of Infinities."

analogy similar to that by which we have seen him conclude properties of the infinite from those of the finite. In this respect his work is a good indication of the looseness of thought at the time. More significantly, his extension of the rule to irrational powers indicates a tendency to break away from the persistent idea, derived from Pythagorean geometry, that irrational magnitudes are not numbers in the strict sense of the word. This was in line with his declaration of the independence of arithmetic from geometry, a freedom which was necessary for the later elaboration of the limit concept which he was here adumbrating.

The proposition above, on the ratios of the powers of integers, Wallis then applied to problems on quadratures and cubatures. In this respect he may be said to have determined the areas and volumes as limits of infinite sequences, in much the same way as Fermat had found them by means of infinite geometric progressions. In fact, the basis for the concept of the definite integral may be considered fairly well established in the work of Fermat and Wallis, although it was to become confused later by the introduction of the conceptions of fluxions and differentials. However, that neither of these men realized fully the significance of this concept is seen in their lack of clarity. We have seen that Fermat did not fully explain the nature of his symbol E. Wallis confused his work with the infinitesimal, identifying infinitely small rectangles with lines, and writing $\frac{1}{\infty} = 0$—ideas which were to lead to the conception, found in Leibniz, of the integral as a sort of totality, rather than as the limit of a sum.

Wallis was interested in another question which characteristically concerned mathematicians of the time—that of the angle of contact (horn angle) formed by two curves with a common tangent, a figure which Euclid, Jordanus Nemorarius, Cardan, and many others had considered. Discussion on this point may have been instrumental in maintaining the concept of the infinitely small as a valid notion, for it lent plausibility to the idea of an ultimate indivisible, smaller than any assignable magnitude and yet seemingly not the same as absolute zero.[261] The postulate of Archimedes of course excluded such angles

[261] For a full discussion of the history of this subject, see Vivanti, *Il concetto*, or the French summary of this book given in *Bibliothica Mathematica*, N. S., VIII (1894) as "Note sur l'histoire de l'infiniment petit."

as magnitudes, just as it had excluded other infinitesimals; but mathematicians of the seventeenth century regarded them as interesting illustrations of their concepts, and discussed the question as to whether they were zero or not. Galileo, Wallis,[262] and others asserted that such angles were absolutely zero, whereas Hobbes, Leibniz, and Newton, for example, held that they were in some way different from zero. On this point argument was possible, of course, only for two reasons: the general lack of critical definitions during this period (a rigorous definition of the tangent to a curve, although implied by the work of Descartes, Fermat, and others, having not been given at the time), and the failure to distinguish clearly between the geometrical figure and its arithmetical measurement. Both of these shortcomings were to be significant later in exposing the calculus of Newton and Leibniz to severe criticism.

After a century of doubt, clear definitions were formulated and the calculus was established upon arithmetical rather than geometrical conceptions. The work of Wallis was an attempt to bring about such an arithmetization, and in this respect it won the support of his contemporary, James Gregory. The latter, in his *Vera circuli et hyperbolae quadratura* of 1667, viewed the passage to the limit as an independent arithmetical operation, suitable as a means of defining new numbers not belonging to the ordinary irrationals.[263] In connection with this work, he constructed in and circumscribed polygons to the circle and hyperbola and showed that, by doubling the number of sides of these, converging series were obtained in which the difference became smaller and smaller. These series consequently had a limit, which, "if one may speak in this manner," could be considered the last polygon in each series. This consequently would give the area of the curvilinear figure.[264] The areas of the circle and the hyperbola Gregory gave to as many as twenty-six figures, although the limit he recognized as in general incommensurable.[265]

This work on the limit of converging infinite series represented a generalization of the earlier propositions of Gregory of St. Vincent and Tacquet on geometric progressions, with which James Gregory was

---

[262] *Opera mathematica*, Vol. II, *De angulo contactus et semicirculi.*

[263] Wallner, "Über die Entstehung des Grenzbegriffes," p. 258; cf. also Georg Heinrich, "James Gregorys 'Vera circuli et hyperbolae quadratura.' "

[264] *Vera circuli et hyperbolae quadratura*, pp. 15–16.   [265] *Ibid.*, pp. 48, 25.

acquainted.[266] He may also have been familiar with the somewhat similar arithmetic work of Roberval, for in another connection he used geometric transformations resembling the so-called Robervallian lines.[267] However, whereas Roberval and Wallis had been led to their arithmetization through the method of indivisibles, Gregory preferred to employ in quadratures the indirect method of the ancients, showing that the difference can be made less than any given quantity.[268] Nevertheless, in connection with this work he adopted the newer analytic methods of Descartes. In this respect also he followed the method of Fermat for determining tangents to curves. For example, the tangent to $y^3 = x^2(a + x)$ he found at the point for which $x = b$ as follows:[269] Choose a second point, with abscissa less than x by a vanishing small amount, o,[270] and assume, "if we may do so," that the corresponding ordinate may be taken indifferently as that of the curve or of the tangent. Then set up the suitable proportion, divide by o, and reject the terms in which o or a power of it remain. The subtangent is in this way found as $z = \dfrac{3b^2 + 3ab}{3b + 2a}$.

Although Gregory did not refer to Fermat in this connection, it is obvious that the methods of the two men are identical with the exception that the E of Fermat has been changed to o, a change of notation which was to be adopted a year or two later by Newton, perhaps under the influence of Gregory's work. The type of arithmetical and analytical work of Fermat, Wallis, and Gregory represented the tendency which was to lead to the calculus, but it met with almost immediate opposition, for the spirit of the age was directed toward the solution of problems through geometrical considerations. Such an arithmetization of mathematics was opposed with particular vigor by two Englishmen, the philosopher Thomas Hobbes and the mathematician and theologian Isaac Barrow.[271] Hobbes objected strenuously to "the whole herd of them who apply their algebra to geometry."[272]

[266] *Ibid.*, p. 20; cf. also p. 123.   [267] Galloys, "Réponse à l'écrit de M. David Gregorie."
[268] See *Geometriae pars universalis*, pp. 27–29, 74 ff.
[269] *Ibid.*, pp. 20–22.   [270] "Nihil seu serum o." *Ibid.*
[271] See Cajori, "Controversies on Mathematics between Wallis, Hobbes, and Barrow."
A summary of this article is given in *Bulletin, American Mathematical Society*, XXXV (1929), 13.
[272] Weinrich, *Über die Bedeutung des Hobbes für das naturwissenschaftliche und mathematische Denken*, p. 91.

He maintained that they mistook the study of symbols for that of geometry, and characterized the *Arithmetica infinitorum* as a "scurvy book."[273] He referred to the arithmetization represented by Wallis as absurd and as "a scab of symbols."[274]

This attitude toward algebra and analytic geometry was probably the result not only of the general predilection in the seventeenth century for geometrical rather than arithmetical methods, but also of Hobbes' exaggerated view of mathematics as an idealization of sensory perception, rather than as a branch of abstract formal logic. Greek thought had accepted mathematics as derived from the experience of the senses by the abstraction from concrete objects of irrelevant properties. Hobbes, however, was unwilling to regard lines as deprived of all breadth, or surfaces of all thickness.[275] Consequently, the infinitely small was for him merely the smallest possible line, plane, or solid—a view of infinitesimals held by the school of mathematical atomists in antiquity and not unlike that of Cavalieri.

Hobbes' view of number was comparable to his attitude toward the geometrical elements. He adopted the Pythagorean notion of number as a collection of units, and he interpreted ratios only in terms of geometrical considerations.[276] This attitude would not only operate against the free use of arithmetic processes which Wallis carried out on the basis of analogy or induction, but would, in fact, when combined with Hobbes' naïve view of geometrical magnitude, make the distinction between rational and irrational number, and the consequent introduction of the limit concept, logically superfluous. Hobbes' ideas in this direction were, as a result, of little consequence in the development of the concepts of the calculus. There is, however, another aspect of his thought which may have been more significant.

We have mentioned that Aristotle thought of motion as the fulfillment of the potential, thus centering attention not on the mathematical aspect of motion, but rather on the metaphysical, attributing to a body in motion a striving toward a goal. It could not of itself, however, reach its goal, since, in order that it do so, the constant application of a force was necessary. In the fourteenth century this Peripatetic theory was questioned, because it could not "save the phenomena." In its place,

[273] *English Works*, VII, 283.
[275] *Ibid.*, VII, 67, 200 ff., 438.
[274] Cf. *ibid.*, VII, 187, 361 ff.
[276] Cf. *Opera omnia*, IV, 27, 36.

Jean Buridan had substituted the doctrine of impetus (or inertia), or the tendency of a body to remain in motion.[277]

This Scholastic view of impetus gave to motion a so-called intensive characteristic, for it centered attention upon the act of moving, rather than on change of position or extension. Such a shift of emphasis made acceptable the notion of motion at a point, an idea which Aristotle had specifically rejected.[278] This was followed immediately by the quantitative treatment of instantaneous velocity found in Calculator, Hentisbery, and Oresme. The idea of impetus was familiar to Nicholas of Cusa, Leonardo da Vinci, and others of the fifteenth and sixteenth centuries, although it was often associated with, and obfuscated by, accretions of Neoplatonic mysticism and of vitalistic and teleological thought, such as are to be found in the works of Nicholas of Cusa, Paracelsus, and Kepler.[279] In 1638 the idea of impetus or inertia culminated in the famous clarification of the laws of dynamics given by Galileo; and at the same time the concept of instantaneous velocity was likewise successfully applied in geometry by Torricelli, Roberval, and Descartes.

It must be kept in mind, of course, that no definition of instantaneous velocity had at the time been given, nor could it have been framed before the development of the differential calculus. However, the concept of inertia was to make the idea of motion at a point intuitively and scientifically acceptable—as well as philosophically interesting and geometrically useful—until such time as it could be made mathematically rigorous.

Philosophers have generally displayed an interest in the calculus because of its associations with problems of motion and variation which present many intriguing metaphysical aspects. Hobbes was especially concerned, inasmuch as he wished to make motion basic in his philosophical scheme. His views were mathematically too naïve to allow of his adding to the discoveries being made in the new analysis, although he may have had a somewhat broader influence on the later

[277] Duhem, Études sur Léonard de Vinci, III, vii–viii.   [278] Physica VI. 234a.

[279] The great German astronomer, for example, in his Mysterium cosmographicum and his Astronomia nova, ascribed a vital faculty to the sun and the planets. In the 1621 edition of the former work, however, he said that one can substitute the word force for soul. See Opera omnia, I, 174; II, 270; III, 176, 178–79, 313. See also Lasswitz, Geschichte der Atomistik, II. 9–12; Duhem, Études sur Léonard de Vinci, II, 199–223.

thought and interpretation of the calculus. Hobbes was much impressed with the success of Galileo's conception of the laws of motion, in terms of inertia and changing velocity. Realizing that physics owes its development to mathematical representation,[280] he wished to describe and explain these ideas in geometrical and metaphysical terms. To this end, he introduced the concept of the *conatus*,[281] a beginning of motion, analogous to the concept of the point as the beginning of geometrical extension. Hobbes thus tried to emphasize, as had the men of the fourteenth century, the idea of motion at a point, rather than that of change of position. Not realizing, as Aristotle had discerned nineteen hundred years before, that instantaneous motion is an intellectual—not an empirical—concept, he attempted to formulate a definition in terms of his ingenuous nominalism, speaking of it as the motion in an infinitely small interval—an interval less than any given interval—that is, through a point.[282]

Hobbes was not so fortunate, as Newton and Leibniz were later, in formulating mathematically his concept of *conatus*. He did not understand the relation between number and spatial quantity, nor did he appreciate that instantaneous velocity is a purely numerical notion. His views, however, were perhaps significant in their influence upon the inventors of the calculus. The excessive nominalism of Hobbes was to lead mathematicians away from a purely abstract view of the concepts of mathematics, such as Wallis had displayed, and to induce them to seek, during more than a century, for an intuitively, rather than a logically, satisfactory basis for the calculus. It was largely on this account that both Newton and Leibniz sought to explain the new analysis in terms of the percipient notion of the generation of magnitude, rather than in terms of the logical conception of number only.[283] This idea of generation is more immediately apparent in the method of fluxions of the English empiricist, Newton; but the German philosopher, Leibniz, justified his differential method also in terms of the analogous idea of continuity. Whereas Newton used the physical idea of the "moment" of growing magnitudes, there grew up in Germany a more metaphysical form of this in the notion of *intensive* magnitude as opposed to *extensive* quantity. Upon this idea of a "tendency" or a

---

[280] *Opera omnia*, II, 137.   [281] *Opera omnia*, I, 177.   [282] *Ibid.*
[283] Cf. Vivanti, *Il concetto d'infinitesimo*, pp. 31–32.

"becoming," the mathematical infinitesimal throve, with the result that philosophers have been reluctant to abandon it, even though modern mathematics has shown that the basis of the calculus is to be found in the derivative rather than in the differential.

There is, furthermore, an inclination on the part of philosophers (and occasionally of mathematicians) to regard Hobbes' *conatus*, or the derived concept of the *intensive* in motion, as the answer to Zeno's paradoxes, inasmuch as even when the time interval has disappeared, the tendency toward motion remains.[284] This attitude adds nothing to the explanation of the paradoxes, for it fails to recognize that the conception of motion at a point, which is the crux of the situation, is not a scientific notion but a mathematical abstraction. As such, the logical difficulties involved have all been cleared up by the calculus and the mathematical continuum. Of course, there is nothing lost (or gained) by calling the derivative (which is, after all, only a number) the intensity at the point, for this is only a change of name; but the answer to the paradoxes must remain that given by mathematics, rather than any made in an attempt to satisfy intuition.

Wallis' arithmetization (as well as Descartes' analytic geometry) was criticized also by the contemporary mathematician, Isaac Barrow. The latter wished to return to the Euclidean view, and maintained that mathematical number has no real existence proper to itself and independent of continuous geometrical quantity. Numbers like $\sqrt{3}$, he felt, cannot, even in thought itself, be abstracted from all magnitude. Such "surd" numbers are "inexplicable," and, "having no merit of their own, they are wont to be banished from arithmetic to another science (which yet is no science), viz., algebra." Barrow held that arithmetic is to be included in geometry, but that algebra is to be included in logic rather than mathematics.[285]

This view would, of course, lead directly away from the limit concept which requires, for its effective use, as well as its logical definition, a conception of number not based on the geometrical interpretation of continuous magnitude. The fact that Barrow advocated a return to the classic conception of number and geometry may have in-

[284] See Lasswitz, *Geschichte der Atomistik*, II, 30; Vivanti, *Il concetto d'infinitesimo*, pp. 31–32.

[285] See Barrow, *Mathematical Works*, pp. 39, 45–46, 51–53, 56, 59.

fluenced his student, Newton, to seek to establish the calculus on the idea of continuous variation found in motion and geometry, and to avoid as far as possible the arithmetical notion of limit. Barrow's distrust of algebraic methods may well have been responsible also for the fact that he did not develop his geometrical discoveries into an effective analytical tool, but in this respect he was not followed by Newton.

Although Barrow did not accept readily the algebra which Italian mathematics had developed, he was impressed by the possibilities offered by the conception of motion at a point, which had been used to such advantage in geometry by Torricelli and others. Time he regarded as a mathematical quantity measurable by, although not dependent on, motion,[286] and upon the suggestion of sensory evidence he thought of it as a continuous magnitude, "passing with a steady flow." This brought Barrow to the problem of the nature of the continuum and the definition of instantaneous velocity. In his treatment of these there is a complete lack of the limit concept and, instead, an attempted blending of atomistic and kinematic views which may have been in part the result of the influence of the Cambridge Platonists. "To every instant of time, or indefinitely small particle of time, (I say instant or indefinite particle, for it makes no difference whether we suppose a line to be composed of points or of indefinitely small linelets; and so in the same manner, whether we suppose time to be made up of instants or indefinitely minute timelets); to every instant of time, I say, there corresponds some degree of velocity, which the moving body is considered to possess at the instant."[287] This passage shows clearly that Barrow's views are essentially infinitesimal and not much clearer than those of Plato, Oresme, Galileo, or Hobbes. In fact, in his argument that the area under a velocity-time curve represents distance, his thought closely parallels that implied by Oresme and expressed by Galileo. "If through all points of a line representing time are drawn. . . parallel lines, the plane surface that results as the aggregate of the parallel straight lines, when each represents the degree of velocity corresponding to the point through which it is drawn, exactly corresponds to the aggregate of the degrees of velocity, and thus most conveniently can be adapted to represent the space traversed also."[288]

[286] Gunn, *The Problem of Time*, p. 57.
[287] Barrow, *Geometrical Lectures*, p. 38.   [288] *Geometrical Lectures*, p. 39.

Barrow admitted that it may be contended rightly that very narrow rectangles should be substituted for the lines, but maintained that "it comes to the same thing whichever way you take it."[289]

In pointing out that "time has many analogies with a line," Barrow again suggested atomistic conceptions, saying that these magnitudes could be considered as constituted either from the continuous flow of one instant or point, or as an aggregation of instants or points.[290] Of all the ways in which a continuous magnitude may be generated, Barrow believed, with Cavalieri, that the one in which it is regarded as composed of indivisibles, "in most cases is perhaps the most expeditious of all, and not the least certain and infallible of the whole set."[291] We have seen that Tacquet had attacked the method of indivisibles, substituting for it a sort of limit of infinitely small quantities, and that Pascal followed him in this. Barrow, on the other hand, defended Cavalieri's method against the valid criticisms of Tacquet.[292]

In spite of the lack of clarity and precision in his views on the continuum, the geometrical results of Barrow represent a remarkably close approach to those of the calculus. They include not only numerous theorems on quadratures and tangents, but also perhaps the clearest recognition up to that time of the significance of the relationship between these two types of problems.[293] All of his propositions, however, are cast in geometrical forms which involve intricate and unnatural constructions, instead of in the analytical symbolism of Descartes, Fermat, and Wallis. If they were recast in terms of the calculus, they would be equivalent to many of the standard rules and theorems on differentiation and integration, including the fundamental theorem of the calculus.[294] Any such attempt to interpret them in terms of the present analytical notations, however, would be misleading. It would suggest, by implication, the possession on the part of Barrow of concepts equivalent to those of the derivative and the integral. It would furthermore give to his work an analytic character which the original was far from exhibiting.

With respect to the concepts involved in Barrow's geometry, we have seen that his views indicate a return to the vague indivisible of

[289] *Ibid.*   [290] *Ibid.*, p. 37.   [291] *Ibid.*, p. 43.   [292] *Ibid.*, pp. 44–46.
[293] See Child, "Barrow, Newton and Leibniz, in Their Relation to the Discovery of the Calculus."
[294] See *Geometrical Lectures*, pp. 30–32.

Cavalieri, rather than progress toward the development of the limit concept fostered by Wallis. As to the form of his work, it is clear that Barrow himself failed to appreciate it as a new type of analysis, sufficiently significant in itself to be developed into an algorithmic procedure. He seems to have realized that he was indicating a new method for finding tangents and areas, but by presenting it in synthetic form he gave it the appearance of being an expansion of the classical geometry of the ancients. "These matters seem not only to be somewhat difficult compared with other parts of Geometry, but also they have not been as yet wholly taken up and exhaustively treated (as the other parts have)."[295]

There is one point in Barrow's *Geometrical Lectures*, nevertheless, at which there is an indication that he may have used an analytic method to arrive at his results, recasting these later into the synthetic form in which he presented them. Also, in this connection, he constructed a diagram which was to become significant in the calculus of Leibniz as the familiar differential triangle. However, similar figures had appeared in the geometry of Torricelli, Roberval, Pascal, and Fermat. At the close of Lecture X, Barrow said:

We have now finished in some fashion the first part, as we declared, of our subject. Supplementary to this we add, in the form of appendices, a method for finding tangents by calculation frequently used by us. Although I hardly know, after so many well-known and well-worn methods of the kind above, whether there is any advantage in doing so. Yet I do so on the advice of a friend [later shown to be Newton[296]]: and all the more willingly, because it seems to be more profitable and general than those which I have discussed.[297]

If Barrow had believed that he was inventing a new subject, it seems likely that it is this method which he would have put forward, instead of his classical proofs.

Let *AP*, *PM* be two straight lines given in position, of which *PM* cuts a given curve in *M*, and let *MT* be supposed to touch the curve at *M*, and to cut the straight line at *T* [fig. 22].

In order to find the quantity of the straight line *PT*, I set off an indefinitely small arc, *MN*, of the curve; then I draw *NQ*, *NR* parallel to *MP*, *AP*; I call *MP* = *m*, *PT* = *t*, *MR* = *a*, *NR* = *e*, and other straight lines, determined by the special nature of the curve, useful for the matter in

[295] *Geometrical Lectures*, p. 66.
[296] See More, *Isaac Newton*, p. 185, n. [297] *Geometrical Lectures*, p. 119.

hand, I also designate by name; also I compare MR, NR (and through them, MP, PT) with one another by means of an equation obtained by calculation; meantime observing the following rules.

Rule 1. In the calculation, I omit all terms containing a power of a or e, or products of these (for these terms have no value).

Rule 2. After the equation has been formed, I reject all terms consisting of letters denoting known or determined quantities, or terms which do not contain a or e (for these terms, brought over to one side of the equation, will always be equal to zero).

Rule 3. I substitute m (or MP) for a, and t (or PT) for e. Hence at length the quantity of PT is found.[298]

From this passage one observes that the "method for finding tangents by calculation" used by Barrow resembles very closely the pro-

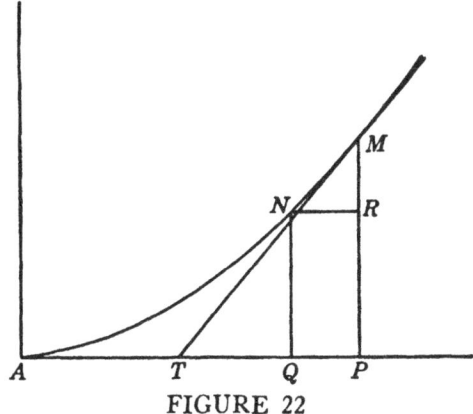

FIGURE 22

cedure now employed in the differential calculus, the letters a and e being equivalent to the customary symbols $\triangle y$ and $\triangle x$. This is an elaboration of Fermat's method, in which, however, only one infinitely small quantity, E, had been used. Barrow's method constitutes an improvement upon that of Fermat in that it makes more convenient the application of the method to implicit functions. Barrow apparently did not know directly of Fermat's work, for he nowhere mentioned his name. Nevertheless, the men to whom he referred as the sources of his ideas include Descartes, Huygens, Galileo, Cavalieri, Gregory of St. Vincent, James Gregory, and Wallis,[299] and it is possible that Fermat's method became known to Barrow through them. Huygens and James Gregory, in particular, made frequent use of Fermat's

[298] *Geometrical Lectures*, p. 120.   [299] See *Geometrical Lectures*, p. 13.

characteristic procedure; and Newton, at any rate, recognized that Barrow's rule was but an improvement upon this method of drawing tangents.[300]

Although the tangent method of Barrow resembles the process of differentiation even more closely than does that of Fermat, one is not at liberty to impute to him the concepts implied by our symbols $\triangle y$ and $\triangle x$. He was quite evidently thinking in terms of geometrical problems and infinitesimals, rather than of functions and symbols for continuous variables. His statement that a and e "have no value" is equivalent in this respect to Fermat's neglect, at the end of the calculation, of all terms involving E. Neither Barrow nor Fermat justified the neglect of these terms, for neither one had a clear conception of a limit. Fermat, perhaps, came nearer to such an idea in thinking of his pseudo-equality as a rigorous equality only when E was zero. Barrow did not tell why the higher powers are to be neglected. The first and third rules given by Barrow in his method are, of course, logically to be justified only in terms of limits. Barrow undoubtedly thought of the triangle MRN as rigorously congruent to MTP only when it was infinitesimally small, for he said here and elsewhere, that "if the arc MN is assumed to be indefinitely small, we may safely substitute instead of it the small bit of the tangent."[301] Passages such as these reappear in the work both of Newton and Leibniz, showing how difficult it was at that time for mathematicians to think arithmetically in terms of limits.

Of all the mathematicians who anticipated portions of the differential and the integral calculus, none approached more closely to the new analysis than did Fermat and Barrow. The former invented analytic methods of procedure equivalent both to differentiations and to integrations, but he appears not to have realized fully the significance of the interrelation between the two types. On the other hand, Barrow appears to have discovered the fundamental inverse relationship, but because he did not develop fully the possibilities of the analytic representation of the operations involved, he was unable to make effective use of it.[302] He systematically reduced inverse-tangent prob-

---

[300] More, *Isaac Newton*, p. 185, n.   [301] *Geometrical Lectures*, p. 61; cf. also pp. 120–21.
[302] See Zeuthen, "Notes," 1897, pp. 565–606; or his *Geschichte der Mathematik im XVI. und XVII. Jahrhundert*, pp. 345–57, for excellent analyses of this portion of Barrow's work.

lems to quadratures, but he did not convert the latter, by means of his inversion theorem, into considerations derived from tangent determinations—that is, he did not express them in terms of the antiderivative, as is generally done in the calculus. Barrow saw no advantage in doing this, because he had not reduced his tangent method to a simple algorithmic form, as did Newton and Leibniz shortly thereafter. Had he done so, he would without doubt have forestalled these men as the founder of the calculus.[303]

This, however, Barrow did not even attempt to do. After preparing his lectures for publication, he turned them over to Newton and Collins for final revision, and gave up mathematics for the study of theology. Newton displayed the analytical knowledge required to develop the geometrical ideas of Barrow into an algorithm, and in fact was already in possession of his methods of the calculus several years before 1670, the year in which Barrow's *Geometrical Lectures* were published. In fact, there had appeared earlier in the Low Countries a number of rules which in application so closely resemble parts of the calculus that they may be thought of as forming a transition from the infinitesimal procedures, developed during this century of anticipation, to the methods of fluxions and differentials. Such rules for tangents and for maxima and minima as those formulated by Sluse, Hudde, and Huygens did not involve new basic conceptions. They were simply canonical forms derived from earlier methods, particularly those of Fermat and Barrow.

Sluse, for example, was perhaps the first mathematician to give a general algorithm for writing down, without following through the analytic calculation required by such methods as those of Barrow and Fermat, the tangent to a curve whose equation is rational in x and y.[304] His rule, which he seems to have formulated in 1652[305] but which was not published until 1673, may be stated in the following terms. Let the equation be $f(x, y) = 0$. Then the subtangent will be the quotient obtained by placing in the numerator all the terms containing y, each multiplied by the exponent of the power of y appearing in it; and in the denominator all the terms containing x, each multiplied by the ex-

---

[303] Child would categorically attribute the invention to Barrow, because, using modern analytical notations instead of the synthetic form of the original, he reads into this work algebraic methods of procedure, rather than geometric proofs of propositions.

[304] Rosenfeld, "René François de Sluse et le problème des tangentes."

[305] Le Paige, "Correspondance de René François de Sluse."

ponent of the power of x appearing in it and then divided by x.[306] This is, of course, equivalent to forming the quotient $\dfrac{y\,f_y(x,\,y)}{f_x(x,\,y)}$, but Sluse was not thinking in terms of derivatives of functions, as we do. He did not give a general demonstration of his rule, but such is easily supplied by the methods of either Fermat or Barrow.

Johann Hudde, in 1659, gave an exactly analogous rule,[307] and this was rediscovered a few years later by Christiaan Huygens.[308] Hudde stated also a rule for writing down, without carrying out the work of Fermat's method, the maximum or minimum value of a rational function of one variable.[309] This is equivalent to setting equal to zero the derivative of the function as found by the rule for quotients given in calculus textbooks.

The almost simultaneous appearance of such rules and formulas indicates that shortly after the middle of the seventeenth century infinitesimal considerations were so widely employed and had developed to such a point that, given a suitable notation, a unifying analytic algorithm was almost bound to follow. Even Huygens, who in his earlier work had scrupulously followed the methods of the ancients, was after 1655 *au courant* with the new points of view and made frequent use of them. He attempted a generalization of the methods of the ancients, in the manner of Valerio;[310] he repeated the infinitesimal demonstrations of Galileo and Torricelli on falling bodies;[311] he showed the influence of Cavalieri in speaking of lines as elements of surfaces;[312] and his work includes frequent applications of Fermat's methods of tangents and maxima and minima.[313] Huygens, however, was a mathematical classicist, and it remained for his two younger friends, Newton and Leibniz, to bind all of this work into what represents probably the most effective instrument for scientific investigation that mathematics has ever produced.[314]

[306] Sluse, "A Method of Drawing Tangents to All Geometrical Curves," p. 38.

[307] Huygens, *Œuvres complètes*, XIV, 446–47.

[308] *Ibid.*, XIV, 442–48, 504–17.

[309] See the second letter by Hudde in *Geometria, a Renato des Cartes anno 1637 Gallice edita*, I, 507–16.

[310] *Œuvres complètes*, XIV, 338–39. [311] *Ibid.*, XVI, 114–18.

[312] *Ibid.*, XI, 158; XII, 5; XIII, 753; XIV, 192, 337.

[313] *Ibid.*, XI, 19; XVI, 153; XIV, *passim*.

[314] Cf. Bell, *The Queen of the Sciences*, p. 8.

V. Newton and Leibniz

FEW NEW branches of mathematics are the work of single individuals. The analytic geometry of Descartes and Fermat was certainly not the result of their investigations only, but was the outgrowth of several mathematical trends which converged in the sixteenth and seventeenth centuries. It was the result of the influence of Apollonius, Oresme, Viète, and many others.

Far less is the development of the calculus to be ascribed to one or two men. We have followed the long and uneven flow of thought which led from the philosophical speculations and the mathematical demonstrations of the ancients to the remarkably successful heuristic methods of the seventeenth century. We have indicated that the procedures invented by Fermat, for example, are almost identical with those found in the calculus, and that the new propositions discovered by Barrow include the geometrical equivalent of the basic theorem of the subject.

The time was indeed ripe, in the second half of the seventeenth century, for someone to organize the views, methods, and discoveries involved in the infinitesimal analysis into a new subject characterized by a distinctive method of procedure. Fermat had not done this, largely because of his failure to generalize his methods and to recognize that the problems of tangents and quadratures were two aspects of a single mathematical analysis—that the one was the inverse of the other. Barrow was unable to establish the new subject for, although the first to recognize clearly the unifying significance of this inverse property,[1] he failed to realize that his theorems were the basis for a new subject. Being unsympathetic with the Cartesian mathematical analysis and the algebraic trend, he implied that his results were to be considered as rounding out the geometry of the ancients.[2]

The traditional view, therefore, ascribes the invention of the calculus to the more famous mathematicians, Isaac Newton and Gottfried Wilhelm von Leibniz. From the point of view of the development of the concepts involved, the aspect which concerns us chiefly here,

[1] *Geometrical Lectures*, p. 124.　　　　　　　　　　[2] *Ibid.*, p. 66.

it might be far better to speak of the evolution of the calculus. Nevertheless, inasmuch as Newton and Leibniz, apparently independently, invented algorithmic procedures which were universally applicable and which were essentially the same as those employed at the present time in the calculus, and since such methods were necessary for the later logical development of the conceptions of the derivative and the integral, there will be no inconsistency involved in thinking of these men as the inventors of the subject. In doing so, however, we are not to consider or to imply that they are responsible for the ideas and definitions underlying the subject at the present time; for these basic notions were to be rigorously elaborated only after two centuries of further effort in this direction. Furthermore, inasmuch as we are here more concerned with ideas than with rules of procedure, we shall not discuss the shamefully bitter controversy[3] as to the priority and independence of the inventions by Newton and Leibniz.[4] Both men owed a very great deal to their immediate predecessors in the development of the new analysis, and the resulting formulations of Newton and Leibniz were most probably the results of a common anterior, rather than a reciprocal coincident, influence.

Attempts have been made by historians of the calculus[5] to trace two distinctly different threads of development: one, the kinematic, leading to Newton through Plato, Archimedes, Galileo, Cavalieri, and Barrow; and the other, the atomistic, tending toward Leibniz through Democritus, Kepler, Fermat, Pascal, and Huygens. There is, however, a complete lack of recognition of such a cleavage by the mathematicians involved, nor can we now distinguish the views and methods of the one "group," throughout the seventeenth century, from those

    [3] See Hathaway, "The Discovery of Calculus," pp. 41–43; and "Further History of the Calculus," pp. 166–67, 464–65, for accusations against Leibniz of "the foundation of a plot to deprive Newton of all credit . . . , with typical German propaganda," and of inaugurating "that system of espionage on scientific work in foreign countries by which the usefulness and credit of as much of that work as possible might be transferred to Germany."
    [4] See Sloman, *The Claim of Leibnitz to the Invention of the Differential Calculus;* Leibniz, *The Early Mathematical Manuscripts* (ed. by Child) for the statement of the suspicions directed against Leibniz. For his defense see Gerhardt's two works, *Die Entdeckung der Differentialrechnung durch Leibniz,* and *Die Entdeckung der höheren Analysis;* and also Mahnke's two articles, "Neue Einblicke in die Entdeckungsgeschichte der höheren Analysis," and "Zur Keimesgeschichte der Leibnizschen Differentialrechnung." For an extensive chronological bibliography on the subject of the controversy, see De Morgan, *Essays on the Life and Work of Newton.*
    [5] See, e. g., Hoppe, "Zur Geschichte der Infinitesimalrechnung," pp. 175–76.

of the other. Galileo, Cavalieri, Torricelli, and Barrow used both fluxionary and infinitesimal considerations, and the procedures of Fermat, Pascal, and Huygens were perhaps as well known to Newton and the English mathematicians as to Leibniz. The geometric developments leading to the fluxionary calculus of Newton were not essentially other than those pointing the way toward the differential calculus of Leibniz. However, after the methods of procedure of the subject had been established, and the logical and metaphysical bases of these were brought into question, the contrast between the points of view and the modes of presentation became heightened by the contrasting scientific and philosophic tastes of the inventors, as well, perhaps, as by the priority controversy in which blind loyalties prevented their successors from appreciating the advantages and disadvantages of the two systems. We shall attempt, therefore, to point out here not only the origins of the work of Newton and Leibniz, but also the nature of the interpretations which they later gave and the significance of these for the development of the fundamental notions of the calculus.

Isaac Newton was the student, at Cambridge, of Isaac Barrow and so came strongly under the influence of the latter, whose *Geometrical Lectures* he helped to prepare for publication. Now Barrow was familiar with the work of Cavalieri and the two views of the generation of geometrical magnitudes there presented—that of indivisibles, and that of flowing quantities. He thought of a tangent to a curve not only as the prolongation of one of the infinitely many lineal elements of which the curve might be assumed to be composed, but also as the direction of motion of a point which, by moving, generated the curve. These views were almost certainly familiar to Newton through his attendance at Barrow's lectures. Barrow, however, lacked an appreciation of the analytical methods of Descartes and Fermat, and failed to realize the significance of Wallis' arithmetization. These newer views, however, had been presented in 1655 by the works of Wallis, to which we have already referred and with which Newton became acquainted during the period of his early mathematical training.[6] Newton, in fact, acknowledged that he had been led to his first discoveries in analysis and fluxions by the *Arithmetica infinitorum* of

[6] Sloman, *The Claim of Leibnitz to the Invention of the Calculus*, pp. 1–7.

Wallis,[7] and the principles of induction and interpolation which Wallis there employed may have been instrumental also in leading Newton to the discovery of the binomial theorem.[8] Newton's conception of number resembles that of Wallis rather than that of Barrow— less a collection of units than an abstract ratio of any quantity to another, a definition which also includes irrational ratios as numbers.[9] Newton in this respect went beyond Wallis and Descartes in regarding negative ratios as numbers in the true sense of the word[10]—a generalization of the geometrical representation of Descartes.

Another element adding to the effectiveness of Newton's presentation of the method of fluxions was the use of infinite series. The Scholastic philosophers had studied infinite series in connection with the geometric representation of variability, and Gregory of St. Vincent, Tacquet, and Fermat had made use of infinite progressions. However, the earliest investigations of general arithmetic infinite series were largely the work of English mathematicians such as Wallis and James Gregory. Incidentally, the important work in this connection by the latter appeared only a year or two before Newton composed his first treatise on the calculus—one in which he employed infinite series in connection with the binomial theorem. The use of such series did indeed make for a universality of application of the method of fluxions and aided in freeing it from geometrical prejudices, but there has been a tendency on the part of historians to focus attention upon Newton's use of infinite series, rather than upon other more essential aspects of his work.[11]

Newton tells us that he was in possession of his fluxionary calculus as early as 1665–66,[12] that is, at some time during the period in which he had heard Barrow's lectures and had discovered the binomial theorem. The first notice of his calculus was given, however, in 1669, in *De analysi per aequationes numero terminorum infinitas*.[13] This was not published until 1711, but it circulated among his friends. In this

[7] More, *Isaac Newton*, p. 184.

[8] Merton, "Science, Technology and Society in Seventeenth Century England," p. 472, n.

[9] Newton, *Opera omnia*, I, 2.

[10] *Ibid.*, I, 3. Cf. Schubert, "Principes fondamentaux de l'arithmétique," pp. 35–37.

[11] Cf. Zeuthen, "Notes," 1895, pp. 194 ff., and a review of these by Tannery in *Bulletin de Darboux*, 2d series, XX (1896), 24–28.

[12] *Opera omnia*, I, 333.   [13] *Opera omnia*, I, 257–82; *Opuscula*, I, 3–28.

monograph he did not explicitly make use of the fluxionary notation or idea. Instead, he used the infinitely small, both geometrically and analytically, in a manner similar to that found in Barrow and Fermat, and extended its applicability by the use of the binomial theorem. In this paper Newton employed the idea of an indefinitely small rectangle or "moment" of area and found the quadratures of curves as follows: Let the curve be so drawn that for the abscissa x and the ordinate y the area is $z = \left(\dfrac{n}{m+n}\right) ax^{\frac{m+n}{n}}$. Let the moment or infinitesimal increase in the abscissa, following the notation of James Gregory, be o. The new abscissa will then be $x + o$ and the augmented area $z + oy = \left(\dfrac{n}{m+n}\right) a(x+o)^{\frac{m+n}{n}}$. If in this expression we apply the binomial theorem, divide throughout by o, and then neglect the terms still containing o, the result will be $y = ax^{\frac{m}{n}}$. That is, if the area is given by $z = \dfrac{n}{m+n} ax^{\frac{m+n}{n}}$, the curve will be $y = ax^{\frac{m}{n}}$. Conversely, if the curve is $y = ax^{\frac{m}{n}}$, the area will be $z = \dfrac{n}{m+n} ax^{\frac{m+n}{n}}$.[14]

Here we have an expression for area which was arrived at, not through the determination of the sum of infinitesimal areas, nor through equivalent methods which had been employed by Newton's predecessors from Antiphon to Pascal. Instead, it was obtained by a consideration of the momentary increase in the area at the point in question. In other words, whereas previous quadratures had been found by means of the equivalent of the definite integral defined as a limit of a sum, Newton here determined first the rate of change of the area, and then from this found the area itself by what we should now call the indefinite integral of the function representing the ordinate. It is to be noted, furthermore, that the process which is made fundamental in this proposition is the determination of rates of change. In other words, what we should now call the derivative is taken as the basic idea and the integral is defined in terms of this. Mathematicians from the time of Torricelli to Barrow had in a sense known of such a relationship, but Newton was the first man to give

[14] *Opera omnia*, I, 281; *Opuscula*, I, 26.

a generally applicable procedure for determining an instantaneous rate of change and to invert this in the case of problems involving summations. Before this time the tendency had been rather in the opposite direction—to reduce problems, whenever possible, to the determination of quadratures. With this step made by Newton, we may consider that the calculus has been introduced.

Newton applied this method to the quadrature of numerous curves, such as $y = x^2 + x^{\frac{3}{2}}$ and $y = \dfrac{a^2}{b + x}$. A few years later, in sending these results to Collins, he described in addition a number of propositions in maxima and minima and in tangents, which he had obtained by means of his methods. It is this letter to Collins of December 10, 1672, which became significant in the controversy as to whether Leibniz made his discoveries independently of Newton. In this letter Newton pointed out frankly that his rules are analogous to those of Sluze and Hudde, although more general;[15] and in another place he admitted that he had gotten the hint for his procedure from the method of Fermat, improved by Gregory and Barrow.[16] The change from the E of Fermat to the o of Gregory and Newton is, of course, trivial. It has from time to time been interpreted as a substitution of zero for E, a view which would reduce Newton's method to a meaningless manipulation of zeroes, somewhat in the manner of Bhaskara.[17] Newton distinctly regarded his symbol as the letter o and not the cipher zero, and in this respect it is entirely comparable to Fermat's E. The significance of Newton's work lay first of all in the fact that he applied the method "directly and invertedly," as he said.[18] In the second place Newton regarded it, in connection with the use of infinite series, as a universal algorithm, whereas that of Fermat, as well as the modifications of this by Sluze, Hudde, and Huygens, availed only in the case of rational algebraic functions.

It will be noticed that although the work of Newton contains the essential procedures of the calculus, the justification of these is not

[15] *Opera omnia*, IV, 510; cf. also *Mathematical Principles* (Cajori), pp. 251–52.

[16] More, *Isaac Newton*, p. 185, n.

[17] The entire interpretation of Newton given by Hoppe, "Zur Geschichte der Infinitesimalrechnung," is vitiated by the fact that he has followed certain older historians in this mistake. Cf. Gerhardt, *Die Entdeckung der höheren Analysis*, p. 80; Weissenborn, *Die Principien der höheren Analysis*, p. 25, n.; Gerhardt, "Zur Geschichte des Streites," p. 131.

[18] More, *loc. cit.*

clear from the explanation he gave. Newton did not point out by what right the terms involving powers of o were to be dropped out of the calculation, any more than Fermat justified omitting the powers of E, or Barrow those of e and a. His contribution was that of facilitating the operations, rather than of clarifying the conceptions. As Newton himself admitted in this work, his method is "shortly explained rather than accurately demonstrated."

In his demonstration above that the area of $y = ax^{\frac{m}{n}}$ is given by $z = \dfrac{n}{n+m} ax^{\frac{n+m}{n}}$, however, we can see some hint of the thought in his mind. The ordinate y seems to represent the velocity of the increasing area, and the abscissa represents the time. Now the product of the ordinate by a small interval of the base will give a small portion of the area, and the total area under the curve is only the sum of all of these moments of area. This is exactly the infinitesimal conception of Oresme, Galileo, Descartes, and others, in their demonstrations of the law of falling bodies, except that these men had found the area as a whole through the addition of such elements, whereas Newton found the area from its rate of change at a single point. It is difficult to tell in exactly what manner Newton thought of this instantaneous rate of change, but he very likely accepted it as similar to the conception of velocity which Galileo had made so familiar but had not defined rigorously. A thorough-going empiricist for whom mathematics was a method rather than an explanation,[19] Newton apparently considered any attempt to question the instantaneity of motion as linked with metaphysics, and so avoided framing a definition of it. Nevertheless, he accepted this notion and made it the basis of his second and more extensive exposition of the calculus, as given in the *Methodus fluxionum et serierum infinitarum*,[20] which was written about 1671,[21] but not published until 1736.

In this book Newton introduced his characteristic notation and conceptions. Here he regarded his variable quantities as generated by the continuous motion of points, lines, and planes, rather than as aggregates of infinitesimal elements, the view which had appeared in

[19] Burtt, *Metaphysical Foundations*, pp. 208–10.

[20] *Opuscula*, I, 31–200.

[21] See Zeuthen, "Notes," 1895, p. 203; cf. also Newton, *Opera omnia*, II, 280.

De analysi. Just as Barrow found the chief characteristic of time in its even flow, so also his pupil Newton, although he did not "formally consider time,"[22] was influenced to make continuous motion fundamental in his system. This concept Newton seems to have felt was sufficiently impelling and so clearly known through intuition as to make further definition unnecessary. The rate of generation Newton called a *fluxion*, designating it by means of a letter with a dot over it, a "pricked letter"; the quantity generated he called a *fluent*, employing in this connection the terms which had appeared earlier in the work of Calculator. Thus if x and y are the fluents, then their fluxions are $\dot{x}$ and $\dot{y}$. Incidentally, Newton in other places[23] proceeded to point out that one may consider the fluxions $\dot{x}$ and $\dot{y}$, in turn, as fluents of which the fluxions are represented by $\ddot{x}$ and $\ddot{y}$, and so on. The fluents of which x and y are the fluxions Newton represented by $\overset{|}{x}$ and $\overset{|}{y}$; the fluents of which these latter quantities are the fluxions were written $\overset{||}{x}$ and $\overset{||}{y}$, and so on.

In the *Methodus fluxionum* Newton stated clearly the fundamental problem of the calculus: the relation of quantities being given, to find the relation of the fluxions of these; and conversely.[24] In conformity with this problem and the new notation, Newton then gave examples of his method. These may be represented by the determination of the fluxion of $y = x^n$. His approach in this case is but slightly different from the earlier exposition in *De analysi*. If o is an infinitely small interval of time, then $\dot{x}o$ and $\dot{y}o$ will be the indefinitely small increments, or moments, of the flowing quantities x and y. In $y = x^n$ one then substitutes $x + \dot{x}o$ for x and $y + \dot{y}o$ for y, expands as before by the binomial theorem, cancels the terms not containing o, and divides throughout by o. Since, moreover, o was assumed to be infinitely small, the terms containing this—that is, the moments of quantities—can be considered as zero in comparison with the others, and are to be neglected.[25]

The result, $\dot{y} = nx^{n-1}\dot{x}$, is, of course, the same as that obtained by Newton previously in the *De analysi* without the use of fluxions.

[22] *Opuscula*, I, 54.
[23] Cf. *Opera omnia*, I, 338; *Opuscula*, I, 208.   [24] *Opuscula*, I, 55, 61.
[25] *Opuscula*, I, 60. The illustration, $y = x^n$, which we have presented is not here given in this form by Newton, but is taken as representative, in form and argument, of the exercises he gave.

It is to be noted that the introduction here of the conception of a fluxion is not an essential modification of the earlier work. The infinitely small enters as persistently as in the 1669 exposition, but in the dynamic form of Galileo's *moment* or the *conatus* of Hobbes rather than in the static form of Cavalieri's *indivisible*. This change serves only in intuition to remove the harshness (as Newton expressed it) from the doctrine of indivisibles.[26] In thought the justification of the neglect of infinitely small terms is to be made on precisely the same basis, whether it be written E, e, a, o, or $o\dot{x}$. Newton himself seemed to feel here some need for the limit concept, for he pointed out that fluxions are never considered alone, but always in ratios.[27] Later, when Newton sought escape from the clutches of the infinitely small, he emphasized this fact much more strongly.

This third stage in his thought appears clearly in *De quadratura curvarum*,[28] which was written in 1676 but not published until 1704. In this treatise Newton sought to remove all traces of the infinitely small. Mathematical quantities were not to be considered as made up of moments or very small parts, but as described by continuous motion. In determining the fluxion of x^n, Newton proceeded much as in the *Methodus fluxionum*, replacing x by $(x + o)$. In conformity with the fluxionary symbolism it would be expected that the increment in x should be designated $o\dot{x}$ instead of o, but inasmuch as Newton is here dealing with only a single variable, the fluxion of this may conveniently be taken as unity. On expanding $(x + o)^n$ by the binomial theorem, and subtracting x^n, the result is, of course, the change in x^n corresponding to the change o in x. Instead, now, of completing the argument by a doubtfully justified neglect of terms, Newton formed the ratio of the change in x to the change in x^n: that is, 1 to $nx^{n-1} + n\left(\dfrac{n-1}{2}\right)ox^{n-2} + \ldots$; and in this he allowed o to approach zero—to vanish. The resulting ratio, 1 to nx^{n-1}, we should speak of as the *limit* of the ratio of the changes, but Newton called it the *ultimate* ratio of the changes—a terminology which was later to lead to some confusion in thought. This ultimate ratio of "evanescent increments" is the same as the prime or first ratio of

[26] *Opera omnia*, I, 250; II, 39.

[27] *Opuscula*, I, 63–64.   [28] *Opera omnia*, I, 333–86; *Opuscula*, I, 203–44.

the "nascent augments." It is likewise the ratio of the fluxions at the point in question.[29]

In the above demonstration, the essential elements of the derivative are more clearly present than in any other part of Newton's work: the emphasis upon a function of one variable, rather than upon an equation in several; the formation of the ratio of the changes in the independent variable and in the function; and, finally, the determination of the limit of this ratio as the changes approach zero. Incidentally, the ratio as expressed by Newton is commonly inverted in the modern derivative. There are in Newton's thought, moreover, certain elements which have since been discarded as adscititious: his appeal to time as an auxiliary independent variable is now considered gratuitous; and the limiting ratio is now regarded as a single number, rather than as the quotient of two rates of change. Had Newton devoted more of his time to clarifying the elements of thought in his demonstration by ultimate ratios, the calculus might have been established upon the concept of the derivative a century before the time of Cauchy. In his first published account of his new analysis Newton suggested this type of argument; but in his illustrations of the method of fluxions in this work he unfortunately resorted to the infinitesimal terminology of his earlier accounts.

Newton's discovery of the calculus dates back to the years 1665 and 1666, as he says in *De quadratura*. Within the following decade he had written out, as we have seen, three accounts of his methods, but had published nothing. By 1676 he became aware that Leibniz was working on similar problems, and on October 24 of this year he sent a letter to Leibniz, through Oldenburg, in which he gave in the form of an anagram a statement of the fundamental problem of his calculus. This seems to have been his only effort to assure his claim to priority in the invention of the calculus. Upon transposing the letters of this anagram and translating, it read: "Given in an equation the fluents of any number of quantities, to find the fluxions and vice versa."[30] Similar statements of the problem of the calculus had been included in the *Methodus fluxionum* and the *De quadratura* which he had already composed.[31]

[29] *Opera omnia*, I, 334.
[30] *Opera omnia*, IV, 540 ff.; cf. also Leibniz, *Mathematische Schriften*, I, 122–47.
[31] *Opera omnia*, I, 339, 342; *Opuscula*, I, 55, 61.

In this letter he also admitted his indebtedness to Wallis, James Gregory, Sluse, and others, but did not give an exposition of his methods. The first published account of his calculus appeared somewhat incidentally, more than ten years later, in the famous *Principia mathematica philosophiae naturalis* of 1687. The propositions in this book, concerned as they are with velocities, accelerations, tangents, and curvatures, are largely those handled now by the methods of the calculus, but Newton presented them in the form of synthetic geometrical demonstrations with an almost complete lack of analytical calculations. Nevertheless, Newton at several points in the work gave indications of more general points of view.

In a series of lemmas in the first book, he expressed the type of argument appearing in *De quadratura curvarum*. "Quantities, and the ratios of quantities, which in any finite time converge continually to equality, and before the end of that time approach nearer to each other than by any given difference, become ultimately equal."[32] This, of course, is the sort of general limit proposition which Stevin, Valerio, Gregory of St. Vincent, Tacquet, Wallis, and others had attempted to substitute for the Greek method of exhaustion. In fact a passage in Gregory of St. Vincent,[33] in which the word "terminus" was used to designate the limit of a progression, may have been the origin of the term "ultimate ratio" which Newton was to use so frequently. Newton's view of a limit, like that of these earlier workers, was bound up with geometric intuitions which led him to make vague and ambiguous statements. Thus he said, "The ultimate ratio of the arc, chord, and tangent, any one to any other, is the ratio of equality."[34] and a little later he spoke of the similarity of the "ultimate forms of evanescent triangles."[35]

These remarks imply that Newton was not thinking arithmetically, as we do now, of the limit of the sequence of numbers representing the ratios of the (arithmetical) lengths of the geometrical quantities involved, as these become indefinitely small, but that he also was influenced by the infinitesimal views of the seventeenth century to think of ultimate geometrical indivisibles. It is true that he never used the expressions ultimate arcs, chord, tangents, or triangles, but only those of ultimate *ratios* and *forms*, expressions which allow of

[32] *Opera omnia*, I, 237; II, 30.   [33] *Opus geometricum*, p. 55.
[34] *Opera omnia*, I, 242; II, 34.   [35] *Ibid.*, I, 243.

rigorously correct abstract interpretations, but which strongly suggest others in terms of the intuitively more attractive view afforded by infinitesimals. That Newton realized the difficulties involved in a naïve view of infinitesimals is indicated, however, in his further statement in the *Principia* that "Ultimate ratios in which quantities vanish, are not, strictly speaking, ratios of ultimate quantities, but limits to which the ratios of these quantities decreasing without limit, approach, and which, though they can come nearer than any given difference whatever, they can neither pass over nor attain before the quantities have diminished indefinitely."[36] This is the clearest statement Newton gave as to the nature of ultimate ratios, but we shall find that, in continuing this argument in a lemma in the second book of his *Principia*, his exposition again took on more strongly the dependence upon the idea of infinitely small quantities, with the concept of limit somewhat hazily implied as basic. It is precisely this lack of arithmetical clarity which led, in the following century, to controversial discussions, not only on the validity of Newton's fluxions, but also as to what Newton really meant by the above statements and others similar to them.

Inasmuch as the *Principia* is written in the old synthetic geometric manner, references to the method of fluxions are not numerous. In the second book, however, there appeared the first publication of "the foundation of that general method."[37] Here one finds the statement of the fundamental principle, "The moment of any genitum is equal to the moments of each of the generating sides multiplied by the indices of the powers of those sides, and by their coefficients continually." Newton proved this first for the product AB as follows: Let AB represent a rectangle and let the sides A and B be diminished by $\frac{1}{2}a$ and $\frac{1}{2}b$ respectively. The diminished area will then be $AB - \frac{1}{2}aB - \frac{1}{2}bA + \frac{1}{4}ab$. Now let the sides of AB be increased by $\frac{1}{2}a$ and $\frac{1}{2}b$ respectively. The area of the enlarged rectangle will then be $AB + \frac{1}{2}aB + \frac{1}{2}bA + \frac{1}{4}ab$. Subtracting the smallest rectangle from the largest, one obtains $aB + bA$ as the moment of the original rectangle, corresponding to the moments a and b of A and B; which proves the proposition for this product. If $A = B$, the moment of A^2 is determined in turn as $2aA$.

[36] *Opera omnia*, I, 251.   [37] *Opera omnia*, II, 277–80.

By the use of the decrements $\frac{1}{3}a$ and $\frac{1}{2}b$ and the increments $\frac{1}{3}a$ and $\frac{1}{2}b$, instead of the increments a and b, Newton here avoided the necessity of dropping the infinitely small term ab. Newton thus made explicit use of infinitely small quantities of first order only—in which respect his work is to be contrasted with that of Leibniz—but his procedure in the proposition above was later justly criticized as implying the omission of infinitesimals of second order.

To find the moment of ABC Newton let $AB = G$ and, by applying the first part of the theorem, obtained $cAB + bCA + aBC$ as the result. Letting $A = B = C$, the moment of A^3 is in turn determined as $3aA^2$. By similar procedures the moment of A^n for positive integral powers is found to be naA^{n-1}. This same result is seen to hold for negative powers also. This is apparent from the following considerations. Let m be the moment of $\frac{1}{A}$. Then from $\frac{1}{A} \cdot A = 1$ and the moment of a product, one obtains $\frac{1}{A} \cdot a + A \cdot m = 0$, or $m = -\frac{a}{A^2}$. This argument is easily generalized to include all negative integral powers, and, with slight modifications, it is applicable to all products of rational powers of variables.

Newton said[38] that this is the foundation of his method of tangents and quadratures; and, in fact, this rule, combined with the use of infinite series, is sufficient for arriving at the essential results of the method of fluxions. However, because Newton supplied here such an unfortunately brief exposition of his procedure and its justification, and inserted this short account in the second book in the unobtrusive form of a lemma[39] to other propositions, some doubt has been expressed as to the seriousness with which it was put forward.[40]

The basis of the calculus as thus first published in the *Principia*, is, of course, to be found in the nature of Newton's moments; but it is just here that Newton was very far from clear in his language. He said on this point, "Finite particles are not moments, but the

---

[38] *Ibid.*   [39] *Opera omnia*, II, 277–80.

[40] Moritz Cantor (Vorlesungen, III, 192) would put but little emphasis upon the lemmas in the *Principia*, whereas Zeuthen ("Notes," 1895, p. 249; cf. also *Geschichte der Mathematik im XVI. and XVII. Jahrhundert*, pp. 382–84) stresses their significance in the calculus.

very quantities generated by the moments. We are to conceive them as the just nascent principles of finite magnitudes." Perhaps realizing that this statement made his moments as vague as the infinitesimals of Cavalieri, Fermat, and Barrow, he justified himself by adding, "Nor do we in this Lemma regard the magnitude of the moments, but their first proportion as nascent."[41] This looks like an attempt to bring in the doctrine of limits, which he had formulated in Book I, in which he regarded the *ratio* as ultimate, without specifying that the quantities entering it were so. Nevertheless, it is difficult to see just how one is to think of the limit of a ratio in determining the moment of *A B*. We have to deal here with two variables and are faced with the equivalent of partial differentiation, unless we have recourse to time as a single independent variable, as Newton next suggested. Perhaps realizing the difficulties in the way of interpreting the proposition in terms of the ratio or proportion of infinitesimals, Newton added another interpretation. "It will be the same thing, if, instead of moments, we use either the velocities of the increments and decrements (which may also be called the motions, mutations, and fluxions of quantities), or any finite quantities proportional to those velocities."[42]

To summarize the above, we see that Newton first had in mind infinitely small quantities which are not finite nor yet precisely zero. "Ghosts of departed quantities" they were fittingly called by the critics of the method in the following century. These offer too great difficulty of conception, so Newton next focused attention on their ratio, which in general is a finite number. Knowing this ratio, one may now substitute for the infinitesimal quantities forming it any other easily conceived finite magnitudes having the same ratio, such as quantities which are thought of as the velocities or fluxions of those entering into the equation. Newton thus offered in the *Principia* three modes of interpretation of the new analysis: that in terms of infinitesimals (used in his *De analysi*, his first work); that in terms of prime and ultimate ratios or limits (given particularly in *De quadratura*, and the view which he seems to have considered most rigorous); and that in terms of fluxions (given in his *Methodus fluxionum*, and the one which appears to have appealed most strongly to his

[41] *Opera omnia*, II, 278.　　　　　　　　　　　[42] *Ibid.*

imagination). The fact that Newton could thus present all three views as essentially equivalent shows us how far he was from viewing his method as quite distinct from the somewhat equivalent methods of his predecessors and contemporaries. In *De quadratura*, after saying that he there considered quantities as described by a continued motion and used the method of prime and ultimate ratios, he asserted that his method is consonant to the geometry of the ancients;[43] and in the *Principia* he also admitted that Leibniz had a similar method for considering the generation of magnitudes—an admission which was, however, omitted from later editions.[44] In fact the method of fluxions is dependent upon some other method, such as limits or infinitesimals, for the determination of the basic relations between the fluxions. Although Newton apparently preferred to link his method of fluxions with the idea of a limiting ratio, he so often used infinitesimals for dispatch that we shall find many of his successors later interpreting the fluxions themselves as infinitely small quantities, confusing them with moments.

Newton himself frequently used the concept of the infinitely small throughout his early work, but tended to become wary of it in his later expositions. In a portion of *De quadratura* which appeared in Wallis's *Algebra* of 1693, Newton had said that terms multiplied by *o* he omitted as infinitely small, thus obtaining the result.[45] In the 1704 publication of the work, on the other hand, he said clearly, that "errors are not to be disregarded in mathematics, no matter how small."[46] The conclusion was to be reached, not by simply neglecting infinitely small terms, but by finding the ultimate ratio as these terms become evanescent. However, even after this he did not abjure the infinitesimal completely, but continued to speak of moments as infinitely little parts. Furthermore, Newton added to the confusion in the thought of his contemporaries on fluxions by failing sometimes to multiply the fluxions by *o* when he wished to represent moments. Although he said that wherever pricked letters represent moments and are without the letter *o*, this letter is always to be understood, very many English mathematicians began to associate fluxions with the infinitely small

[43] *Opera omnia*, I, 338.   [44] *Ibid.*, II, 280.
[45] See De Morgan, "On the Early History of Infinitesimals in England," p. 324. See also Cajori, "Newton's Fluxions," p. 192, and Raphson, *The History of Fluxions*, p. 14.
[46] *Opera omnia*, I, 338.

differentials of Leibniz.[47] Newton's final view of the basis of the subject, however, seems to be that shown in his remark in *De quadratura:* "I have sought to demonstrate that in the method of fluxions it is not necessary to introduce into geometry infinitely small figures.[48]

We have seen that most of Newton's work on the calculus was written in the period from 1665 to 1676, but none of it had been published during that time. It has been suggested[49] that Newton's long delay in the publication of his three chief works on the calculus was occasioned by the fact that he was dissatisfied with the logical foundations of the subject. In the meantime, however, other mathematicians were looking for the general principle needed to solve the problems of tangents, maxima and minima, and quadratures. The methods of Fermat had already been modified by Huygens, Hudde, Sluze, and others. These men were contemporaries of the most versatile genius of the period, Gottfried Wilhelm von Leibniz, who, like Newton, was to develop rules and a literal symbolism for putting all of the infinitesimal considerations together under an algorithmic procedure. Although interested primarily in law and logic, Leibniz had in his early days written a little on arithmetic and mechanics. In 1672, however, he met Huygens in Paris and was urged by him to make a deeper study of mathematics. On a visit to London, in 1673, he met a number of mathematicians, learned much about infinite series, purchased a copy of Barrow's *Lectures*, and may have known, through Collins, of Newton's *De analysi*. After his return to Paris in the same year, he studied the mathematical works of Cavalieri, Torricelli, Gregory of St. Vincent, Roberval, Pascal, Descartes, Wren, James Gregory, Sluse, Hudde, and others.[50] The background of Leibniz in infinitesimal analysis was thus not greatly different from that of Newton, for the results of these men were well known in England as well as on the Continent. The early mathematical reading of Leibniz was thus largely on geometry, but he had other interests also which may have been

[47] Montucla (*Histoire des mathématiques*, II, 373) misinterprets Newton's occasional omission of these letters as an indication of confusion in Newton's thought between velocity and increment.

[48] *Opera omnia*, I, 333; *Opuscula*, I, 203.

[49] Merz, *A History of European Thought in the Nineteenth Century*, II, 630.

[50] See Gerhardt, *Die Entdeckung der Differentialrechnung durch Leibniz*, p. 31 and *passim*.

decisive in shaping his analysis. His first mathematical paper had been on combinatorial analysis, and he always retained a strong arithmetical tendency. One of the first fruits of his study of problems in quadratures was the "Arithmetical Tetragonism," in which he found the area of a unit circle to be given by four times the infinite series $1 - \frac{1}{3} + \frac{1}{5} - \frac{1}{7} + \ldots$.[51] These formalistic and arithmetical considerations were now to combine in an interesting way with the geometry which Leibniz had begun to master.

During his study at this time Leibniz was working on the problem of tangents, as well as that of quadratures, and had reached a solution based upon the "characteristic triangle"—the differential triangle which had appeared in various forms, particularly in the works of Torricelli, Fermat, and Barrow. It is difficult to determine the filiation of events leading to the invention by Leibniz of the differential calculus, but he himself, in a letter written thirty years later, attributed the inspiration for his use of the differential triangle to a figure (fig. 18 above) he had run across, about 1673, in Pascal's *Traité des sinus du quart de cercle*.[52] Leibniz said that on the reading of this example in Pascal a light suddenly burst upon him and that he then realized what Pascal had not—that the determination of the tangent to a curve depended upon the ratio of the *differences* in the ordinates and abscissas, as these became infinitely small, and that the quadrature depended upon the *sum* of the ordinates, or infinitely thin rectangles, for infinitesimal intervals in the abscissas. Moreover, the operations of summing and of finding differences were mutually inverse. Barrow had, in a sense, realized this also, for his *a* and *e* method for tangents involved the differences of ordinates and abscissas, his quadratures were effected by the summation of infinitesimals, and his inversion theorem showed the relationship of the two problems; but he had never developed these into a unified procedure. Leibniz, on the other hand, continued studying the characteristic triangle, encouraged by

[51] See *Mathematische Schriften*, V, 88; cf. also Leibniz, *Early Mathematical Manuscripts*, p. 163.

[52] Gerhardt and Zeuthen stress the obligation of Leibniz to Pascal. Child feels that inasmuch as Leibniz would have found in Barrow anticipations of the differential triangle much clearer than that in Pascal, he may have failed to mention Barrow either through a desire not to point out his indebtedness or through having forgotten the influence Barrow had had upon him. See especially Leibniz, *Early Mathematical Manuscripts*, pp. 15–16, for a discussion of the diagrams involved on this point; cf. also the works of Gerhardt and Zeuthen cited in the bibliography.

Huygens, and related this work to his former interest in combinatorial analysis.[53]

In the harmonic triangle and in the arithmetic triangle of Pascal, there are striking relationships. For example, in the arithmetic tri-

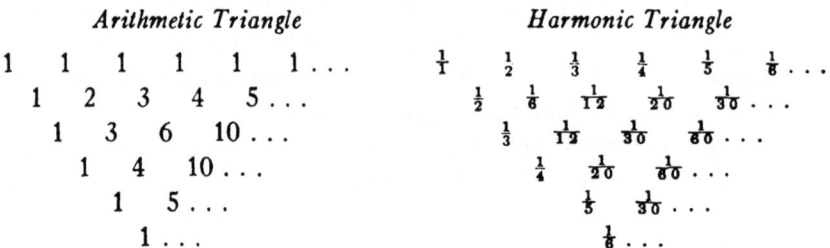

Arithmetic Triangle   *Harmonic Triangle*

angle any element is the sum of all of the terms in the line above it and to the left, and it is also the difference between the two terms directly below it. Similarly, in the harmonic triangle any element is the sum of the terms in the line below and to the right, and it is also the difference between the two terms just above it.

That is, in Pascal's arithmetic triangle, if we designate by x the numbers in any one line, the numbers of the first line below will be the sum of all the x's up to this point, reading from left to right; those in the second line below will be the sum of the sums of all the x's, and so on. Conversely, the lines above represent the differences, the differences of the differences, and so on. Similarly, in the harmonic triangle, with respect to the elements of any line, those below represent the differences, the differences of the differences, and so on; those above are the sums, the sums of the sums, and so on—reading from the right, however. Thus in these triangles we see that sums and differences are the inverses of each other,[54] just as the problem of tangents, which depends on the differences of ordinates, is the inverse of that of quadratures, which depends on the sum of all the ordinates, in the sense of Cavalieri. However, whereas the differences between elements in the arithmetic and harmonic triangles are finite, those between the ordinates of a curve are infinitesimal, and the formulas applicable in the former case no longer hold in the case of curves.

[53] See *Mathematische Schriften*, V, 108, 404–5; see also Newton, *Opera omnia*, IV, 512–15, and Gerhardt, *Die Entdeckung der Differentialrechnung durch Leibniz*, pp. 54–56.

[54] See Leibniz, *Early Mathematical Manuscripts*, p. 142, and *Mathematische Schriften*, V, 397.

It was therefore necessary for Leibniz to develop a method of procedure for determining sums and differences of infinitesimals. This he appears to have done by about 1676, the time at which Newton composed *De quadratura*. He had, about a year before, adopted his characteristic notation. He employed $\int x$, or later $\int x dx$, for the "sum" of all the x's—or the "integral" of x, as he called it later, on the suggestion of the Bernoulli brothers.[55] For the "differences" in the values of x, he wrote dx, although he had at first used $\frac{x}{d}$ for this, in order to imply that finding the "difference" involved a lowering of the dimension of the quantity.

Just as in the *Principia* Newton had begun by finding the moment of the product AB, so Leibniz determined the "difference" of the product xy. Although at the outset Leibniz was uncertain about his method and hesitated as to whether or not $d(xy)$ is the same as $dxdy$ and whether $d\left(\frac{x}{y}\right)$ is equal to $\frac{dx}{dy}$,[56] he in the end answered these questions correctly, determining that $d(xy) = xdy + ydx$ and $d\left(\frac{x}{y}\right) = \frac{ydx - xdy}{y^2}$. These values he found by allowing x and y to become $x + dx$ and $y + dy$ respectively. Upon subtracting the original value of the function from the new one and observing that $dxdy$ is infinitely small in comparison with the terms xdy and ydx, the results are obtained.

Having established the rules for differences and quotients, Leibniz was then able to extend these to all integral powers of a variable, the difference of x^n being $nx^{n-1} dx$. Because summation is the inverse of determining a difference, the integral of x^n is, of course, $\frac{x^{n+1}}{n+1}$.[57]

We have had occasion to observe the derivation of the latter result in various forms in the methods of previous investigators, but only in the work of Newton and Leibniz was it obtained as the inverse of another fundamental operation. A comparison of the derivations by

[55] See James Bernoulli, "Analysis problematis antehac propositi," p. 218.

[56] Leibniz, *Early Mathematical Manuscripts*, p. 102; Gerhardt, *Die Entdeckung der Differentialrechnung durch Leibniz*, pp. 24, 38.

[57] Cf. *Mathematische Schriften*, V, 226 ff.

the methods of fluxions and differences with those by quadratures as given by Cavalieri, Torricelli, Roberval, Pascal, Fermat, and Wallis will convince one of the enormous operational facility to be gained by such a method of procedure. As Newton made the rules for fluxions basic in his method, so also Leibniz looked upon the operation of finding "differences" as fundamental in his "differential and summatory" calculus. Ever since Newton and Leibniz invented their methods and combined them in this manner with the discovery of the fundamental inverse property, this point of view has continued in the elementary calculus. Differentiation is in general the fundamental operation, integration being regarded simply as the inverse of this.

There has been retained also in the calculus a certain element of confusion in terminology which is the result of the somewhat different attitudes of Newton and Leibniz with respect, not to the determination of the integral, but to its definition. Newton defined the fluent as the quantity generated by a given fluxion—that is, as the quantity having a given magnitude as its fluxion, or as the inverse of the fluxion. In keeping with this emphasis upon the indefinite integral, Newton included in both the *Methodus fluxionum* and *De quadratura* what amounts to a table of integrals. Leibniz, on the other hand, defined the integral as the sum of all the values of a magnitude,[58] or the sum of an infinite number of infinitely narrow rectangles, or—as modern mathematics would express it—as the limit of a certain characteristic sum. These two points of view have been perpetuated in the elementary calculus, in which there are two integrals: the indefinite integral and the definite integral. The origins of these, in the history of the subject, are even now sometimes brought vividly to mind in referring to the former as "the integral in the sense of Newton" and to the latter as "the integral in the sense of Leibniz."[59] Too much stress should not be put upon such a distinction, however, because Newton and Leibniz were both well aware of the two aspects of the integral.[60]

[58] "Seu data differentia dy invenire terminum y, est invenire summas omnium differentiarum seu dy." See Leibniz, "Isaaci Newtoni tractatus duo." Cf. also Gerhardt, *Die Entdeckung der Differentialrechnung durch Leibniz*, p. 45.

[59] Cf. Saks, *Théorie de l'intégrale*, p. 122.

[60] Hoppe ("Zur Geschichte der Infinitesimalrechnung," pp. 186–87), misled by Newton's symbol *o*, has completely misinterpreted the facts on this point. He would have Newton make the summation of indivisibles fundamental and have Leibniz oppose this point of view by emphasizing the operation of differentiation. The situation is, on the contrary, if anything quite the reverse of this.

Newton and Leibniz are known as the founders of the calculus largely because they established, in the periods 1665–66 and 1673–76 respectively, the methods and relationships outlined above. There was another aspect of their work which the inventors felt carried great weight—the generality of their methods. Both men pointed out that, unlike the anticipatory procedures, their methods were applicable even in the case of radicals. The justification for such an assertion was made by Newton largely upon the basis of infinite series. If $(x + o)^n$ is to be expanded by the binomial theorem, the number of terms will be infinite for values of n which are not positive integers. No conclusion can in general be drawn from an application of the theorem in this case unless the series is convergent, but neither Newton nor his successors for a century later fully appreciated the need for investigations into the question of convergence. Leibniz in this respect had perhaps even less caution than many of his contemporaries, for he seriously considered whether the infinite series $1 - 1 + 1 - 1 + 1 - \ldots$ was equal to $\frac{1}{2}$.[61] He had as well fewer scruples than Wallis in the wide generalization of rules demonstrated only for a small number of special cases. Although he had indicated a proof of his rule for the "difference" of x^n for integral values of n only, he announced that this would hold for all values, and entitled the first printed account of the calculus "A New Method for Maxima and Minima, as Well as Tangents, Which Is Not Obstructed by Fractional or Irrational Quantities."[62]

This first published treatise of the calculus, a memoir of six pages appearing in the *Acta eruditorum* of 1684, three years earlier than Newton's first account,[63] must have repelled most readers seeking an introduction to the new method. Even to the Bernoulli brothers, who did so much to popularize the subject in its early stages, it was "an enigma rather than an explication."[64] In the first place, it contained many misprints.[65] Secondly, Leibniz here imitated the bareness and

[61] Cf. *Mathematische Schriften*, V, 382–87.

[62] "Nova methodus pro maximis et minimis, itemque tangentibus, que nec fractas nec irrationales quantitates moratur, et singulare pro illis calculi genus." *Acta eruditorum*, 1684, pp. 467 ff. See *Mathematische Schriften*, V, 220 ff. For an English translation of this see Raphson, *History of Fluxions*, pp. 19–27.

[63] See *Mathematische Schriften*, V, 220 ff.

[64] See Leibniz, *Mathematische Schriften*, III (Part 1), 5, n.

[65] Eneström, "Über die erste Aufnahme der Leibnizschen Differentialrechnung."

simplicity of the *Géométrie* of Descartes, although in later articles he attempted fuller explanations. His 1684 work contained, besides the definition of the "difference" or differential of a quantity, the rules—without proof—for the differentials of sums, products, quotients, powers, and roots, with a few applications to tangents and to problems in maxima and minima and points of inflection. It is interesting that among these examples is one in which Leibniz derived the law of refraction, using Fermat's principle. This suggests the influence of Fermat, who had himself given a similar demonstration and whose work was repeated later by Huygens; but Leibniz did not refer to him here. Although quadratures were not included in this first paper, Leibniz developed the applications to such problems two years later, in another paper in the *Acta eruditorum*.[66] In subsequent articles in this and other journals,[67] Leibniz gave further developments and applications of his calculus—such as the determination of the differentials of logarithms and exponentials, and of osculating figures.

In all of this work Leibniz realized that he was creating a new subject. It has been suggested that it was only after the method of Leibniz had achieved marked success that Newton came to regard the method of fluxions as constituting a new subject and an organized mode of mathematical expression,[68] rather than simply as a useful modification of some earlier rules. This is belied by the fact that Newton had by 1676 written out three accounts of his method; but it remains true, nevertheless, that Leibniz expressed himself more vigorously on this point than did Newton. He said that his analysis was to be compared with the methods of Archimedes in much the way that the work of Viète and Descartes had been to the geometry of Euclid,[69] in that it dispensed with the necessity of imagination. In order to popularize it, he announced explicitly all the rules of operation, even the simplest,[70] presenting these as though they were rules of algebra, and pointing to the reciprocal relation of powers and roots as analogous to that subsisting between his "sums" and "differences," or integrals and differentials.[71]

[66] "De geometria recondita et analysi indivisibilium atque infinitorum." *Acta eruditorum*, 1686, pp. 292–300; see also Leibniz, *Mathematische Schriften*, V, 226 ff.

[67] See Leibniz, *Mathematische Schriften*, V, for these papers.

[68] De Morgan, *Essays on the Life and Work of Newton*, pp. 32–34.

[69] *Mathematische Schriften*, II, 123.

[70] Zeuthen, "Notes," 1895, p. 236.   [71] *Mathematische Schriften*, V, 231, 308.

This didactic spirit of Leibniz is in contrast with the reticence which Newton displayed on the subject of his method of fluxions, perhaps from a morbid fear of opposition. With reference to the logical and philosophical justifications of his procedures, on the other hand, Leibniz was less emphatic than Newton. He did not make a really serious effort on this point, because he felt that the calculus, as a *modus operandi*, brought its demonstrations with it. He did not wish to make of the infinitely small a mystery, as had Pascal; nor did he turn to geometrical intuition for clarification. In appealing only to intelligence, he stressed rather the algorithmic nature of the method, as he himself spoke of it. In this sense he may justly be considered one of the founders of formalism, as opposed to intuitionism, in mathematics. He had confidence that if he formulated clearly the appropriate rules of operation and that if these were properly applied, something reasonable and correct would result,[72] howsoever doubtful might be the meaning of the symbols involved. This attitude reflects well the corresponding difficulties experienced at the time with imaginary numbers. Leibniz, unlike Aristotle, seemed to feel that his position was to be justified by an appeal to the principle of sufficient reason to determine, in this connection, the transition from possibility to actuality.[73] The importunateness of his contemporaries, however, made it necessary for him now and then to attempt a further clarification of the basic conceptions of his differential calculus. In this respect he was neither lucid nor consistent.[74]

From his earliest to his latest work Leibniz made use of the principle that in a relationship containing differentials of various orders, those of lowest order only are to be retained, because all the others will be infinitely small with respect to these. This is, in a new algebraic form, essentially the same as the doctrine which Roberval and Pascal had employed when they had held that a line is as nothing in comparison with a square. Leibniz had carried over into analysis the idea of infinitely small quantities of different orders, based upon a principle of homogeneity which geometrical considerations had suggested. Whereas Fermat, Barrow, and Newton had made use only of

[72] Klein, *Elementary Mathematics from an Advanced Standpoint*, p. 215.

[73] Enriques, *Historic Development of Logic*, p. 77.

[74] Hoppe ("Zur Geschichte der Infinitesimalrechnung," p. 184) has maintained that the thought of Leibniz was much deeper and more accurate than that of Newton, but his view is based upon a misinterpretation of the work of the latter.

first order infinitesimals, Leibniz conceived of an infinite number of such orders, corresponding in a sense to the infinite ranks in the system of monads found in his philosophical scheme. However, in his definition of the differential of the first order, Leibniz vacillated; and for those of higher order he was very far from giving a satisfactory explanation.

In the first published account of the calculus, Leibniz gave a singularly satisfactory definition of his first-order differentials. He said that the differential dx of the abscissa x is an arbitrary quantity, and that the differential dy of the ordinate y is defined as the quantity which is to dx as the ratio of the ordinate to the subtangent. Barrow's rule for tangents had, in a sense, implied a similar definition, for it had required the substitution of the ordinate for a and the subtangent for e; but Barrow's a and e were vague infinitesimals. In the definition of Leibniz given above, the differentials are finite, assignable quantities, entirely comparable with those defined in the calculus of today. This fact has led to the assertion that "Leibniz from the beginning of the new calculus defined the differential absolutely as did Cauchy."[75]

In a sense this is true, but such a statement is quite misleading for two reasons. In the first place, the definition of Leibniz presupposes logically a satisfactory definition of the tangent line, just as Cauchy's differential depends upon the notion of the derivative. In each case it would be expected that the explanation should be in terms of limits. Leibniz, however, unlike Cauchy, defined the tangent as a line joining two infinitely near points of the curve, these infinitely small distances being expressible by means of differentials or differences between two consecutive values of the variable.[76] This constitutes a *petitio principii* which indicates that the avoidance of infinitely small quantities in the thought of Leibniz was only superficial. Of course, it is possible that Leibniz intended his language to be interpreted in the precise sense of limits, as we do when we speak of the tangent as the line through two consecutive or coincident points; but a further consideration of Leibniz's writings will make such a view appear a misinterpretation of his whole thought. Leibniz, throughout his work, regarded the

[75] Mansion, *Résumé du cours d'analyse infinitésimale*, Appendix, "Esquisse historique," p. 221; cf. also editorial note, *Mathesis*, IV (1884), 177.

[76] *Mathematische Schriften*, V, 220 ff.; Gerhardt, *Die Entdeckung der Differentialrechnung durch Leibniz*, p. 35.

differential as fundamental, as a recent study has shown.[77] Modern mathematics agrees with Cauchy, however, in making this notion subordinate to that of a limit by defining it in terms of the derivative. The reason for this change of view is to be found in the failure of attempts made by Leibniz and others to give a satisfactory definition of the differential independent of the method of limits.

The attempts of Leibniz to give satisfactory definitions of differentials of higher orders were unsuccessful. He said that ddx or d^2x is to dx as dx is to x,[78] making no distinction between the differentials of independent and dependent variables. Somewhat similarly he said that if $dx : x = dh : a$, where a is a constant and dh is a constant differential, then $d^2x : dx = dh : a$ or $d^2x : x = dh^2 : a^2$; and in general $d^e x : x = dh^e : a^e$, where e may even be fractional.[79] Perhaps realizing that this definition could not be consistently applied, he later gave a geometrical interpretation which, although lacking precision of statement, can be correctly interpreted in terms of derivatives. Given any curve, let dx be at every point an assignable quantity and let dy be such that the ratio of dy to dx is that of the ordinate to the subtangent. If then for every point on the curve we plot on the same axes a new point whose ordinate is proportional to dy, the result will be a new curve, the differentials of which will be the "differentio-differentials," or "second differentials," of the original curve.[80] This geometric representation is in general equivalent to that which would be obtained by plotting the values of the ratio $\dfrac{dy}{dx}$ for every point. The second differential d^2y would then be determined from the derivative of the new curve—that is, from the second derivative of the original curve. Leibniz, however, did not regard derivatives as fundamental, so that his remarks here cannot be regarded as constituting a satisfactory definition of d^2y, any more than can that of dy in terms of the tangent. The lack of such suitable definitions led to bizarre uses of the differential symbolism. In 1695 John Bernoulli, in a letter to Leibniz, wrote such expressions as: $\sqrt[3]{d^6y} = d^2y$ and $\dfrac{d^3y}{d^2x} = d^3yd^{-2}x = d^3y\int^2x$.[81]

---

[77] Petronievics, "Über Leibnizens Methode der direkten Differentiation."

[78] *Mathematische Schriften*, V, 325; cf. also III (Part 1), 228.

[79] *Ibid.*, III (Part 1), 228.   [80] "Addenda ad schediasma proximo," p. 370.

[81] Leibniz, *Mathematische Schriften*, III (Part 1), 180.

In the absence of satisfactory definitions, Leibniz resorted frequently to analogies to clarify the nature of his infinitely small differentials. At one point he made use of the imagery of Newton and spoke of his differentials as the momentary increments or decrements of quantities.[82] Again he applied the thought of Hobbes and said that the *conatus* is to motion as a point to space or as one to infinity.[83] The infinitely small he considered the study of the vanishing or incipiency of magnitudes, as opposed to quantities already formed.[84] He applied this analogy to second-order differentials as well. If in nature a motion is thought of as pictured by a line, then the impetus or velocity is represented by an infinitely small line, and the acceleration by a line doubly infinitely small.[85]

Appealing again to geometrical intuition, Leibniz said that as a point added nothing to a line, because it is not homogeneous or comparable, so differentials of a higher order in his method may likewise be neglected.[86] Pursuing this thought further, he said that if one thinks of geometrical magnitudes as represented by the ordinary quantities of algebra, then the first differentials refer to tangents, or the directions of lines, and the higher differentials refer to osculations or curvatures.[87] In this sense the differentials were to be compared with the Euclidean angle of contact which was less than any assignable magnitude and yet not zero.[88]

In the absence of rigorous definitions Leibniz continued to multiply analogies. In a somewhat less critical vein he said that the differential of a quantity can be thought of as bearing to the quantity itself a relationship analogous to that of a point to the earth or of the radius of the earth to that of the heavens.[89] In another place he said that as the earth is infinite with respect to a ball held in the hand, so the distance of the fixed stars is doubly infinite with respect to the ball;[90] and this analogy he repeated later, substituting a grain of sand for the ball.

John Bernoulli, the pupil of Leibniz, with great naïveté pointed out other analogies which the work of Galileo and of Leeuwenhoek

[82] *Mathematische Schriften*, VII, 222.
[83] *Philosophische Schriften*, IV, 229. Cf. also *Mathematische Schriften*, III, 536 ff.
[84] *Philosophische Schriften*, VI, 90. [85] "Testamen de motuum coelestium," p. 86.
[86] *Mathematische Schriften*, V, 322. [87] *Ibid.*, V, 325–26; cf. also p. 408.
[88] *Mathematische Schriften*, V, 388. [89] "Testamen de motuum coelestium," p. 85.
[90] *Mathematische Schriften*, V, 350, 389.

had made possible. He compared the orders of the infinitely great with the relations of the stars to the sun, to the planets, to the satellites of these, to the mountains on the last named, and so on. The infinitely small quantities in the same way resembled the apparently number-less grades of animalcules which the microscope had disclosed.[91] Leibniz, while appreciating the comparison, cautioned Bernoulli that the latter were of finite size, whereas the differentials were infinites-imals.[92]

It is interesting to notice that whereas Newton had used the infinitesimal conception in his early work, only to disavow it unequiv-ocally later and to attempt to establish the idea of fluxions on the doctrine of prime and ultimate ratios of finite differences—that is, in terms of limits—we shall find that with Leibniz the tendency is somewhat in the other direction. Beginning with finite differences, he was to be confirmed in his use of infinitesimal conceptions by the operational success with which his differential method met, although he seems to have remained largely in doubt as to its logical justifica-tion. The divergent views of these two men were perhaps less the result of dissimilar mathematical traditions than of varying tastes. Newton, the scientist, found in the notion of velocity a basis which to him appeared to be satisfactory; Leibniz, the philosopher, who was, as well, perhaps as much a theologian as a scientist,[93] preferred to find this in the differential, the counterpart in thought of the monad, which was to play such a large part in his metaphysical system.

Although Leibniz continued to employ the conceptions and the methods of infinitesimals, his justification of the infinitely small was not at first the result of serious effort, inasmuch as the notion had been employed by geometers with more or less indulgence throughout the century. Although Huygens and some few others did not readily accept the new calculus, they had not opposed it. In 1694, however, the Dutch physician and geometer, Bernard Nieuwentijdt, who was in this respect to be succeeded by a long line of mathematicians, opened an attack upon the lack of clarity in Newton's work and upon the validity of the higher differentials of Leibniz. Although he admitted in general the correctness of the results of the new methods,

[91] Leibniz and Bernoulli, *Commercium philosophicum et mathematicum*, I, 410 ff.
[92] *Ibid.*; cf. also *Mathematische Schriften*, III (Part 2), 518, 524.
[93] Mach, *Science of Mechanics*, p. 449.

he felt that they entailed some obscurity and said that they not infrequently led to absurdities. The tangent procedure of Barrow he criticized because *a* and *e* were taken as zero.[94] Newton's evanescent quantities he regarded as too vague, and he said that he could not follow the reasoning in the lemmas on limits, by which quantities, which in a given time tend to equality, are in the end equal.[95] In the analysis of Leibniz he questioned the manner in which the sum of infinitesimals might be a finite quantity.[96]

In a further attack in the following year Nieuwentijdt said that Leibniz had not made the nature of infinitely small quantities any clearer than had Newton and Barrow, nor could he explain in what way the differentials of higher order differed from those of the first order.[97] Nieuwentijdt tried to develop a method which would solve the problems of Leibniz without using infinitesimals of higher order; but the result was a failure.[98] The infinite and infinitesimal Nieuwentijdt defined unsatisfactorily, as had Nicholas of Cusa, as quantities respectively greater than and less than any given magnitude.[99] Thus, whereas the work of Newton and Leibniz had tended, in certain respects, toward the limit concept, the view of Nieuwentijdt represented a reversion to a less critical manipulation of the static infinite and the infinitesimal, which in the following century was boldly extended to the infinitely large and the infinitely small of higher orders, particularly by Fontenelle. This was unfortunate, as a serious effort to establish a secure foundation for the calculus was to be desired at the time.

Leibniz answered Nieuwentijdt in 1695 in the *Acta eruditorum*,[100] defending himself from "overprecise" critics, whom he likened to the Skeptics of long before. He maintained that we should not be led by excessive scrupulousness to reject the fruits of invention. His method, he maintained, differed from that of Archimedes only in the expressions used, being in this respect more direct and better adapted to the art of discovery.[101] After all, he held, the phrases "infinite"

[94] Nieuwentijdt, *Considerationes circa analyseos ad quantitates infinite parvas applicatae principia*, p. 8
[95] *Ibid.*, pp. 9–15. [96] *Ibid.*, p. 34; cf. also pp. 15–24.
[97] *Analysis infinitorum seu curvilineorum proprietates ex polygonorum natura deductae.*
[98] See Weissenborn, *Die Principien der höheren Analysis*, Section 10.
[99] *Analysis infinitorum*, p. 1.
[100] See *Mathematische Schriften*, V, 318 ff. [101] *Mathematische Schriften*, V, 350.

and "infinitesimal" "signify nothing but quantities which one can take as great or as small as one wishes"—somewhat as Aristotle had regarded the infinitely small as indicating a potentiality only—"in order to show that an error is less than any which can be assigned—that is, that there is no error."[102] Nevertheless, he reiterated that one can use these as "ultimate things"—that is, as actually infinite and infinitely small quantities—as a tool, much "as algebraists retain imaginary roots with great profit."[103] This Janus-faced appearance of the differential recurs frequently in the work of Leibniz. Although he commonly spoke of it as infinitely small, he often used instead the term incomparably small.[104] He held that "if one preferred to reject infinitely small quantities, it was possible instead to assume them to be as small as one judges necessary in order that they should be incomparable and that the error produced should be of no consequence, or less than any given magnitude."[105]

At another point Leibniz said that the differential was "less than any given quantity" and compared the neglect of differentials to the work of Archimedes, who "assumed, together with all those following him, that quantities which did not differ by a given quantity were in fact equal."[106] He here seemed to feel that his method constituted but a cryptic form of expression of the method of exhaustion, so that further justification was superfluous. However, the meaning of the symbols themselves was still not clear.

The hesitation which Leibniz displayed as to whether the differentials were to be regarded as assignables or inassignables is perhaps most neatly illustrated by a procedure he suggested in his *Historia et origo calculi differentialis*, written a year or two before his death. In order to avoid working with quantities which are not really infinitesimals but which are treated as such—amphibia between existence and nonexistence, as Leibniz called the imaginary numbers—he resorted to a bit of sleight of hand. Leibniz here used the symbols $(d)x$ and $(d)y$ to represent finite assignable differences; then, after the calculation had been completed, he replaced them by the inassignable

[102] *Philosophische Schriften*, VI, 90.
[103] Leibniz, *Early Mathematical Manuscripts*, p. 150.
[104] *Mathematische Schriften*, V, 407; cf. also p. 322.
[105] "Testamen de motuum coelestium," p. 85.
[106] "Responsio ad nonnullas difficultates, a Dn. Bernardo Nieuwentijt," p. 311.

infinitesimals or differentials dx and dy, "as a kind of fiction," because, after all, "$dy : dx$ can always be reduced to a ratio $(d)y : d(x)$ between quantities which are without any doubt real and assignable."[107]

Just how this jump from assignables to inassignables and back again is to be justified, Leibniz did not make plain. It appears from this argument that he realized that it was not the individual differentials, but only their ratio, that constituted the important consideration. Newton likewise understood that the significance of his method lay in the *ratio* of fluxions, so that for fluxions one could substitute other finite quantities in the same proportion. That the limit concept does not stand out clearly in their work is probably the result of the fact that they, and their contemporaries, were thinking always of a ratio as the quotient of two numbers, rather than as a single number in its own right.[108] Only after the development of the general abstract concept of real number was the way clear to interpret both the fluxionary and the differential calculus in terms of the limit of an infinite sequence of ratios or numbers; but this interpretation did not become generally acceptable for another century. The meanings of the terms "evanescent quantities" and "prime and ultimate ratio" had not been clearly explained by Newton, his answers being equivalent to tautologies:

But the answer is easy: for by the ultimate velocity is meant that, with which the body is moved, neither before it arrives at its last place, when the motion ceases, nor after; but at the very instant when it arrives. . . . And, in like manner, by the ultimate ratio of evanescent quantities is to be understood the ratio of the quantities, not before they vanish, nor after, but that with which they vanish.[109]

This sounds very much like the quotient of two infinitesimals, although Newton added that "those ultimate ratios with which such quantities vanish are not truly the ratios of ultimate quantities but the limits to which the ratios of quantities, decreasing without end, always converge." In other words, it is the *ratio* in which Newton was interested, not the evanescent quantities themselves; but he failed to define this ratio unequivocally.

[107] See Leibniz, *Early Mathematical Manuscripts*, p. 155; Weissenborn, *Die Principien der höheren Analysis*, p. 104; Gerhardt, *Die Entdeckung der Differentialrechnung durch Leibniz*, p. 31.
[108] Pringsheim, "Nombres irrationnels et notion de limite," pp. 143–44.
[109] *Opera omnia*, I, 250–51; II, 40–41.

Leibniz had a somewhat similar idea. Just as Newton never calculated a single fluxion but always a ratio, so also Leibniz realized that it was the ratio of, or relationship between, the differentials which was significant. These could therefore be regarded as any finite quantities, the ratio of which was that of the ordinate to the subtangent. For pragmatic reasons, however, Leibniz retained the infinitely small, justifying this by saying that if one desired rigor, he could for the "inassignables" substitute "assignables" having the same ratio. However, just as Newton did not clearly explain how ratios of evanescent quantities become, or are related to, "prime" or "ultimate" ratios, so Leibniz was unable to explain the transition from finite to infinitesimal magnitudes. Leibniz admitted that one could not prove or disprove the existence of infinitely small quantities.[110]

Furthermore, Leibniz felt that the justification for his calculus lay in the ordinary mathematical considerations already known and used, and that it was "not necessary to fall back upon metaphysical controversies such as the composition of the continuum."[111] Nevertheless, when called upon to explain the transition from finite to infinitesimal magnitudes, he resorted to a quasi-philosophical principle known as the law of continuity. We have seen previous applications made of this doctrine by Kepler and by Nicholas of Cusa. The latter may have influenced Leibniz in this respect, as well as in the philosophical doctrine of monads.[112]

Leibniz, however, gave to the doctrine of continuity a clarity of formulation which had previously been lacking and perhaps for this reason looked upon it as his own discovery. This "postulate" Leibniz expressed, in a letter to Bayle in 1687, as follows: "In any supposed transition, ending in any terminus, it is permissible to institute a general reasoning, in which the final terminus may also be included."[113] Thus in his manipulations in the calculus, "the difference is not assumed to be zero until the calculation is purged as far as is possible by legitimate omissions, and reduced to ratios of nonevanescent quantities, and we finally come to the point where we apply our result

[110] Leibniz and Bernoulli, *Commercium philosophicum et mathematicum*, I, 402 ff.; cf. also *Mathematische Schriften*, III (Part 2), 524 ff.

[111] Leibniz, *Early Mathematical Manuscripts*, pp. 149–50.

[112] See Zimmermann, "Der Cardinal Nicolaus Cusanus als Vorläufer Leibnizens."

[113] Leibniz, *Early Mathematical Manuscripts*, p. 147; see also *Philosophische Schriften*, III, 52; *Mathematische Schriften*, V, 385.

to the ultimate case"[114]—ostensibly by virtue of the law of continuity. Thus even in the work of Leibniz the idea of a limit was implicitly invoked, although the logical order was reversed. Leibniz justified the limiting condition by the law of continuity, whereas mathematics has since shown that the latter must itself first be defined in terms of limits. In this manner of thinking Leibniz seems still to be striving to make use of a vague idea of continuity which we feel we possess and which had bothered thinkers since the Greek period.

Newton, of course, had tacitly ensconced part of his difficulty with the continuum in the comforting notion of continuous motion, although he too, in his prime and ultimate ratio, was implicitly invoking the law of continuity of Leibniz. While Newton sought to avoid the limit concept by notions acceptable to scientific empiricism, Leibniz had recourse to ideas of ultimate form, suggested by metaphysical idealism. Even when the quantities—such as differentials—involved in a relationship become inassignable, nevertheless he felt that the ultimate form remained. A point, he held, was not that whose part is zero, but whose extension is zero.[115] In a similar connection Leibniz had asked Wallis, "Who does not admit a figure without magnitude?"[116] The characteristic triangle was for Leibniz one in which the form of the triangle remained after all magnitude had been abstracted from it,[117] much as Newton had spoken of the ultimate forms of evanescent triangles.[118]

Leibniz had intended to write a volume on the subject of the infinite. This book, which would have constituted a definitive expression of his views, did not appear. However, the attitude which Leibniz expressed in connection with his law of continuity seems to have been that which, with some vacillations and modifications, he held until his death. In his *Théodicée* he had said, in speaking of infinite and infinitely small quantities, "but all that is nothing but fictions; every number is finite and assignable, as is every line likewise."[119] Nevertheless, several years later, in writing to Grandi, he said, "Meanwhile, we conceive the infinitely small not as a simple

[114] Leibniz, *Early Mathematical Manuscripts*, pp. 151–52.
[115] *Philosophische Schriften*, IV, 229.
[116] *Mathematische Schriften*, IV, 54; cf. also IV, 63.
[117] Cf. Freyer, *Studien zur Metaphysik*, p. 10.
[118] *Opera omnia*, I, 243.   [119] *Philosophische Schriften*, VI, 90.

and absolute zero, but as a relative zero, (as you yourself well remark), that is, as an evanescent quantity which yet retains the character of that which is disappearing."[120]

Leibniz here clearly has in mind the law of continuity. This principle, the origin of which lay in the nature of the infinite, he felt to be absolutely necessary in geometry, as well as useful in physics.[121] Accordingly, he regarded an equality as a particular case of an inequality, and an infinitely small inequality as becoming an equality.[122] Acceptance of the law of continuity would therefore justify the omission of differentials of higher order, and it appears that it is upon this basis that Leibniz would have his calculus justified. A rather inexact tradition would impute to Leibniz a belief in actually infinitesimal magnitudes.[123] However, Leibniz himself, in a letter written about two months before his death, said emphatically that he "did not believe at all that there are magnitudes truly infinite or truly infinitesimal."[124] These conceptions he regarded as "fictions useful to abbreviate and to speak universally."[125] The link between these fictions and reality he undoubtedly felt would be found in his law of continuity, which he had taken as basic in all his later work in the calculus.[126]

We have traced the development of the calculus through the invention of the methods of Newton and Leibniz, and have found that the concepts involved had not yet been clarified. Newton gave three interpretations of his procedure, and although indicating a preference for the notion of prime and ultimate ratios as the most rigorous, he did not elaborate any one into a careful logical system. Leibniz displayed a similar lack of decision, for although he employed the infinitesimal method throughout, he wavered in his attitude toward differentials, considering them variously as inassignables, as qualitative zeros, and as auxiliary variables.

It has very misleadingly been said that Leibniz developed his method logically, while Newton was changeable in his conceptions, using limiting ratios to mask his infinitesimals.[127] The converse can, with

[120] *Mathematische Schriften*, IV, 218 f.
[121] *Philosophische Schriften*, III, 52. [122] *Ibid.*, II, 105.
[123] Cf. Klein, *Elementary Mathematics from an Advanced Standpoint*, p. 214.
[124] *Opera omnia* (Dutens), III, 500. [125] *Ibid.*
[126] Scholtz, *Die exakte Grundlegung der Infinitesimalrechnung bei Leibniz*, p. 39.
[127] Hoppe, "Zur Geschichte der Infinitesimalrechnung," p. 184; Moritz Cantor, "Origines du calcul infinitésimal," p. 24.

perhaps more justice, be asserted. The position of Leibniz was more difficult to maintain than had been that of the users of indivisibles, for he was required to explain infinitesimals not only of first order, but of all orders. Newton had avoided a multiplicity of infinitesimals through the method of fluxions, which required only a single increment—that in the time, o. His method was consequently well adapted to an interpretation in terms of limits, assuming that the idea of fluxion has first been properly defined or is recognized as a primary, undefined notion. Leibniz, on the other hand, had no quantity to serve as an independent variable, so that although he recognized that it was not the single differentials but only their ratio which was significant, he could not indicate how this fact was to be applied in his work. Newton had experienced this same difficulty when, in his *Principia* demonstration, he abandoned the fluxionary method and resorted to a use of infinitesimals in connection with a function of two variables.

Furthermore, the notation of Leibniz concealed, perhaps more effectively than that of Newton, the logical basis of the calculus. Leibniz had developed—as the result of painstaking investigation, patient experimentation, and frequent correspondence on the subject with other mathematicians[128] (particularly the Bernoulli brothers)— a symbolism which was remarkably felicitous when applied to the solution of problems. Because it is so convenient as to be almost automatic, this notation has been maintained to the present day. Nevertheless, this very success operated to mislead Leibniz as to the rigorous formulation of the subject. His system of notation caused him to think of his differential ratios as quotients, and his integrals as sums—of what, he could not say—rather than as limiting values of certain characteristic functions. His view seems to have been that first-order differentials were incomparables (a useful fiction), which by the law of continuity had the same ratio as that of the ordinate to the subtangent. Second-order differentials were the inassignables which were to those of first order as the latter were to unity. Such a definition of course confused $\dfrac{d^2y}{dx^2}$ with $\left(\dfrac{dy}{dx}\right)^2$, and this unsatisfactory state of affairs was not cleared up until the infinitesimal calculus was explained later in terms of limits.

[128] Cajori, *A History of Mathematical Notations*, II, 180–81.

Newton's notation likewise offered difficulties in the way of logical formulation. Not only did the infinitely small insinuate itself into the method of fluxions in the form of a "moment," but Newton occasionally failed to exercise care in distinguishing clearly between fluxions and moments. Furthermore, in the terminology of "prime and ultimate ratios" there was the idea, which was to persist for a century, that the limiting quantities were ratios, rather than single numerical values. Newton recognized that the ultimate ratio was a ratio of fluxions and not of moments. However, he was not clear in his explanations on this point and failed to point out the analogy between this ratio and the so-called sum of an infinite series. As the Euclidean view of number had been based upon the ratio of geometrical magnitudes, so here Newton emphasized the ratio of two velocities, rather than the single limiting value of a function, as brought out in the nineteenth-century definitions of the derivative.

With such indecision and lack of clarity on the part of the inventors of the calculus, it is hardly surprising to discover confusion among their followers as to the nature of the subject. This was intensified by the fact that many mathematicians failed completely to distinguish between the two systems and so misunderstood the arguments of their authors. This lack of distinction was due to some extent to Newton and Leibniz themselves. In the *Acta eruditorum* of 1705[129] an account of Newton's *De quadratura* was given, probably by Leibniz, in which it was stated that in this book the differentials of Leibniz were merely replaced by fluxions, and it is implied that these are essentially the same. In England, the committee of the Royal Society investigating the priority claims of Newton and Leibniz reported in the *Commercium epistolicum* of 1712[130] that "The *differential method* is one and the same with the *method of fluxions*, excepting the name and mode of notation; Mr. Leibniz calling those quantities *differences* which Mr. Newton calls *moments* or *fluxions*, and marking them with the letter *d*, a mark not used by Mr. Newton."[131] Thus even Newton's own countrymen did not realize that although he had used infinitely

[129] See Leibniz, "Isaaci Newtoni tractatus duo," p. 34; cf. also Bertrand, "De l'invention du calcul infinitésimal."
[130] See Newton, *Opera omnia*, IV, 497–592, for a reprint of this. See also the review (probably by Newton), *Philosophical Transactions*, XXIX (1714–16), pp. 173–224.
[131] See *Opera omnia*, IV, 588–89.

small moments in his early work, he professed to have nothing to do with them in his later expositions in terms of fluxions and of prime and ultimate ratios. This is more easily understood when we recognize that Newton himself never admitted a change in his view, and that his prime and ultimate ratio, unless interpreted under the limit concept or the principle of continuity, will involve the infinitely small.

The first published volume devoted expressly to the history of the calculus—*The History of Fluxions* by Joseph Raphson—appeared in 1715 and illustrated well the prevailing confusion of thought. To the author of this volume and to Halley, Newton had in 1691 planned to entrust the preparation for publication of his *De quadratura*.[132] However, a change of heart led Newton to withhold publication until 1704, by which time he had, as has been noted, emphatically renounced infinitesimals. Raphson, nevertheless, failed to recognize any change in the point of view and continued to confuse fluxions and moments. The small letters, which in the *Principia* Newton had employed to designate moments, Raphson interpreted as symbols of fluxions, and these he identified (all too uncritically) with the infinitesimals of Barrow and Nieuwentidjt, and with the differentials of Leibniz.[133] Raphson went so far as to regard fluents as infinitely great with respect to the finite quantities from which they were obtained; and fluents of higher orders were in similar manner interpreted as of higher degrees of infinity.[134]

In all fairness it is to be remarked that the object of Raphson's work was not to furnish a clarification of the basic notions of the calculus. The book contains, in fact, no formal definitions for such terms as fluxion, moment, or fluent. The purpose was partly to present the various methods in such a way as to make as easy as possible the application of the rules of procedure to specific problems, but "chiefly to do justice to their authors in point of chronology."[135] In both respects Raphson vigorously maintained the superiority of the work of Newton. With respect to priority his views were essentially those expressed in the report in the *Commercium epistolicum* of three years earlier, suggesting the possibility of plagiarism on the part of Leibniz.[136]

---

[132] See Raphson, *The History of Fluxions*, pp. 2–3.   [133] *Ibid.*, p. 4.
[134] *Ibid.*, p. 5.   [135] *Ibid.*   [136] Cf. *Ibid.*, pp. 8, 19, 61, 92.

As to the Leibnizian method and notation, Raphson most unjustly characterized this as "less apt and more laborious" and as a "far-fetched symbolizing" of "insignificant novelties."[137] This unfortunate attitude, which excessive deference to the great reputation of Newton had engendered, was to prevail among British mathematicians for just about a century following the appearance of Raphson's work, with the result that the differential notation made little headway in England before 1816. This period was characterized likewise by the same confusion of fluxions and the infinitely small as found in the exposition given by Raphson. As it was, a large number of the textbooks on the method of fluxions regularly interpreted fluxions as infinitely small quantities,[138] thus adding to the prevailing confusion in the interpretation of the conceptual bases of the calculus, which furnished the provocation for a century of criticism and controversy anent the subject.

[137] *Ibid.*, p. 19.
[138] Cajori, *A History of the Conceptions of Limits and Fluxions*, Chap. II; De Morgan, "On the Early History of Infinitesimals in England."

VI. The Period of Indecision

THE FOUNDERS of the calculus had clearly stated the rules of operation which were to be observed, and the astonishing success of these when applied to mathematical and scientific problems by Euler, Lagrange, Laplace, and a host of others led men to overlook somewhat the highly unsatisfactory state of the logic and philosophy of the subject. Throughout the whole of the eighteenth century there was general doubt as to the nature of the foundations of the methods of fluxions and the differential calculus. In England Newton's lack of clarity and his inconsistency in notation was followed by a confusion of fluxions with moments. On the Continent the metaphysical rationalism of Leibniz was neglected by his followers, who freely attempted to interpret the differentials as actual infinitesimals or even as zeros, and who criticized Leibniz for his hesitancy in this respect.

Such a state of affairs could not long continue unchallenged. Nieuwentijdt had earlier questioned the validity of differentials of higher orders. Gassendi had referred to the vanity of mathematical demonstrations based upon the infinitesimal methods,[1] and Bayle had used the difficulties of the infinite in the service of a general skepticism.[2] However, the most general and significant attack upon the structure of the new analysis was launched in 1734 by the philosopher and divine, George Berkeley, in a tract called *The Analyst*.[3]

Berkeley had previously attacked Newton's cosmology in his *Essay towards a New Theory of Vision*, but the motive prompting his animadversions in *The Analyst* was as largely that of supplying an apology for theology as it was of inflicting upon the proponents of the new calculus a rebuke for the weak foundations of the subject. This we gather from the subtitle of the tract: *Or a Discourse Addressed to an Infidel Mathematician*. [This referred to Newton's friend, Edmund Halley.] *Wherein It Is Examined Whether the Object, Principles, and Inferences of the Modern Anaylsis Are More Distinctly Conceived, or*

---

[1] See Bayle, *Dictionnaire historique et critique*, XV, 63.
[2] Cohn, *Geschichte des Unendlichkeitsproblems*, pp. 193–97.
[3] *The Works of George Berkeley*, Vol. III.

More Evidently Deduced, than Religious Mysteries and Points of Faith. "First Cast the Beam Out of Thine Own Eye; and Then Shalt Thou See Clearly to Cast Out the Mote Out of Thy Brother's Eye."

Berkeley in this work did not deny the utility of the new devices nor the validity of the results obtained. He merely asserted, with some show of justice, that mathematicians had given no legitimate argument for their procedure, having used inductive instead of deductive reasoning.[4] His objections were not to the mathematician as an artist and computist, but as "a man of science and demonstration" and "in so far forth as he reasons." After giving an account of Newton's method of fluxions which is eminently fair, except for the irrelevant ridicule of successive fluxions, Berkeley pointed out his specific objections. Since Newton had admitted to no change in his point of view, Berkeley legitimately took advantage of this fact to criticize a demonstration in the *Principia* in which the author had made use of infinitely small quantities in determining the moment of a product.[5] Newton had found the moment of the rectangle AB by giving to A and B the decrements and increments $\frac{1}{2}a$ and $\frac{1}{2}b$, then subtracting the diminished rectangle from the augmented. Berkeley objected that we must take the whole amounts a and b as increments or decrements to find the moment of AB, and in this case we would have to omit from the final calculation the infinitely small quantity ab. Berkeley with perfect right criticized leaving out the infinitely small quantity ab, if the result is to be rigorously correct, and cited a passage in *De quadratura*, in which Newton professed not to neglect anything, no matter how small.

Newton had not gone into the justification of his demonstration, but had indicated that one could find this in the method of prime and ultimate ratios. Berkeley therefore attacked this type of reasoning, as applied by Newton in one of his leading demonstrations: that of finding the fluxion of x^n in *De quadratura*, in which Newton had sought to avoid the infinitely small.[6] Here it will be recalled that Newton had given x an increment o, had expanded $(x + o)^n$ by the binomial theorem, had subtracted x^n to obtain the increment in x^n, had divided by o to find the ratio of the increments of x^n and of x, and had then

[4] Cf. *ibid.*, p. 30.
[5] *Ibid.*, pp. 22–23.   [6] *Ibid.*, III, 24–25.

let o become evanescent, thus determining the ultimate ratio of the increments (or of the fluxions). Berkeley averred that Newton here disregarded the law of contradiction, assuming first that x has an increment and then, in order to reach the result, allowing the increment to be zero, i. e., assuming that there was no increment. Berkeley maintained that the supposition that the increments vanish destroys the supposition that they were increments. Such an interpretation of Newton's meaning, which of course results in the consideration of the indeterminate ratio $\dfrac{0}{0}$, is not unjustified, inasmuch as Newton did not sufficiently explain the terms "evanescent quantity" and "prime and ultimate ratio," upon which the reasoning depends. The modern interpretation in terms of limits, of course, considers the infinite sequence formed by the ratios as the increment approaches zero, and this has no last term, although it is defined as having a limit. Newton's expression "ultimate ratio" is misleading, to say the least, and in all events indicates a lack of appreciation of the subtle difficulties involved in the concepts of infinity, continuity, and real number—difficulties not resolved until the second half of the nineteenth century.

Although Berkeley's arguments were directed chiefly against the British method of fluxions, the method of differentials, as used on the Continent by L'Hospital and others, also came in for criticism.[7] He explained that in finding the tangent by means of differentials, one first assumes increments; but these determine the secant, not the tangent. One undoes this error, however, by neglecting higher differentials, and thus "by virtue of a twofold mistake you arrive, though not at science, yet at the truth." This interpretation of the validity of the results of the calculus, as due to a compensation or errors,[8] we shall find advanced again by Euler, Lagrange, and Carnot, proponents of the differential method who wished to clarify its bases.

Berkeley's criticism of Newton's propositions was well taken from a mathematical point of view, and his objection to Newton's infinitesimal conceptions as self-contradictory was quite pertinent. On the other hand, his objections to some of Newton's quantities—such as fluxions, nascent and evanescent augments, moments as increments *in statu nascenti*, prime and ultimate ratios, infinitesimals, ultimate

[7] *Ibid.*, III, 20, 29.   [8] *Ibid.*, III, 32.

forms of evanescent triangles—on the grounds of their putative incomprehensibility or inconceivability, although just as much called for, were really misdirected. Assuming that the symbols involved have been clearly and logically defined (except for the primary undefined elements of the subject), it is of no mathematical consequence whether one can conceive of them in some manner corresponding to physical perception. Thus Berkeley argued that the concept of velocity depends upon space and time intervals and that it is consequently impossible to conceive of an instantaneous velocity, i. e., of a velocity in which these intervals are zero.[9] His argument is of course absolutely valid as showing that instantaneous velocity has no physical reality, but this is no reason why, if properly defined or taken as an undefined notion, it should not be admitted as a mathematical abstraction. It is interesting to notice that just as the arch-materialist Hobbes, being unable to conceive of lines without thickness, denied them to geometry, so also Berkeley, the extreme idealist, wished to exclude from mathematics the "inconceivable" idea of instantaneous velocity. This is in keeping with Berkeley's early sensationalism, which led him to think of geometry as an applied science dealing with finite magnitudes which are composed of indivisible "minima sensibilia."[10]

In line with this idea, Berkeley suggested that it would be better in the calculus to consider increments than velocities. At any rate, he warned, increments and velocities should not be confused as Newton had upon occasion done.[11] He accepted Cavalieri's indivisibles, but insisted that they were finite in number, saying that divisibility to infinity is only a fiction, and that infinitely small magnitudes are inconceivable, for they imply the existence of extension without perception by the mind through the senses. Berkeley was unable to appreciate that mathematics was not concerned with a world of "real" sense impressions. In much the same manner today some philosophers criticize the mathematical conceptions of infinity and the continuum, failing to realize that since mathematics deals with relations rather than with physical existence, its criterion of truth is inner consistency rather than plausibility in the light of sense perception or intuition.

Although those of Berkeley's arguments which are based upon the

[9] *Ibid.*, III, 19.　　[10] Johnston, *The Development of Berkeley's Philosophy*, pp. 82–86.
[11] *The Works of George Berkeley*, III, 46–47.

inconceivability of the notions involved lose their force in the light of the modern view of the nature of mathematics, it is clear that there was an obvious need for a logical clarification of many of the terms Newton had used. Berkeley's animadversions, although those of a nonmathematician, were successful in making this fact appreciated. As a result, there appeared within the next seven years some thirty pamphlets and articles which attempted to remedy the situation. The first appeared in 1734, a pamphlet by James Jurin, *Geometry No Friend to Infidelity: or, A Defence of Sir Isaac Newton and the British Mathematicians, in a Letter Addressed to the Author of the Analyst by Philalethes Cantabrigiensis*. This defense is weak in the extreme. Jurin maintained categorically that fluxions are clear to those versed in geometry. With respect to Berkeley's specific criticism on Newton's determination of the moment of AB, Jurin gave two answers: the moment ab in Berkeley's explanation he held to be as a pin's head to the globe of the earth or of the sun or of the orb of the fixed stars; the procedure of Newton, however, he defended by asserting that the moment is the arithmetic mean of the increment and decrement! In answer to Berkeley's objections to the determination of the fluxion of x^n as given by Newton in *De quadratura*, Jurin ingenuously said that one is not to let the increment in this case be nothing, but to let it "become evanescent" or "be upon the point of evanescence," affirming that "there is a last proportion of evanescent increments."[12] Jurin's response shows that he had no adequate appreciation either of Berkeley's arguments or of the nature of the limit concept.

Berkeley answered Jurin in 1735, in *A Defence of Freethinking in Mathematics*,[13] and justly asserted that the latter was attempting to defend what he did not understand.[14] In this work Berkeley again appealed to the divergence in Newton's views—as presented in *De analysi*, the *Principia*, and *De quadratura*—to show a lack of clarity in the ideas of moments, fluxions, and limits. Jurin's reply in the same year, in *The Minute Mathematician*, was again evasively tautological. He explained that "A nascent increment is an increment just beginning to exist from nothing, or just beginning to be generated, but not yet

---

[12] See *Geometry No Friend to Infidelity*, pp. 35, 52 ff. See also *The Minute Mathematician*, p. 74.

[13] *Works*, III, 61 ff. [14] *Ibid.*, III, 78.

arrived at any assignable magnitude how small soever."[15] By Newton's ultimate ratio he understood literally "their ratio at the instant they vanish."[16] Instead of explaining Newton's lemma on the moment of a product in terms of limits, Jurin allowed himself to become involved in the tangle of infinitesimals and was forced to resort to the idea of inassignables of Leibniz, saying that the magnitude of a moment is nothing fixed or determinate, but is "a quantity perpetually fleeting and altering till it vanishes into nothing; in short it is utterly unassignable."[17]

Berkeley, whose *Analyst* "marks a turning-point in the history of mathematical thought in Great Britain,"[18] now dropped out of the controversy,[19] but the unsatisfactory nature of Jurin's arguments was pointed out by Benjamin Robins in *A Discourse Concerning the Nature and Certainty of Sir Isaac Newton's Methods of Fluxions and of Prime and Ultimate Ratios*, as well as in articles in current journals.[20] As indicated in the title of his book, Robins distinguished, not three views in Newton's work, but two: that of fluxions and that of prime and ultimate ratios. The former he considered to be the more rigorous, saying that Newton used the latter only to facilitate demonstrations.[21] He added that the method of fluxions is established without recourse to the method of limits.[22] Robins admitted that Newton's use of moments in the lemma of the second book of the *Principia* was such as to allow interpretations resembling the language of infinitesimals; but he said that Newton thought it sufficient to indicate once for all that this can be made conformable to the method of prime and ultimate ratios, as presented in the lemmas of the first book.[23]

Although Jurin had denied the possibility of infinitely small constants, he had somewhat hazily espoused infinitely small variables, or

---

[15] See *The Minute Mathematician*, p. 19.   [16] *Ibid.*, p. 30; cf. also p. 56.

[17] *Ibid.*, p. 56.

[18] Cajori, *A History of the Conceptions of Limits and Fluxions*, p. 89.

[19] He had also answered the tract, *A Vindication of Sir Isaac Newton's Principles of Fluxions Against the Objections Contained in the Analyst*, by J. Walton, but neither the *Vindication*—a tautological paraphrase of Newton—nor Berkeley's answer contains any significantly new views. See *The Works of George Berkeley*, III, 107.

[20] See Robins, *Mathematical Tracts*, Vol. II.   [21] *Mathematical Tracts*, II, 86.

[22] See Gibson, "Berkeley's Analyst," pp. 67–69.

[23] Cf. *Mathematical Tracts*, II, 68 ff.

vanishing quantities. Robins was more emphatic in his disavowal of infinitesimals of any kind and said that Newton's statements involving moments are to be interpreted in terms of prime and ultimate ratios. For example, whereas Jurin had said that $Ab + Ba$ was equal to $Ab + Ba + ab$ when a and b banish, Robins said that $Ab + Ba$ was as much of the increment of AB as is necessary for expressing the ultimate ratio. This indicates that Robins realized the more clearly that the logical basis is to be found in the method of limits, although he was not clear, inasmuch as the product AB involves two independent variables, as to exactly how this was to be applied to the case in hand.

The limit conception of Robins represents a formulation of ideas which Valerio and Tacquet had expressed somewhat vaguely a century earlier—ideas to which Robins referred.[24] It indicates, as well, the dependence of this notion, in his thought, upon geometric intuition, for Robins spoke not only of the limits of "ratios of vanishing quantities" but also of the limits of the "forms of changing figures," giving as an illustration the circle as the limit of the inscribed regular polygon, as the number of sides is indefinitely increased.[25] This confusion of the arithmetical with the geometrical had been responsible for much of the vagueness in the work of Newton and of Leibniz, and was to persist during the following century. Nevertheless, Robins realized more clearly than did Jurin the nature of the limit concept. He recognized that the phrase "the ultimate ratio of vanishing quantities" was a figurative expression, referring, not to a last ratio, but to a "fixed quantity which some varying quantity, by a continual augmentation or diminution shall perpetually approach, . . . provided the varying quantity can be made in its approach to the other to differ from it by less than by any quantity how minute soever, that can be assigned,"[26] . . . "though it can never be made absolutely equal to it."[27]

Robins realized, as Jurin did not, that the varying quantity need not be considered as finally reaching the fixed quantity as its last value, although this latter "is considered as the quantity to which the varying quantity will at last or ultimately become equal."[28] In the controversy between Robins and Jurin, the question as to whether a

[24] *Ibid.*, II, 58.   [25] *Ibid.*, II, 54.

[26] *Mathematical Tracts*, II, 49.   [27] *Ibid.*, II, 54.   [28] *Ibid.*

variable was to be considered as necessarily reaching its limit played a large part. Robins upheld the negative side; Jurin insisted that there are variables which reach their limits and vigorously accused his opponent of misinterpreting Newton's true meaning. It is difficult to judge from Newton's words exactly what he meant. The phrase "ultimate ratio" certainly favors Jurin's interpretation, but to avoid the logical difficulties inherent in questions of infinitesimals and the meaning of $\frac{0}{0}$, it was necessary at the time to accept the more logical view of Robins that the variable need not attain its limit.

Robins has recently been criticized[29] for defining his limit in such a way that the variable never reaches its limit, on the grounds that although this has certain pedagogical advantages, it involves a less general conception of limit than Jurin's, and that under this conception of a limit Achilles could not overtake the tortoise. This criticism has been followed by the assertion that one can assume a time rate of doubling of the number of sides of a polygon inscribed in a circle, such that the circumference (the limit) is reached by the polygon (the variable). Such an argument is entirely beside the point. Besides confusing the numerical concept of limit with a geometrical representation, it merely substitutes an infinite time series for Zeno's distance series. This it does under the assumption, apparently, that this is intuitively more impelling, because of our vague idea of the relentless flow of time, which in the Achilles is subordinated to the static idea of distance. The question as to whether the variable S_n reaches the limit S is furthermore entirely irrelevant and ambiguous, unless we know what we mean by *reaching* a value and how the terms "limit" and "number" are defined independently of the idea of reaching. Definitions of number, as given by several later mathematicians, make the limit of an infinite sequence identical with the sequence itself. Under this view, the question as to whether the variable reaches its limit is without logical meaning. Thus the infinite sequence .9, .99, .999 . . . , *is* the number one, and the question, "Does it ever *reach* one?" is an attempt to give a metaphysical argument which shall satisfy intuition. Robins could hardly have had

[29] See Cajori, *A History of the Conceptions of Limits and Fluxions*, Chap. IV.

such a sophisticated view of the matter, but he apparently realized, as Jurin does not appear to have done,[30] that any attempt to let a variable "reach" a limit would involve one in the discussion as to the nature of $\frac{0}{0}$. Thus he is hardly to be criticized for his restriction.

We shall find that the question as to whether a variable reaches its limit or not has no significance, in the light of modern definitions. At that time, however, it was important as an indication that mathematicians still felt that the calculus must be interpreted in terms of what was intuitively reasonable, rather than of that which was logically consistent. This is apparently the reason why Robins considered the method of fluxions more satisfactory than that of prime and ultimate ratios. Everyone assumed he had a clear idea of instantaneous motion, although logically this is defined, as Robins failed to realize, in terms precisely equivalent to those needed to make the idea of prime and ultimate ratio rigorous.

That the work of Robins was not fully appreciated in England is seen in the fact that although Berkeley's *Analyst* was frequently discussed in the flood of texts on fluxions appearing in 1736 and 1737, the Jurin-Robins controversy was not referred to.[31] Mathematicians were still not satisfied with the method of limits and, although the discussion on infinitesimals may have discouraged the use of the idea of moments, it did not banish them. In 1771 the article on fluxions in the *Encyclopedia Britannica* read: "The fluxion of any magnitude at any given point is the increment that it would receive in any given time, supposing it to increase uniformly from that point; and as the measure will be the same, whatever the time be, we are at liberty to suppose it less than any assigned time."[32] The old confusion between fluxions and moments had not yet ended, and Robins' clarification of Newton's method of prime and ultimate ratios was not sufficient to establish this as the basis until after continental influences, in the early nineteenth century, had brought about a turning point in British mathematics.

In the meantime, however, numerous attempts, some noteworthy

---

[30] See Jurin, "Considerations upon Some Passages of a Dissertation Concerning the Doctrine of Fluxions Published by Mr. Robins," pp. 68 ff.

[31] Cajori, *A History of the Conceptions of Limits and Fluxions*, p. 179.

[32] *Ibid.*, p. 240.

and others insignificant, were made to find new and more satisfactory forms and arguments in which to present Newton's method. By far the ablest and most famous of these was made by Colin Maclaurin. In his *Treatise of Fluxions*, in 1742, he aimed, not to alter the conceptions involved in Newton's fluxions, but to demonstrate the validity of his method by the rigorous procedures of the ancients[33]— to deduce the new analysis from a few "unexceptional principles."[34] Maclaurin professed in the preface of this work that the *Analyst* controversy had given occasion to his treatise. Therefore he proceeded with extreme circumspection, omitting the notation of fluxions until toward the end of the long two-volume treatise. Like Robins, he banished the infinitely small as inconceivable and as "being too bold a Postulatum for such a Science as Geometry."[35] He did not, however, see any objection to introducing into geometry the idea of an instantaneous velocity, for he felt that there can be no difficulty in conceiving velocity wherever there is motion.[36] In fact, the mathematical sciences, Maclaurin said, included velocity and motion, as well as the properties of figures.[37] Time, he conceived with Barrow, to flow in a uniform course that serves to measure the changes of all things.[38]

Barrow, however, had defined velocity as the power by which a certain space may be described in a certain time,[39] somewhat as Aristotle had considered motion a manifestation of a potentiality. Maclaurin, on the other hand, tried to define instantaneous velocity in a manner recalling the attempt of Oresme: "The velocity of a variable motion at any given term of time is not to be measured by the space that is actually described after that term in a given time, but by the space that would have been described if the motion had continued uniformly from that term."[40] Maclaurin realized that if an instantaneous speed is "susceptible of measuring, it is only in this sense."[41] He recognized, that is, that science deals only with actual intervals; but he failed to see that as a mathematical notion instantaneous velocity could be defined by extrapolating beyond sense impressions, through the limit of an average rate of change as the intervals approach zero.

Although Maclaurin considered his explanation of fluxions as a

[33] *A Treatise of Fluxions*, I, 51 ff. [34] *Ibid.*, I, Preface. [35] *Ibid.*, I, iv.
[36] *Ibid.*, I, iii. [37] *Ibid.*, I, 51. [38] *Ibid.*, I, 53.
[39] *Ibid.*, I, 54. [40] *Ibid.*, I, 55. [41] *Ibid.*

criticism of the method of differentials, his interpretation of fluxions, in terms of the intervals which would be generated if the motion were to continue uniformly, left the way open to an explanation of Newton's fluxionary procedures in terms of finite differences and limits, by which the differential calculus of Leibniz was also to be explained. Brook Taylor had recognized the importance of this type of exposition and had some years previously composed a book on the subject—his *Methodus incrementorum directa et inversa*.

Whereas Robins and Maclaurin emphasized in Newton's work the interpretations in terms of fluxions, Taylor said that Newton had founded his method on prime and ultimate ratios.[42] Newton had recognized that the limit of the ratio of the moments was the same as the ratio of the corresponding fluxions or velocities. Whereas Robins and Maclaurin had assumed that everyone had a clear idea of instantaneous velocity, Taylor felt that it was easier to conceive of moments and to obtain the ratios of fluxions from these.[43] In this respect, however, his work at times approached closely to an unclear manipulation of zeros, resembling the later procedures of Euler. Taylor said that the relation of fluxions was to be obtained from that involving finite differences. In this respect his view resembled that of modern mathematicians, although Leibniz said at the time that it was "putting the cart before the horse."[44] However, Taylor, like Leibniz, was not clear with respect to the transition from finite differences to fluxions, for he held that to bring this about one simply wrote zero for the "nascent increments." Ultimate ratios, he thought, are those in which the quantities are already evanescent and are made zero,[45] an attitude which was to appear on the Continent in the work of Euler.

The view of Taylor on instantaneous velocities was shared by Thomas Simpson, author of a popular textbook on the method of fluxions which appeared first in 1737 and in an enlarged edition in 1750. Simpson felt that by taking fluxions as mere velocities, the imagination is confined to a point, and, without proper care, insensibly involved in metaphysical difficulties.[46] Maclaurin recognized this difficulty also and said that fluxions *were measured by* the quantities

[42] Taylor, *Methodus incrementorum directa et inversa*, Preface.   [43] *Ibid.*
[44] *Mathematische Schriften*, III (Part 2), 963.
[45] *Methodus incrementorum directa et inversa*, Preface, and p. 3.
[46] *The Doctrine and Application of Fluxions*, pp. xxi–xxii.

they would generate, if they were to continue uniformly. Simpson went beyond Maclaurin and *identified* fluxions of accelerated quantities with the increments which would be generated in a given portion of time if the "generating celerity" were to continue uniformly.[47] Simpson therefore followed Taylor in reverting to Newton's use of moments. He employed v in place of Newton's o, and as Newton had in his *De analysi* omitted terms containing o, so Simpson dropped out powers of v when the points coincided, because these powers he felt were due to the acceleration of the motion,[48] whereas, according to his definition, the generating celerity was to be regarded as continuing uniformly.

The views of Taylor and Simpson are similar to those which Jurin expressed in opposing Robins and serve to indicate the continued recurrence and widespread use of infinitesimal quantities in England during the century following the time of Newton, and the confusion of these with fluxions. There was also another element in British thought which operated against an early clarification of the bases of the calculus. The method of differences of Taylor failed to exert a decisive influence not only because of the novelty of his notation and his lack of clarity of expression, but also because it was essentially arithmetical and so involved a degree of abstraction which seems to have been unwelcome in British mathematics at the time. The basis of the method of fluxions remained essentially geometrical in its conception of quantity, in spite of Newton's work in infinite series; and the tenacity with which English mathematicians clung to the idea of velocity was probably due as much to a desire for an intuitively satisfying conceptual background as to loyalty to their great predecessor. This is shown particularly in the treatise of Maclaurin, which represents the high point in the rigorous interpretation of the calculus in terms of geometrical and mechanical notions. This work, however, was as little read as it was widely praised, and it consequently had probably no more influence than had Robins' treatises, which it resembles in its general ideas.

Much of the *Analyst* controversy and the confusion in the interpretation of the limit concept was due to the lack of a clear distinction between questions of geometry and those of arithmetic, and to the

[47] *Ibid.*, p. 1; cf. also p. xxii.   [48] *Ibid.*, pp. 3–4.

absence of the formal idea of a function. These weaknesses in the English view are brought out particularly in Newton's determination of the moment of a product, which was the object of some of Berkeley's most pertinent strictures. Newton had indicated vaguely that this was to be made rigorous by the application of the limit method, which is essentially arithmetical, but he had interpreted the product geometrically as the area of a rectangle. As a consequence, there was no indication as to whether the product was to be considered a function of one or of two variables. Time was, in a sense, made to take the place of an independent variable, but this was merely to aid in the conceptual representation and not to reduce the problem to one expressible in terms of a function of a single independent variable, as is necessary for differentiation.

We shall find that on the Continent there was a growing tendency to link the calculus with the formal concept of a function, instead of with the intuitional conceptions of geometry. The ideas developed as a result of this trend were not significantly influential in changing British views until the beginning of the following century, but meanwhile there were in England occasional abortive attempts to substitute arithmetical devices and ideas for those of geometry and dynamics.[49] Here and there the idea was expressed that the operations of the new analysis should proceed along the line afforded by the ordinary methods of arithmetic and algebra, and that the introduction of the doctrine of motion was unwarranted and unnecessary. However, the efforts to establish an algebraic calculus were characterized by lack of rigor in exposition.

One of the best known and least objectionable of the efforts to base the calculus only upon principles received in algebra and geometry, "without the aid of any foreign ones relating to an imaginary motion or incomprehensible infinitesimals," was made in 1758 by John Landen in what he called *The Residual Analysis*. Instead of computing the quotient of fluxions, or of differentials, Landen calculated "the value of the quotient of one residual divided by another."[50] By a residual was to be understood an expression of the form $x - x$, or $x^n - x^n$. Landen's method thus was based upon an uncritical manipulation of

[49] See Cajori, *A History of the Conceptions of Limits and Fluxions*, Chap. IX.
[50] *The Residual Analysis*, p. v.

indeterminate forms. Given a function $F(x)$, Landen found "we shall frequently have occasion to assign the quotient of $F - F$ divided by $x - x$.[51] For example, the fluxion, or residual quotient, of $x^{\frac{m}{n}}$ he found by writing

$$\frac{x^{\frac{m}{n}} - v^{\frac{m}{n}}}{x - v} = x^{\frac{m}{n} - 1} \left\{ \frac{1 + \dfrac{v}{x} + \left(\dfrac{v}{x}\right)^2 + \left(\dfrac{v}{x}\right)^3 + \cdots}{1 + \left(\dfrac{v}{x}\right)^{\frac{m}{n}} + \left(\dfrac{v}{x}\right)^{\frac{2m}{n}} + \left(\dfrac{v}{x}\right)^{\frac{3m}{n}} + \cdots} \right\}$$

and then taking $v = x$.[52]

The method of Landen has been characterized as making use of the limit of D'Alembert supposed to be attained instead of a terminus which can be approached as closely as desired.[53] Such a judgment is most charitable. If Landen possessed the limit concept, he certainly hid it most effectively under a misleading notation and terminology at a time when there was need for a clear and open recognition of its fundamental importance. Although we shall find that D'Alembert was at the time urging upon continental mathematicians that the logical basis of the differential calculus was to be found in the notion of limits, there was in England no strong leader to propagate this doctrine, unhampered by geometrical and mechanical superfluities. When it finally imposed itself upon British mathematics, it was as the result of European developments, to which we must now turn.

While the English mathematicians were so greatly occupied with arguments as to the validity of the views involved in the method of fluxions, the differential calculus was rapidly gaining in popularity on the Continent. The algorithmic essentials of the differential calculus of Leibniz had appeared in the *Acta eruditorum* for 1684, and those of the "calculus summatorius," or integral calculus, followed in 1686. Inasmuch as Leibniz, unlike Newton, corresponded extensively with numerous mathematicians on the subject of the new analysis, seeking for the most suitable forms of notation and presentation, there grew up a group of enthusiastic admirers of the subject who were soon able to make contributions of their own.

[51] *Ibid.*, p. 5.
[52] *Ibid.*, pp. 5–6; see also *A Discourse Concerning the Residual Analysis*, pp. 5, 41.
[53] See Cajori, *A History of the Conceptions of Limits and Fluxions*, pp. 238–39.

In Switzerland, for example, there were John and James Bernoulli, the former of whom in 1691–92 wrote a little treatise on the differential calculus, although this was not published until 1924.[54] The first published textbook on the subject appeared in 1696. This was by a French disciple of Leibniz, the Marquis de l'Hospital, and the title is characteristic of his approach—*Analyse des infiniments petits pour l'intelligence des lignes courbes.* This work is based, at least in part, on the earlier work of John Bernoulli.[55] Although l'Hospital did not in this book discuss the nature of the basic concepts of the calculus, he played a significant rôle in the popularization of the new subject, both through the fact that his text appeared in numerous editions[56] and through the influence he exerted in the *Journal des savants.*[57] Through this journal and the *Acta eruditorum* was created an atmosphere of enthusiasm for the differential calculus which led to a disregard on the Continent of the method of fluxions. It is interesting to notice that this parallels the comparative lack of regard for Newtonian science, in favor of that of Descartes, until the former was popularized in France by Voltaire. In Germany the differential calculus was popularized in the philosophical works of Christian Wolff, as well as by the mathematical work of the Bernoulli brothers. In Italy an enthusiastic interest in the differential calculus was manifested by Guido Grandi, another of the correspondents of Leibniz and author of a number of works on the calculus during the early eighteenth century.

In spite of the popularity which the calculus of Leibniz enjoyed, there was a total lack of clarity and agreement as to the basis of the analysis. Voltaire called the calculus "the Art of numbring and measuring exactly a Thing whose Existence cannot be conceived."[58] The indecision which Leibniz had displayed was shared also by his followers. John Bernoulli's little volume on the differential calculus begins with the paradoxical postulate that a quantity which is diminished or increased by an infinitely small quantity is neither diminished

---

[54] See *Die Differenzialrechnung von Johann Bernoulli.*

[55] See Eneström, "Sur la part de Jean Bernoulli dans la publication de l'Analyse des infiniment petits"; cf. Rebel, *Der Briefwechsel zwischen Johann (I.) Bernoulli und dem Marquis de l'Hospital,* pp. 9 ff.

[56] Paris, 1715, 1720, and 1781; Avignon 1768.

[57] Sergescu, "Les Mathématiques dans le 'Journal des Savants,' 1665–1701."

[58] See *Letters Concerning the English Nation,* p. 152.

nor increased.[59] In other words, Bernoulli made fundamental the omission of differentials of higher orders,[60] rather than the limit concept. Similarly, in the integral calculus the figure bounded by an infinitely small piece of a curve, the ordinates of its end points, and the corresponding difference in the abscissas he considered as a parallelogram.[61] Although he thought of a surface as the sum of such differentials of area, he did not define the integral as such a sum, as had Leibniz, but rather as the inverse of the differential, with the addition of a suitably chosen constant[62]—a definition which persisted throughout the next century.

We have seen that Leibniz generally considered his differentials as only indefinitely or imcomparably small, but John Bernoulli boldly asserted in a letter to Leibniz in 1698 that inasmuch as the number of terms in nature is infinite, the infinitesimal exists *ipso facto*.[63] This assertion he attempted to clarify in a manner recalling the exposition of Pascal in which the latter had pointed out that, through the reciprocal relationship, the existence of the indefinitely small was implied by that of the indefinitely great. Bernoulli sought to apply this type of argument to the case of actual infinitesimals: Let the infinite series $\frac{1}{2}$, $\frac{1}{4}$, $\frac{1}{8}$, . . . be given. Then if there are ten terms, one-tenth exists; if there are a hundred, then a hundredth exists; . . . if the number of terms is infinite, as is here supposed, the infinitesimal exists.[64] Leibniz, in answering, wisely cautioned him that arguments concerning the finite need not hold for the infinite, and that furthermore the infinite and the infinitesimal may be imaginary, even though they determine real relationships.[65] John Bernoulli, nevertheless, disregarded this *caveat* and persisted in views with respect to the infinitely large and the infinitely small which call to mind the earlier work of Wallis.

Of the two famous Bernoulli brothers, John had the greater originality and imagination, but James was superior in critical power.[66]

[59] John Bernoulli, *Die Differentialrechnung*, p. 11.

[60] Cf. Leibniz, *Mathematische Schriften*, III (Part 1), 366.

[61] John Bernoulli, *Die Differentialrechnung*, p. 11.

[62] John Bernoulli, *Die erste Integralrechnung*, pp. 3, 8, 11–12.

[63] Leibniz, *Mathematische Schriften*, III (Part 2), 555.

[64] *Ibid.*, pp. 563 ff.; see also Leibniz and John Bernoulli, *Commercium philosophicum et mathematicum*, I, 400–31.

[65] *Commercium philosophicum et mathematicum*, I, 370.

[66] Mach, *The Science of Mechanics*, pp. 427–28.

This fact is well illustrated by their respective attitudes toward the calculus. Whereas John expressed the positive attitude of Leibniz with reference to infinitesimals, his brother James put forth Leibniz's more cautious view. He found the use of the infinite not sufficiently convincing and too far from the opinion of the ancients.[67] He held that the infinitely small was not to be thought of as a determined quantity, but as a fiction of the spirit—"a perpetual fluxion toward nothing." As a consequence, "the ratio $\dfrac{2yy + dy^2}{4yy - dy^2}$ is always variable and does not become fixed unless dy is perfectly zero."[68] This view of the differential as a variable would associate the calculus with the method of limits, but James Bernoulli was unable to express this notion clearly because he failed, as had Leibniz, to distinguish between independent and dependent variables. In other words, the function concept had not yet become primary.

Although James Bernoulli attempted to avoid the pseudo-infinitesimal and maintained that a quantity smaller than any given magnitude is zero,[69] he vacillated in his attitude. On occasion he asserted that the Euclidean axiom "If equals are taken away from equals, the results are equal," need not be absolutely true when incomparably small quantities are involved.[70] For this reason he gave warning that in the calculus of the infinitely small, one must proceed with caution to avoid paralogisms.[71]

Wolff, a follower of Leibniz in mathematics and philosophy whose works enjoyed a wide distribution, adopted a modification of the views of James Bernoulli which the writings of Leibniz had suggested. He thought of the infinitely large and the infinitely small as impossibilities or as convenient geometrical fictions, useful for discovery, which result from a figurative manner of speaking.[72] By calling something infinitely large, he said, one simply means that it can exceed any number. Similarly, the infinitely small is not really a quantity in the strict sense of the word, but rather some sort of imaginary symbolism as Leibniz had sometimes held.[73]

If some mathematicians followed Wolff in denying the reality of the

[67] Leibniz, *Mathematische Schriften*, V, 350.
[68] Leibniz, *Mathematische Schriften*, III (Part 1), 52–56.
[69] *Opera*, I, 379.　　　　　　　　[70] *Opera*, II, 765.　　　　　　[71] *Ibid.*
[72] *Philosophia prima, sive ontologia*, pp. 597–602.　　　　　　　　　　[73] *Ibid.*

infinitely small, others took the opposing view. In Italy Guido Grandi upheld the existence of absolutely infinite and infinitesimal magnitudes of various orders. Those of first order he defined as quantities which bear to any finite magnitude of the same kind a ratio respectively greater and less than any assignable number.[74] Those of higher order were similarly defined in terms of those of lower order.[75] Quantities differing by less than any assignable magnitude he considered equal, because this, he felt, represented only a short way of saying what Euclid and Archimedes had meant in their work with in and circumscribed figures.[76] As an example of the addition of differentials to give a finite magnitude, Grandi referred to the paradoxical result $1 - 1 + 1 - 1 + \ldots = 0 + 0 + \ldots = \frac{1}{2}$. This, he suggested to Leibniz, could be compared with the mysteries of the Christian religion and with the creation of the world by which an absolutely infinite force created something out of absolutely nothing.[77]

In France, after a period of hesitation, views more bold even than those expressed by Grandi were to appear. L'Hospital, the first student of John Bernoulli,[78] had presented the views of his teacher in his *Analyse des infiniments petits;* and a little later Pierre Varignon, Bernoulli's "best friend in France,"[79] worked for the new analysis. However, as Nieuwentijdt had opposed the calculus of Leibniz and Berkeley that of Newton, so there arose in 1700, in the French Académie des Sciences, a lively dispute as to the validity of the infinitesimal methods.[80] In this discussion Rolle maintained that the new methods led to paralogisms, while John Bernoulli maintained in an *argumentum ad hominem* that Rolle did not understand the calculus.[81] Varignon attempted to clarify the situation by showing indirectly that the infinitesimal methods could be reconciled with the geometry of Euclid.[82]

In 1727, however, the French author Bernard de Fontenelle, a friend of Varignon, could boast that there were no longer two parties

[74] *De infinitis infinitorum et infinite parvorum ordinibus,* pp. 22–23.
[75] *Ibid.,* pp. 26 ff.   [76] *Ibid.,* p. 39.
[77] Cf. Leibniz, *Mathematische Schriften,* IV, 215–17; Reiff, *Geschichte der unendlichen Reihen,* p. 66.
[78] Fedel, *Der Briefwechsel Johann (i) Bernoulli-Pierre Varignon,* p. 3.   [79] *Ibid.,* p. 2.
[80] See a note in *Histoire de l'Académie des Sciences,* 1701, pp. 87–89.
[81] Leibniz, *Mathematische Schriften,* III (Part 2), 641–42.
[82] Fedel, *Der Briefwechsel Johann (i) Bernoulli-Pierre Varignon,* p. 25.

in the Académie. He evidenced this fact by publishing his *Élémens de la géométrie de l'infini*, in which there is no semblance of doubt expressed on the subject. The work of Fontenelle displays an absolute dogmatism with respect to the infinite. Recognizing that geometry is entirely intellectual and independent of the actual description and existence of figures,[83] Fontenelle did not discuss the subject from the point of view of science or metaphysics, as had Aristotle and Leibniz. He objected to regarding the infinite as a mystery and protested that Cavalieri was too modest in his treatment.[84] Confidently following Wallis, Fontenelle wrote ∞ as the last term of the infinite sequence 0, 1, 2, 3, . . ., although he realized that the manner in which the series goes from the finite to the infinite is inconceivable.[85] On the basis of this definition of the infinite, Fontenelle went on to include in his calculation not only integral powers of ∞, but even fractional and infinite powers as well, using such symbols as $\infty^{\frac{1}{2}}$ and ∞^{∞^3}, and writing such equalities as $\infty \cdot \infty^{\infty-1} = \infty^{\infty}$.[86] As Wallis had written the infinitely small as $\frac{1}{\infty}$, so Fontenelle derived his orders of infinitesimals as the reciprocals of the powers of infinity. The differentials dy and dx he held to be magnitudes of the order $\frac{1}{\infty}$, although he defined these in terms of the characteristic triangle of Leibniz.[87]

It is interesting to note that we have found three men—Wallis, John Bernoulli, and Fontenelle—who tried in an arithmetic manner to derive the infinitely small as the reciprocal of the infinitely large. (Pascal had in this manner sought only to relate the *indefinitely* large to the *indefinitely* small.) Such attempts lacked all semblance of mathematical rigor because of the lack at that time of satisfactory definitions of either the infinite or the infinitesimal. These efforts were furthermore counter to the general tendency of the time, which found the basis of mathematics in geometrical conceptions. Arithmetic had not become sufficiently abstract and symbolic to free itself of spatial interpretations, for number was still interpreted metrically as a ratio of geometrical magnitudes. Descartes had affirmed the identity of numerical and geometrical calculations. Newton had said that a num-

[83] *Élémens de la géométrie de l'infini*, Preface.   [84] *Ibid.*
[85] *Ibid.*, pp. 30–31.   [86] *Ibid.*, pp. 40 ff.   [87] *Ibid.*, p. 311.

ber was a ratio of quantities, and Wolff wrote that a number was anything which referred to unity in the same way as one straight line to another.[88] The methods of fluxions and differentials were, as a result, naturally considered convenient processes for solving geometrical problems. Although the results were usually expressed in algebraic terminology, the bases were sought in the geometry of the ancients, rather than in arithmetical conceptions. With the greatest mathematician of the first half of the eighteenth century, however, a change of view—anticipated to some extent by Wallis, Bernoulli, and Fontenelle—entered the new analysis.

Leonhard Euler wrote a prodigious number of books and articles promoting the new analysis, organizing it, and putting it on a formalistic basis. Most of his predecessors had considered the differential calculus as bound up with geometry, but Euler made the subject a formal theory of functions which had no need to revert to diagrams or geometrical conceptions.[89] Leibniz had used the word function somewhat in our sense, and had boasted that his infinitesimal method was not limited to algebraic functions, as was that of Descartes, but was applicable to logarithms and exponentials as well. Nevertheless, Euler was the first mathematician to give prominence to the function concept and to make a systematic study and classification of all the elementary functions, together with their differentials and integrals.

The word function, however, meant for Euler not so much any quantity conceived as depending on variables, as an analytic expression in constants and variables which could be represented by simple symbols.[90] Functionality was a matter of formal representation, rather than conceptual recognition of a relationship. The almost automatic development of the calculus during the eighteenth century was largely the result of this formalistic view, to which the notation of Leibniz was so remarkably well adapted. However, the greater the success achieved by the differential calculus, the less constrained Euler felt to justify his procedures. His views on the bases of the subject were elementary in the extreme, resembling somewhat those

---

[88] *Elementa matheseos universalis*, I, p. 21; cf. also Pringsheim, "Nombres irrationnels et notion de limite," p. 144 n.

[89] See the introduction to his *Institutiones calculi differentialis. Opera omnia*, Vol. X.

[90] See Brill and Noether, "Die Entwickelung der Theorie der algebraischen Funktionen in älterer und neuerer Zeit."

of Wallis, Taylor, John Bernoulli, and Fontenelle. He felt that the notions of the infinitely great and the infinitely small did not hide so great a mystery as was commonly thought. An infinitely small or evanescent quantity he held to be simply one which will be zero.[91]

This view might well have served as the basis for an interpretation in terms of limits, in which the differentials are simply variables approaching zero as a limit. Euler, however, did not proceed in this manner. Throughout the development of the calculus, the pandemic infinitesimal had, at various stages, been taken as a constant quantity less than any assignable magnitude. Euler emphatically rejected any such notion of mathematical atomism or monadology, inveighing against this as a "wretched abuse of the principle of sufficient reason."[92] He asserted, as had James Bernoulli, that a number less than any given quantity must of necessity be zero.[93] The differentials dx and dy were therefore simply zero.[94] Although he admitted the existence of an infinite number of infinitesimals, as found in the differentials of higher orders, these he held were all zero.[95] Leibniz had at one point suggested that the differentials could be regarded as qualitative zeros, which nevertheless retained by the law of continuity the character of the relationships of the finite quantities from which they were derived. Euler, in conformity with his formalistic view, held less philosophically that the zeros represented by differentials were to be distinguished through the recognized fact that the ratio $\dfrac{0}{0}$ could, in a sense, represent any ratio of finite numbers, $\dfrac{n}{1}$.[96] Thus for Euler the calculus was simply the determination of the ratio of evanescent increments—a heuristic procedure for finding the value of the expression $\dfrac{0}{0}$.[97]

[91] *Opera omnia*, X, 69.
[92] *Letters to a German Princess*, II, 61; cf. also *Opera omnia*, X, 67.
[93] *Opera omnia*, X, 69–70.　　　　　　　　　　　[94] *Ibid.*, pp. 70–72.
[95] *Ibid.*; cf. also XI, 5.　　　　　　　　　　　　[96] *Ibid.*, X, 70.
[97] Cohen (*Das Princip der Infinitesimalmethode und seine Geschichte*, p. 96) displays a lack of understanding of the limit concept in saying, in connection with Euler's use of zeros: "Es ist offenbar, dass er hierein nur der Grenzmethode folgt, dieselbe aber überbietet: indem er das Inkrement selbst, nicht den Vorgang, als verschwindend (incrementum evanescens) annimmt, als Null."

In justifying the omission in the differential calculus of infinitesimals of higher order, Euler's argument again lacked clarity. He held that, in the expression $dx \pm dx^2$, the infinitely small quantity dx^2 vanishes before dx does, so that for $dx = 0$ the ratio of $dx \pm dx^2$ to dx will be one of equality.[98] At one point he suggested that in omitting terms involving differentials, one made allowance for certain errors,[99] as Berkeley had maintained; and on occasion he employed difference calculus as a practical substitute for the differential calculus.[100] Euler, however, did not justify the transition from one to the other. Leibniz had explained the substitution of inassignables for assignables as validated by the law of continuity, but Euler followed Taylor in simply substituting zero for the increments. For example, in determining the differential of x^2 he allowed x to become $x + \omega$. The ratio of the increments in x and x^2 was then $1 : 2x + \omega$. This is always different from the ratio $1 : 2x$ unless ω vanishes. Euler therefore substituted 0 for ω and thus obtained the evanescent ratio $\dfrac{dx^2}{dx} = \dfrac{2x}{1}$ in much the same manner as Taylor had found the fluxion of x^2.[101]

With respect to the infinite, Euler adopted the views of Wallis and Fontenelle. Inasmuch as the sum of the series $1 + 2 + 3 + \ldots$ can be made greater than any finite quantity, it must be infinite and can be represented by the symbol ∞.[102] At another point he suggested that ∞ was a sort of limit between the positive and the negative numbers, in this respect resembling the number 0. In a similar manner he held that the relationship $\dfrac{a}{0} = \infty$ was to be interpreted as meaning that nothing times infinity can result in a finite magnitude.[103] Furthermore, as $\dfrac{a}{dx}$ will be infinite because $dx = 0$, so $\dfrac{a}{dx^2}$ will be infinite of second order; and, more generally, corresponding to the orders of the differentials, there is an infinite number of grades of infinity.[104] If x is infinite, he held that between 1 and $x^{1,000}$ there are 1,000 grades of

[98] *Opera omnia*, X, 71.
[99] Cf. Weissenborn, *Die Principien der höheren Analysis*, p. 158.
[100] Cf. Bohlmann, "Übersicht über die wichtigsten Lehrbücher der infinitesimalrechnung von Euler bis auf die heutige Zeit."
[101] *Opera omnia*, X, 7.
[102] *Ibid.*, p. 69.   [103] *Ibid.*, p. 73.   [104] *Ibid.*, p. 75.

infinity.[105] The lack of care with which Euler handled the infinite is evidenced also in his use of divergent series. As Leibniz had suggested that $1 - 1 + 1 - 1 + \ldots = \frac{1}{2}$, so Euler held that from $\dfrac{1}{(1 + 1)^2}$ $= \frac{1}{4}$ one could conclude that $1 - 2 + 3 - 4 + 5 - \ldots = \frac{1}{4}$. Under a somewhat different point of view Euler added that $1 - 3 + 5 - 7 + \ldots = 0$. Numerous similar examples of divergent series are to be found in his work.[106]

Inasmuch as Euler restricted himself to well-behaved functions, he did not become involved in those subtle difficulties connected with the notions of infinity and continuity which were later to make such a naïve position untenable. Although his views on the fundamental principles of the calculus lacked all semblance of the precision and rigor which entered mathematics in the following century, the formalistic tendency which his work inaugurated was to free the new analysis from all geometrical fetters.[107] It also made more acceptable the arithmetic interpretation which was later to clarify the calculus through the limit concept which Euler himself neglected.

While Euler, under the influence of Leibniz and the Bernoullis, was working so successfully on the assumption that differentials were zeros, his contemporary, Jean le Rond D'Alembert, was promulgating the doctrine which was to be elaborated ultimately into that accepted at the present time. Although the Newtonian-Leibnizian priority controversy had estranged British and Continental mathematicians, neither group was completely unaware of the views of the other. Thus Robins had, in 1739, criticized Euler for his crude conception of the infinitesimal, saying that the error of his ways was due to following "that inelegant computist" (John Bernoulli) who instructed him.[108]

In like manner the substance of the Berkeley-Jurin-Robins controversy was known to mathematicians on the Continent. Buffon, in the historical introduction to his translation of Newton's *Method of Fluxions*, criticized Berkeley and Robins for taking exception to some of Newton's arguments and warmly espoused Jurin's weak and prolix defense as "solid, brilliant, admirable."[109] Although he felt that

---

[105] *Opera omnia*, XV, 298.   [106] See *Opera omnia*, XIV, 585 ff.
[107] Cf. Merz, *A History of European Thought in the Nineteenth Century*, I, 103.
[108] Cajori, *A History of the Conceptions of Limits and Fluxions*, pp. 139–40.
[109] *La Méthode des fluxions et des suites infinies*, pp. xxvii–xxix.

Robins had criticized Euler and Bernoulli unfairly,[110] Buffon himself opposed Euler's views on the infinite and the infinitesimal. He held that the sequence 1, 2, 3, . . . had no last term and that the infinitely large and the infinitely small were only "privations."[111] Buffon, more interested in the natural than the mathematical sciences, did not elaborate this view; but the doctrine that the terms infinitely large and infinitely small signified only indefinitely large and indefinitely small was expounded more fully by the mathematician D'Alembert, who made this the basis of his theory of limits.

D'Alembert was probably led by two earlier works to regard the method of limits as fundamental in the calculus. These books, mentioned in his article on the *différentiel* in the famous *Encyclopédie*, were Newton's *De quadratura curvarum* of 1704 and De la Chapelle's *Institutions de géométrie* of 1746, a popular text which connected him with the adumbrations of the limit idea by Stevin, Gregory of St. Vincent, and others.[112]

D'Alembert interpreted Newton's phrase "prime and ultimate ratio" not literally, as a first or last ratio of two quantities just springing into being, but as a limit. One quantity he called the limit of another if the second can approach the first nearer than by any given quantity, or so that the difference between them is absolutely inassignable. Properly speaking, however, he felt that the varying quantity never coincides with, or is equal to, its limit.[113] D'Alembert thus agreed essentially with Robins' interpretation of Newton's meaning. He applied the same idea to the differential calculus. It will be recalled that Leibniz had believed that one could think of differentials as inassignable quantities, which by the law of continuity express the ultimate relationship between quantities deprived of magnitude. However, just as D'Alembert had denied the validity of the conception of an ultimate ratio, so also he rejected that of an ultimate relationship which can be determined by differentials. He said that the basis of the differential calculus, like that of the method of fluxions, was to be found in the idea of a limit. "The differentiation of equations consists simply in finding the limits of the ratio of finite differences of

[110] *Ibid.*, p. xxix.
[111] *Ibid.*, p. ix.
[112] Cajori, "Grafting of the Theory of Limits on the Calculus of Leibniz."
[113] Article, "Limite," in *Encyclopédie*. Cf. also Pierpont, "Mathematical Rigor," p. 33.

two variables included in the equation."[114] This D'Alembert believed to be the true metaphysics of the calculus, admitting, incidentally, that this was more difficult to develop than were the rules of application.

Nieuwentijdt had criticized Leibniz's differentials of higher order as nonexistant. D'Alembert, however, said that the distinction is irrelevant, for the differential notation is to be considered merely as a convenient abridgment or manner of speaking, used to avoid the circumlocution necessary in expressing the limit concept. He held that no such thing as an infinitesimal existed in its own right.

A quantity is something or nothing: if it is something, it has not yet vanished; if it is nothing, it has literally vanished. The supposition that there is an intermediate state between these two is a chimera.[115]

Just as the calculus of first order differences was explained in terms of limits, so D'Alembert defined infinitesimals of second and higher orders in a terminology equivalent to that of limiting ratios. His first explanation of these quantities lapsed into a phraseology dangerously resembling the naïve assertion of Leibniz that dx^2 is defined as being to dx as dx is to one, for D'Alembert said: "When one says that a quantity is infinitely small with respect to a quantity which is itself already infinitely small, this signifies merely that the ratio of the first of these quantities to the second is always as much smaller than the second quantity as is the latter than a given quantity." He added immediately, however, the explanation interpretable in terms of limiting ratios, "and that the ratio can be supposed as small as we please in imagining the second quantity sufficiently small."[116] This definition of infinitesimals of higher order indicates the general manner in which these quantities are interpreted at the present time, but it lacks clarity and resolution. Furthermore, it does not unequivocally state that the limit of the ratio must be zero for infinitesimals of higher order. These weaknesses in D'Alembert's explanation were corrected early in the following century by Cauchy. Meanwhile the interpretation of the infinitely small as a fixed infinitesimal continued in spite of D'Alembert's opposition.

[114] Article, "Différentiel," p. 977.
[115] *Mélanges de littérature, d'histoire, et de philosophie*, pp. 249–50.   [116] *Ibid.*

As D'Alembert had interpreted the word infinitesimal as meaning indefinitely small and had defined this in terms of limits, so also did he try to clarify the other concept which had bothered mathematicians since the Greek period—that of the infinite. He asserted—contrary to the view of Fontenelle—that the notion of infinity is really that of the indefinitely large and is only a convenient abridgment for the interpretation in terms of the doctrine of limits. With this understanding, he pointed out that one can have orders of infinitely large quantities analogous to those of infinitesimals. A line is said to be infinite of the second order with respect to another if the ratio of these is greater than any given number; it is of the third order if the ratio of the product of it by any finite number to the square constructed on the other line is greater than any given number.[117]

This interpretation of infinity is, of course, that which Euler had in mind when he said that the logarithm of an infinite number is an infinite number of lower order than that of any root of an infinite number.[118] It corresponds to that which mathematicians use when they speak of orders of infinity with respect to functions and is concerned only with the limits of ratios. It has nothing to do with the doctrine of infinite aggregates, which appeared in the late nineteenth century and which was necessary rigorously to establish the calculus. D'Alembert (like most of his contemporaries) would have been incapable of appreciating the modern notion of the actual infinite. In geometry, D'Alembert explained, there was no need to suppose the existence of an actual infinite, so that the question of its existence did not concern mathematics. Whereas the modern concept of the infinite is based upon arithmetic conceptions, D'Alembert's interpretations were largely in terms of geometry, even though in the chart of the divisions of knowledge, in the preface of the *Encyclopédie*, infinitesimal considerations are entered under algebra rather than geometry.

Because of his geometrical ideology, D'Alembert's elaboration of the limit concept lacked the clear-cut phraseology necessary to make it acceptable as a substitute for the infinitesimal interpretation. Thus to say with D'Alembert that the secant becomes the tangent when the two points are one and that it is therefore the limit of the secant,[119] imposes the necessity of visualizing the process by which two points become one, thus leaving the interpretation open to Zeno's criticisms.

To some mathematicians, at least, D'Alembert's limit concept appeared to be enmeshed in as dark a metaphysics as was that of the infinitely small. Consequently the majority of textbooks on the calculus published on the Continent at that time continued to prefer the explanations of Leibniz. Of twenty-eight publications appearing from 1754 to 1784, fifteen interpreted the calculus in Leibnizian terminology, six in terms of limits, four in terms of Euler's zeros, two in terms of fluxions, and one (that of Lagrange) in terms of a method to be described later.[120] Nevertheless, the limit idea continued from time to time to be put forth as the logical manner of interpreting the calculus. Hyacinth Sigismund Gerdil, for example, in 1760–61[121] followed D'Alembert in saying that the infinite and the infinitesimal are to be excluded from the calculus, their place being taken by limits.[122]

Another supporter was found in A. G. Kästner, the author of a popular textbook, *Anfangsgründe der Analysis des Unendlichen*, which appeared in 1761. In this Kästner said that he made use of Newton's method of first and last ratios, although he availed himself also of the abridged notation which the differential calculus afforded and which he compared, as a sort of ellipsis, with the figures of speech used by poets.[123] Following D'Alembert, he denied the existence of both the infinitely large and the infinitesimal, although the latter insinuated itself into his work nevertheless. In determining differentials, for example, Kästner allowed the independent variable z to be given an increment e, the function Z of z thereby taking on the increment E. "If now e is indefinitely diminished, the limit which the ratio $E : e$ approaches indefinitely is called the ratio of the differentials of Z and z and the infinitely small quantities e and E are called the differentials of z and Z."[124] This lack of that distinction between the increments and the differentials which is essential in the interpretation of the calculus in terms of the derivative, shows clearly how little even professed proponents of the limit concept could avoid using the infinitely small in their explanations. Evidently Kästner did not recognize the

[120] Cajori, "Grafting of the Theory of Limits on the Calculus of Leibniz."

[121] Cf. Lagrange, *Œuvres*, X, 269–70; VII, 598.

[122] See Moritz Cantor, *Vorlesungen*, IV, 643–44.

[123] *Anfangsgründe*, Vorrede, pp. xiii–xiv.   [124] *Anfangsgründe*, p. 10.

significance of the admonition of Newton and D'Alembert that the ultimate ratio was not a ratio of ultimate quantities.

That the differentials, rather than the differential quotient, were fundamental in his thought is brought out in another connection. In determining the differential E of $Z = \dfrac{z^n}{a^{n-1}}$ at the point for which $z = 0$, we should first determine the derivative at this point and then multiply this by e, the differential of z. The result is $E = 0$. Kästner, however, first found the increment E in Z, corresponding to the increment e in z, and then substituted $z = 0$ before determining the limit of the ratio $\dfrac{E}{e}$. His result was $E = \dfrac{e^n}{a^{n-1}}$, a conclusion which Euler and James Bernoulli had reached by their infinitesimal methods.[125]

A number of other mathematicians of the time felt that the infinitely small could not furnish a satisfactory basis for the calculus.[126] Most of these attempted to substitute some form of the limiting idea, but there was one notable exception. Joseph Louis Lagrange displayed a skeptical attitude toward the infinitely small, echoing Bishop Berkeley in saying that the correctness of the results of the differential calculus was to be explained as due to a compensation of errors.[127] However, his attitude toward the limit concept was cool also, for he considered this involved with metaphysical difficulties.[128] He felt that D'Alembert's definition of the tangent as the limit of a secant was unsatisfactory, inasmuch as after the secant has become the tangent, nothing prevents it from continuing as a secant on the other side of the point in question.[129]

Furthermore, the method of fluxions did not appeal to Lagrange because of the introduction of the irrelevant notion of motion. Euler's presentation of dx and dy as 0 failed also to satisfy him because he felt that we have no clear and precise idea of the ratio of two terms which become zero.[130] Lagrange, as a result, sought a simple algebraic

[125] *Ibid.*, p. 13. See also James Bernoulli, *Opera omnia*, II, 1097; Euler, *Opera omnia*, X, 561.
[126] See Moritz Cantor, *Vorlesungen*, IV, Section XXVI.
[127] Lagrange, *Œuvres*, VII, 598. See also IX, 17.
[128] *Œuvres*, VII, 325–26; cf. also III, 443, and IX, 18.
[129] *Œuvres*, VII, 325.   [130] *Œuvres*, IX, 17–18.

method free from the objections found in others. As early as 1759 he seems to have been satisfied that he had found this, for he wrote to Euler in that year that he believed he had developed the true metaphysics of the principles of mechanics and of the differential and integral calculus, as far as was possible.[131] Lagrange probably had in mind the method[132] which he proposed in a paper of 1772, "Sur une nouvelle espèce de calcul relatif à la différentiation et à l'intégration des quantités variables."[133]

Here he recalled that Leibniz had pointed out the analogy—now called Leibniz's Rule—between differentials of all orders of the product of two variables and the powers of the same order of a binomial in these variables, and had also remarked that the same correspondence subsisted between the negative powers and the integrals. Following this suggestion, Lagrange made use of a similar analogy in connection with infinite series. The series $f(x + h) = f(x) + f'(x)h + f''(x)\dfrac{h^2}{2!} + \ldots$ had been known at least from the time of Taylor, whose name it bears. In this series, the coefficients of the powers of h involve the ratios of differentials, or of fluxions. However, the series can be derived without reference to these notions. What would be more natural, therefore, than to define differentials and fluxions in terms of the coefficients of such a series? This procedure would (only on the surface, as we know now) obviate the necessity of introducing either limits or infinitesimals into the work, and the calculus would thus be reduced to simple algebraic operations. This notion of the differential and the integral calculus appeared to Lagrange to be

the clearest and simplest which had yet been given. It is, as one sees, independent of all metaphysics and of any theory of infinitely small or vanishing quantities.[134]

Lagrange proceeded therefore to make Taylor's series fundamental in his work, assuming implicitly that all functions allow of such an expansion. The coefficients p, p', $\ldots q$, q', $\ldots$, of the powers of h, k, $\ldots$ in the Taylor's series for $u(x + h, y + k, \ldots)$ he defined as the "fonctions dérivées" of u. The differential calculus then consisted,

[131] *Œuvres*, XIV, 173; cf. III, 443, and VII, 325–28.
[132] See Jourdain, "The Ideas of the 'Fonctions analytiques' in Lagrange's Early Work."
[133] See *Œuvres*, III, 439–76.   [134] *Œuvres*, III, 443.

for Lagrange, in "finding directly and by simple and facile procedures the derived functions p, p', ..., q, q', ... of the function u"; and the integral calculus consisted, inversely, in "determining by means of these latter functions the function u."[135]

Lagrange's method is based on the unwarranted supposition that every function can be so represented and handled. Moreover, the escape from the infinitely large and the infinitely small, as well as from the limit concept, is only illusory, inasmuch as these notions enter into the critical question of convergence which Lagrange did not adequately consider. Furthermore, his method lacks the operational suggestiveness and facility which the Leibnizian ideas and notation afforded. Logically, however, the Lagrangian definition had an advantage in that it sought, as had the work of Euler, to make fundamental the formalism of the theory of functions, rather than the preconceptions of geometry, mechanics, or philosophy.

Lagrange has been criticized[136] for giving up, in favor of mathematical formalism, the "generative" concept which has frequently been felt to be the basis of the methods of fluxions and differentials. Such criticism is based on a failure to recognize that mathematics is most useful when unencumbered by psychological preconceptions. Euler had attempted misdirectedly to formalize the Leibnizian conceptions by making the differentials zeros. D'Alembert had made an effort to present a satisfactory idea of the notion of limit, but had failed to give the concept a clear and precise formalism which would make it logically unequivocal. Lagrange, therefore, sought still another mode of presentation, based on the function concept which Euler had emphasized and popularized. Incidentally, in so doing he focused attention for almost the first time upon the quantity which is now the central conception in the calculus—that of the derived function, or the derivative, or the differential coefficient (a terminology reminiscent of Lagrange's method, which was, however, introduced later by Lacroix).[137] Lagrange, in this connection gave not only the name from which the word derivative was adopted, but also the notation $f'x$, modifications of which are still conveniently used.

[135] *Ibid.*

[136] Cohen, *Das Princip der Infinitesimalmethode und seine Geschichte*, p. 100.

[137] See *Bibliotheca Mathematica* (3), I (1900), 517 for notes on the origin of these designations.

Newton had emphasized the method of prime and ultimate ratios of increments or of fluxions. Although this ratio can be interpreted as a *single number* or quantity, which we now call the derivative, Newton seems to have had in mind the idea of this quantity as the *ratio* of increments or of fluxions, or of quantities proportional to these. Moreover, although the fluxions themselves can only be defined rigorously as derivatives, Newton does not appear to have realized this.

In the differential calculus likewise it is clear that although Leibniz realized the significance of the ratio of two infinitesimals, he never seems to have thought of this ratio as a single number, but rather as a quotient of inassignables, or of assignables proportional to these. With D'Alembert's insistence that the differential calculus was to be rigorously interpreted only in terms of limits, one comes close to the conception of a derivative, but even here there is the lack of the notion of a single function, or number, obtained as the limit of a single infinite sequence.

Like Newton and Leibniz, D'Alembert appears to have had in mind not a function, but two sides of an equation, the limits of which are equal. Kästner likewise displayed the predominance of the idea of a ratio. Although he recognized the fundamental significance of the limit concept, he nevertheless defined the differential quotient literally as a ratio of differentials.[138] In Lagrange's method the term derivative can for the first time be applied with strict propriety, for his "fonction dérivée" is merely a single coefficient of a term in an infinite series and is completely divested of any idea of ratio, or limiting equality. It is a single quantity or function, and although Lagrange's definition was not that which was to be accepted in the end, the notion of a derived function may well have aided in making the general acceptance of the current definition possible.

Although Lagrange's method appeared in the *Miscellanea Taurinensia* for 1772, it received little recognition—perhaps as a result of the novelty of the ideas and the notation involved—and the search for a satisfactory basis for the calculus continued. To encourage efforts in this direction the Berlin Academy, of which Lagrange at the time was president, in 1784 offered a prize for the best exposition of a clear

[138] *Anfangsgründe*, p. 4.

and precise theory of the mathematical infinity. The prize-winning essay was that of Simon L'Huilier: *Exposition élémentaire des principes des calculs supérieurs*. This was published in 1787 and appeared again (in Latin) in 1795. In this work L'Huilier proposed to show that "the method of the ancients, known under the name of Method of Exhaustion, conveniently extended, suffices to establish with certainty the principles of the new calculus."[139] In conformity with this purpose, he modified the method of exhaustion to interpret it in terms of limits. Making the limit concept basic in his exposition, L'Huilier agreed with D'Alembert that "in the differential calculus it was not necessary to pronounce the name differential quantity."[140]

As in modern textbooks, L'Huilier made the differential ratio or quotient fundamental, defining this as the limit of the ratio of the increment in the function to that in the independent variable. L'Huilier regarded this form of presentation of the calculus as a development of that of Newton and other English authors, but his exposition indicated an advance over their work. He focused attention upon a single number (the derivative) as the limit of a single variable (the ratio of the increments), rather than upon an ultimate ratio of two evanescent quantities, or of two fluxions, or of any two quantities which have the same ratio as these. Although he retained the name "differential quotient" and the symbol $\frac{dy}{dx}$ to represent this quantity, he insisted that the latter was nothing but a symbol which was to be interpreted as a single number.[141] In dealing with differential quotients of higher orders he again warned that the symbol $\frac{d^2y}{dx^2}$ was not to be broken up as a quotient. This is in striking contrast with the work of Newton and Leibniz, in which fluxions and differentials of any order were regarded as having a significance independent of the ratio or equation in which they entered. The differential quotient of L'Huilier was a single number or function, equivalent to Lagrange's derived function, and represented essentially the present conception of the derivative.

Although L'Huilier's definition of the differential quotient is in

[139] *Exposition élémentaire*, p. 6.
[140] *Ibid.*, p. 141.   [141] *Ibid.*, p. 32.

most respects that to be found in present day elementary textbooks on the calculus, he does not seem to have been aware of the possible difficulties involved in the limit concept. He had avoided the mysticism of the infinitesimal, the vagueness of the ultimate ratio, and the inanity of the symbol $\frac{0}{0}$; but he failed to appreciate that the subtlety of the limit concept was to make an extremely careful definition essential. L'Huilier was dealing only with very simple functions, so that he was unaware of the inadequacies of his presentation. His variable was always less than or greater than his limit. "Given a variable quantity always smaller or greater than a proposed constant quantity; but which can differ from the latter by less than any proposed quantity however small; this constant quantity is called the limit in greatness or in smallness of the variable quantity."[142] The variable could not oscillate, as it may under our more general view.

More seriously, L'Huilier fell into an error which is suggested by the vague idea of uniformity expressed in the law of continuity of Leibniz. He said that "if a variable quantity at all stages enjoys a certain property, its limit will enjoy this same property."[143] That this notion persisted also in the nineteenth century is apparent from the statement of William Whewell: "The axiom . . . that what is true up to the limit is true at the limit, is involved in the very conception of a limit."[143a] The falsity of this doctrine is immediately apparent from the fact that *irrational* numbers may easily be defined as the limits of sequences of *rational* numbers, or from the observation that the properties of a polygon inscribed in a circle are not those of the limiting figure—the circle. The mistake of L'Huilier in this connection was probably the result of his failure to see that the limit concept was to be identified with the nature of infinite converging sequences and with the question of the nature of real numbers and the continuum. The inadequacy of his conception of number may be further inferred from the fact that he felt it necessary to distinguish between limiting values and limiting ratios.

[142] *Ibid.*, p. 7.

[143] *Exposition élémentaire*, p. 167. Cajori ("Grafting of the Theory of Limits on the Calculus of Leibniz"), apparently unaware of this statement, has said that this principle was used but not stated in the eighteenth century.

[143a] *History of Scientific Ideas*, I., 152.

Although L'Huilier correctly sought the basis of the calculus in the limit concept, his exposition of this was an oversimplification of what was later realized to be a very difficult question. Following D'Alembert in regarding the infinite from the point of view of magnitude rather than of aggregation, he denied the existence of an actually infinite quantity because he felt that its acceptance would lead to such "contradictions" as $\infty + n = \infty - n$. He consequently maintained that he had shown the calculus to be independent of all idea of the infinite, whether large or small. In this he failed to realize that the whole theory of limits is based, in the last analysis, upon that of infinite aggregates. This fact was not clearly recognized until the following century.

L'Huilier's monograph was not widely read, nor were his views generally accepted at the time. The indecision as to the true basis of the calculus remained as acute as before. As a result, there appeared in 1797 what was perhaps the most famous attempt to clear up the difficulties in the situation: the *Réflexions sur la métaphysique du calcul infinitésimal*, by L. N. M. Carnot, the remarkable soldier, administrator, and mathematician to whom the French Assembly gave the title "L'Organizateur de la Victoire." Carnot's work enjoyed a truly remarkable popularity, appearing in numerous editions and several languages from that time until quite recently.[144]

In view of the lack of clarity and uniformity in the then-current expositions of the calculus, Carnot wished to make the theory rigidly precise. Considering the many conflicting interpretations of the subject, he sought to know "in what the veritable spirit of the infinitesimal analysis consisted."[145] In his selection of the unifying principle, however, he made a most deplorable choice. He concluded that "the true metaphysical principles of the Infinitesimal Analysis . . . are nevertheless . . . the principles of the compensation of errors,"[146] as Berk-

---

[144] The first edition of Paris, 1797, was followed by an enlarged second edition at the same place in 1813. Unless otherwise stated, references in this work are to this second edition. The first edition was translated into English by W. Dickson in the *Philosophical Magazine*, VIII (1800), 222–40; 335–52; IX (1801), 39–56. The second edition appeared in an English translation by W. R. Browell at Oxford, in 1832, as *Reflexions on the Metaphysical Principles of the Infinitesimal Analysis*. Other French editions appeared in Paris in 1839, 1860, 1881, and 1921 (2 vols.) A Portuguese translation appeared at Lisbon in 1798, a German one at Frankfurt a. M. in 1800, and an Italian one at Pavia in 1803.

[145] *Réflexions sur la métaphysique du calcul infinitésimal*, p. 1.   [146] Browell trans., p. 44.

eley and Lagrange had suggested. In his expansion of this view, he
reverted substantially to ideas which Leibniz had expressed. He held
that to be certain that two designated quantities are rigorously equal,
it is sufficient to prove that their difference cannot be a "quantité
designée."[147] Paraphrasing Leibniz, Carnot said further: that for any
quantity one may substitute another which differs from it by an
infinitesimal;[148] that the method of infinitesimals is nothing more than
that of exhaustion reduced to an algorithm;[149] that "quantités inap-
préciables" are merely auxiliaries which are introduced, like imaginary
numbers, only to facilitate the computation, and which are eliminated
in reaching the final result.[150]

Carnot even echoed the favorite explanation of Leibniz in terms of
the law of continuity. He held that one can envisage the infinitesimal
analysis under two points of view, according as the infinitesimals are
taken to be "quantités effectives" or are regarded as "quantités
absolument nulles." (Leibniz, however, had not admitted them as
absolutely, but only relatively, zero.) In the first case, he felt that the
calculus was to be explained upon the basis of a compensation of
errors: "imperfect equations" were to be made "perfectly exact" by
the simple expedient of eliminating the quantities whose presence
occasioned the errors;[151] in the latter case he considered the calculus
an "art" of comparing vanishing quantities with each other in order
to discover from these comparisons the relationships between the proposed quantities.[152] To the objection that these vanishing quantities
either *are* or *are not* zero, Carnot responded that "what are called
infinitely small quantities are not simply any null quantities at all,
but rather null quantities assigned by a law of continuity which
determines the relationship."[153] This explanation is strikingly like that
given by Leibniz about a century earlier.

Along with his establishment of the "true metaphysics" of the
infinitesimal calculus, Carnot proceeded to show that the diverse
views of the subject were essentially reducible to this same basis.

[147] *Réflexions sur la métaphysique du calcul infinitésimal*, p. 31.
[148] *Ibid.*, p. 35.   [149] *Ibid.*, p. 39.
[150] *Ibid.*, pp. 38–39. "Les mathématiques ne sont-elles pas remplies de pareilles énigmes?"
said Carnot, in this connection.
[151] Cf. Dickson's translation, p. 336.
[152] *Réflexions sur la métaphysique du calcul infinitésimal*, p. 185.   [153] *Ibid.*, p. 190.

In demonstrating this, he pointed out that the method of exhaustion made use of analogous systems of known auxiliary quantities. Newton's method of prime and ultimate ratios was similar, except that in it the lemmas freed the work from the need for the argument by a double reductio ad absurdum. Carnot felt that the methods of Cavalieri and Roberval also were admittedly corollaries of the method of exhaustion.[154] Descartes' method of undetermined coefficients, he held, touched the infinitesimal analysis closely, the latter being but "a felicitous application" of the former![155] The method of limits he recognized to be not different from that of first and last ratios, so that it was likewise a simplification of the method of exhaustion.[156] Furthermore, these methods all lead to the results of infinitesimal analysis, but by a difficult and circuitous route.[157] Lagrange's method he likewise saw linked to the infinitesimal calculus, in that it neglects the other terms of the infinite series.[158] The divers points of view he therefore felt, somewhat as had L'Huilier, to be merely simplifications of the method of exhaustion which were effected by reducing this to a convenient algorithm. Inasmuch as the infinitesimal method combined the facility of the procedures of approximate calculation with the exactitude of the results of ordinary analysis, he saw no point in attempting, under the guise of greater rigor, to substitute for it any less natural method.[159]

Although Carnot's work was widely read, it can hardly be said to have led to a clearer understanding of the difficulties inherent in the new analysis. Although he realized that differentials were to be defined somewhat as variables, anticipating, to a certain extent, the view of Cauchy, he was unable to give suitable definitions to these because, like Leibniz, he thought in terms of equations rather than of the function concept which led Cauchy to make the derivative fundamental. Furthermore, Carnot, one of a school of mathematicians who emphasized the relationship of mathematics to scientific practice,[160] appears, in spite of the title of his work, to have been more concerned about the facility of application of the rules of procedure than about the logical reasoning involved. In this respect his work resembles

---

[154] *Ibid.*, pp. 139–40. [155] *Ibid.*, pp. 150–51. [156] *Ibid.*, p. 171.
[157] *Ibid.*, p. 192. [158] *Ibid.*, pp. 194–97. [159] *Ibid.*, pp. 215–16.
[160] Merz, *A History of European Thought in the Nineteenth Century*, II, 100–1.

that of Leibniz, whose explanations of differentials he so largely paraphrased, and whose method he defended with almost polemic warmth.

Mathematicians on the Continent had been in essential agreement that pragmatically the differential calculus offered the best method of procedure; but it was precisely on the *logic* of the matter that they were at variance. On the latter point Carnot was of little assistance, for after pointing out what most men had long realized—that all of the methods of the new analysis were essentially related—he proceeded to make basic the infinitesimal system, the one which was logically, perhaps, the weakest of all. In so doing he pointed toward a view diametrically opposed to that which D'Alembert[161] and L'Huilier had indicated and along which the rigorous development was ultimately to proceed, following the work of Cauchy.[162]

The year 1797, in which the first edition of Carnot's work was published, saw the appearance also of the famous work of Lagrange, *Théorie des fonctions analytiques, contenant les principes du calcul différentiel, dégagés de toute considération d'infiniment petits ou d'évanouissans, de limites ou de fluxions, et réduits à l'analyse algébrique des quantités finies.* This book developed with care and completeness the characteristic definition and method in terms of "fonctions dérivées," based upon Taylor's series, which Lagrange had proposed in 1772. In it the author gave not only an attempted proof of the incorrect theorem that every continuous function may be so expanded, but also the determination of the "fonctions derivées" (or derivatives) of the elementary functions, and numerous applications to geometry and mechanics. Carried along by the authority of Lagrange's great reputation, the method now enjoyed a short-lived period of comparative success. As had been the case with Maclaurin's treatise on fluxions, mathematicians praised the new method highly,[163] although they seldom used it. They explained that inasmuch as the notations were

---

[161] Mansion (*Esquisse de l'histoire du calcul infinitésimal*, p. 290, n.) has misdirectedly characterized Carnot's ideas as a development of those of D'Alembert.

[162] The assertion (See Smith, "Lazare Nicholas Marguérite Carnot," p. 189) that Carnot "paved the way for Cauchy's notable memoir" is justified only in the most general sense.

[163] Cf. the review in *Monthly Review*, London, N. S., XXVIII (1799), appendix, pp. 481–99; also Valperga-Caluso, "Sul paragone del calcolo delle funzioni derivate coi metodi anteriori"; also the set of monographs by Froberg, and others, called *De analytica calculi differentialis et integralis theoria, inventa a cel. La Grange.*

less convenient and the calculations involved more embarrassing than those found in the methods of the ordinary differential and integral calculus, it was sufficient that one were assured through Lagrange's method of the legitimacy of other more expeditious methods. This seems to have been the view of Lagrange himself, for side by side with his method of derived functions he continued throughout his life to employ the notation of differentials.

Most of the objections to the method of Lagrange were based upon the inconveniences of the notation and the operations, but before long doubts began to arise as to the correctness of the principle that all continuous functions could be expanded in Taylor's series. It was pointed out[164] that such an expansion was possible only for the more simple functions, and that consequently the method was of limited applicability. Furthermore, Sniadecki, a Pole, correctly explained that the method was fundamentally identical with the limit method.[165] However, the views of convergence, continuity, and function held at the time were not sufficiently definite to permit a deeper clarification of these ideas.

An interesting but somewhat misdirected tirade against Lagrange's point of view was delivered by another Polish mathematician, Hoëné Wronski. An eager devotee of the differential method of Leibniz and of the transcendental philosophy of Kant, he protested with some asperity against the ban on the infinite in analysis which Lagrange had wished to impose. He criticized Lagrange not so much for the absence of logical rigor in his free manipulation of infinite series— although he did pertinently ask where Lagrange got the series $f(x + i)$ $= A + Bi + Ci^2 + Di^3 + \ldots$ with which he opened his proof of Taylor's series—as for his lack of a sufficiently broad view. Wronski believed that modern mathematics is to be based on the "supreme algorithmic law" $Fx = A_0\Omega_0 + A_1\Omega_1 + A_2\Omega_2 + A_3\Omega_3 + \ldots$, where the quantities $\Omega_0, \Omega_1, \Omega_2, \Omega_3, \ldots$ are any functions of the variable x. Being the supreme law of mathematics, the irrecusable truth of this law he held to be not mathematically derived, but given by transcendental philosophy.[166]

[164] See Dickstein, "Zur Geschichte der Prinzipien der Infinitesimalrechnung"; cf. also Lacroix, *Traité du calcul*, 2d ed., III, 629–30.

[165] Dickstein, "Zur Geschichte der Prinzipien der Infinitesimalrechnung."

[166] *Réfutation de la théorie des fonctions analytiques*, Vol. IV of Wronski's *Œuvres*.

Wronski was, of course, correct in saying that the method of derived functions was too restricted, in that it was limited only to certain functions which could be so expanded. However, his general views on the calculus were far from those accepted at the present time. Whereas Lagrange had attempted to give a formal logical justification of the subject, Wronski asserted that the differential calculus constituted a *primitive algorithm* governing the *generation* of quantities, rather than the laws of quantities *already formed*. Its propositions he held to be expressions of an absolute truth, and the deduction of its principles he consequently regarded as beyond the sphere of mathematics. The explication of the calculus by the methods of limits, of ultimate ratios, of vanishing quantities, of the theory of functions, he felt constituted but an indirect approach which proceeded from a false view of the new analysis. In seeking to examine the principles of the subject by purely mathematical means, Wronski believed geometers were simply wasting time and effort. He called upon them to give up "that servile imitation of ancient geometers in avoiding the infinite and that which depends upon it."[167]

In his own work Wronski made a more uncritical use of the infinite than had even Wallis and Fontenelle. He said, for example, that "the *absolute* meaning of the number π was given by the expression ...

$$\frac{4\,\infty}{\sqrt{-1}}\left\{(1 + \sqrt{-1})^{\frac{1}{\infty}} - (1 - \sqrt{-1})^{\frac{1}{\infty}}\right\}.$$

."[168] Perhaps because of the novelty of his notation, as well as of this bizarre use of the symbol ∞, Wronski did not exert a strong influence upon the development of the calculus. The mathematics of the time was about to accept what he opposed: the rigorous logical establishment of the calculus upon the limit concept. Nevertheless, the work of Wronski represents an extreme example of a view which we shall find recurring throughout the nineteenth century. In regarding the calculus as a means of explaining the growth of magnitudes, followers of this school of thought were to attempt to retain the concept of the infinitely small, not as an extensive quantity but as an intensive magnitude. Mathematics has excluded the fixed infinitely small because it has failed to establish the notion logically; but transcendental philosophy has sought to preserve primitive intuition in this respect by interpreting it as

[167] *Ibid.*, pp. 39–40.   [168] *Ibid.*, p. 69.

having an a priori metaphysical reality associated with the generation of magnitude.

Lagrange's *Théorie des fonctions* was only one, but by far the most important, of many attempts made about this time to furnish the calculus with a basis which would logically modify or supplant those given in terms of limits and infinitesimals. Condorcet, Arbogast, Servois, and others put forth methods similar to that of Lagrange.[169] Condorcet, as early as 1786, had begun a work on the calculus, based on series and finite differences, but this was interrupted by the Revolution and did not appear.[170]

In 1789 L. F. A. Arbogast presented to the Académie des Sciences a "true theory of the differential calculus" along the lines of Lagrange and Condorcet. This work, published in 1800 under the title *Du calcul des dérivations*, sought to establish the principles of the calculus independently of limits and the infinitely small, and with the simplicity and certitude found in ordinary algebra.[171] Arbogast assumed, as had Lagrange, that the function $F(a + x)$ could be expanded in a series of powers of x, and then showed that the coefficients in this could be identified with the more familiar differential quotient.[172]

These new methods resemble the attempts which had been made in England, following the Berkeley-Jurin-Robins dispute, to give arithmetical procedures for the calculus. They serve to indicate, as did the British controversy, a lack of satisfaction, at the time, with the methods of limits, fluxions, and infinitesimals. The full title of Lagrange's *Théorie des fonctions* indicates his discontent with these methods. F. J. Servois, in his *Essai sur un nouveau mode d'exposition des principes du calcul différentiel*, which appeared in 1814,[173] expressed himself more strongly. He called the method of limits a "gothique hypothèse"[174] and that of differentials a "strabisme infinitésimal."[175]

---

[169] For references to this work, see Dickstein, *op. cit.*; Moritz Cantor, *Vorlesungen*, IV, Section XXVI; and Lacroix, *Traité du calcul différentiel et du calcul intégral*, I, 237–48, and preface.

[170] See Lacroix, *Traité du calcul*, I, xxii–xxvi.

[171] See Zimmermann, *Arbogast als Mathematiker und Historiker der Mathematik*, pp. 44–45; cf. also Arbogast, *Du calcul des dérivations*, Preface; and Lacroix, *Traité du calcul*, Preface.

[172] *Du calcul des dérivations*, pp. xii–xiv; cf. also p. 2.

[173] This appeared in the *Annales de Mathématiques*, V (1814–15), pp. 93–141, and was also published separately in Nîmes, 1814. References given in this work are to the Nîmes publication.

[174] *Essai*, p. 56.   [175] *Ibid.*, p. 65.

The introduction of the idea of infinity he objected to as useless; and, with reference to the views which Wronski had expressed on the calculus in his *Réfutation* of Lagrange, he asserted that he "had well foreseen, on reading Kant, that geometers would sooner or later be the object of the cavils of his sect."[176] Servois tried, in his turn, to establish the calculus on a combination of finite differences and infinite series, in which the idea of limits was, as in the case of the method of Lagrange, ostensibly eliminated by a disregard of the essential question of convergence.

Further attempts along this line continued for sometime: in 1821 in *Eine neue Methode für den Infinitesimalkalkul* of Georg von Buquoy; in 1849 in the *Essai sur la métaphysique du calcul intégral* of C. A. Agardh; and as late as 1873 in an article, "Exposition nouvelle des principes de calcul différentiel," by J. B. Brasseur.[177] These invariably depended upon expansions in series to avoid "any metaphysical notions," and in this respect they resembled closely the earlier efforts of Lagrange, Condorcet, Arbogast, and Servois. However, at almost precisely the period of the work of Servois a tendency toward rigor in series was setting in. This made clear the basic weakness of all such attempts to avoid the method of limits through an uncritical formal manipulation of infinite series. The final formulation of the calculus was not destined to be in terms of one of these new methods which were developed, but was to be based on one of the very notions they sought to avoid—that of a limit.

In the year 1797, in which Carnot attempted a concordance of all the systems of the calculus and in which Lagrange tried seriously to establish his method, there was published the first volume of perhaps the most famous and ambitious textbook on the subject which had appeared up to that time—*Traité du calcul différentiel et du calcul intégral* of Lacroix. Lacroix declared in the preface of this work that it had been the new method of Lagrange which had inspired him to compose a treatise on the calculus which should have as a basis the luminous ideas which this method had substituted for the infinitely small. His text, however, indicates well the indecision of the period, for in spite of the professed aim of the work, the foundation of the subject remained somewhat in doubt. Lacroix interpreted the method

[176] *Ibid.*, p. 68.   [177] See Bibliography below, for full citations of these works.

of series of Lagrange in terms of the limits of D'Alembert and L'Huilier. In this respect, however, he obscured the significance of this relationship by speaking of the limit of divergent series, and by following Euler in the study of such infinite series as $1 - 1 + 2 - 6 + 24 - 120 + \ldots$[178] Furthermore, he did not share with Lagrange the suspicion that the method of Leibniz was based on a false idea of the infinitely small. Lacroix admittedly made use of infinitesimals.[179] Although he accepted the metaphysics of Lagrange, he adopted the differential notation of Leibniz.[180] This fact at times confused his thought and led him, as it had Euler, to regard the differential coefficient (as Lacroix called it) as a quotient of zeros.[181]

The lack on the part of Lacroix of critical distinction in dealing with the methods of Newton, Leibniz, and Lagrange gave to his work an appearance somewhat resembling the attempt of Carnot to demonstrate the congruity of the numerous representations of the calculus. The mathematician and astronomer Laplace praised this attitude on the part of Lacroix, saying that such a *rapprochement* of methods served as a mutual clarification, the true metaphysics probably being found in what they had in common.[182] It appears, however, that at that time at least it was deplorable, for it led to a confusion of thought on the subject just when logical precision was most needed. However, in his more popular work of 1802—the *Traité élémentaire*, an abridgment of his larger treatise—Lacroix omitted the method of Lagrange and made the explanation in terms of limits basic, although again with a lack of rigor which made interpretations in terms of infinitesimals possible. The success of this work, which went through many editions (the ninth French edition appearing in 1881) and was translated into several languages, led to other texts of the same type; and it was largely through these that the method of limits became familiar, if not rigorous. It was through such texts that the Leibnizian notation and the doctrine of limits supplanted in England the method of fluxions and interpretations which had already become hopelessly confused with the infinitely small.

The year 1816, in which Lacroix's shorter work was translated into

[178] *Traité du calcul*, III, 389.
[179] *Ibid.*, I, 242. [180] *Ibid.*, p. 243. [181] *Ibid.*, I, 344.
[182] Letter to Lacroix, published in *Le Nouveau Traité du calcul différentiel*, 2d ed., preface, p. XIX.

English, "marks an important period of transition,"[183] because it witnessed the triumph in England of the methods used on the Continent. This particular point in the history of mathematics marks a new epoch for a far more significant reason, for in the very next year there appeared a work by Bolzano—*Rein analytischer Beweis des Lehrsatzes dass zwischen je zwei Werthen, die ein entgegengesetztes Resultat gewähren, wenigstens eine reelle Wurzel der Gleichung liege*—which indicated the rise of the period of mathematical rigor in all branches of the subject.[184] In the calculus the new attitude resulted in the logical establishment of the higher analysis upon the limit concept, and thus brought to an end the period of indecision which had begun with the inventions of the method of fluxions and of the differential calculus.

[183] Cajori, *A History of the Conceptions of Limits and Fluxions*, pp. 270-71.
[184] See Pierpont, "Mathematical Rigor," pp. 32-34.

VII. The Rigorous Formulation

THE OBJECTIONS raised in the eighteenth century to the methods of fluxions, of prime and ultimate ratios, of limits, of differentials, and of derived functions were in large measure unanswered in terms of the conceptions of the time. The arguments were in the last analysis equivalent to those which Zeno had raised well over two thousand years previously and were based on questions of infinity and continuity. The proponents of all methods except that of differentials, however, protested that they had no need to invoke the notion of the infinite and disregarded entirely that of continuity. The advocates of the differential calculus, in turn, although attempting to justify its procedures in terms of these concepts, were quite unable to furnish logically consistent explanations of them. By most mathematicians these were considered to be metaphysical ideas and so to lie beyond the realm of mathematical definition.

It is interesting, in looking back, to see that the methods most adverse to the introduction into mathematics of the notions of infinity and of continuity were precisely those which made this introduction possible. The method of limits, which at that time appeared to lead to neither infinity nor the law of continuity, was to furnish the logical basis for these; and the method of Lagrange, which was developed in order to avoid these difficulties, was to bring up questions which pointed the way toward their solution. We have seen that in the early nineteenth century critics of Lagrange began to question the validity of his principle that a continuous function can always be expressed, by means of Taylor's theorem, as an infinite series. They began to ask what was meant by a function in general and by a continuous function in particular, and to criticize the almost indiscriminate use of infinite series. Lagrange had made a start in the direction of greater care in the use of series when he pointed out that one must consider the remainder in every case. This warning serves to indicate, perhaps, why he thought he had avoided the use of the infinite and the infinitely small. Like Archimedes, he evidently did not consider the series as continued to infinity, but only to a point at which the remainder was sufficiently small. In the nineteenth century, however, the concept of infinity was to become basic in the calculus through the use of infinite series and infinite aggregates.

One of the pioneers in the matter of greater rigor in the fundamental conceptions of the calculus—in its arithmetization and in the careful study of the infinite—was the Bohemian priest, philosopher, and mathematician, Bernhard Bolzano.[1] In 1799 Gauss had given a proof of the fundamental theorem of algebra—that every rational, integral algebraic equation has a root—using considerations from geometry. Bolzano, however, wished a proof involving only considerations derived from arithmetic, algebra, and analysis. As Lagrange had felt that the introduction of time and motion into mathematics was unnecessary, so Bolzano sought to avoid in his proofs any considerations derived from spatial intuition.[2]

This attitude made necessary, in the first place, a satisfactory definition of continuity. The calculus may, indeed, be thought of as having arisen from the Pythagorean recognition of the difficulty involved in attempting to substitute numerical considerations for supposedly continuous geometrical magnitudes. Newton had avoided such embarrassment by appealing to the intuition of continuous motion, and Leibniz had evaded the question by his appeal to the postulate of continuity. Bolzano, however, gave a definition of continuous function which, for the first time, indicated clearly that the basis of the idea of continuity was to be found in the limit concept. He defined a function $f(x)$ as continuous in an interval if for any value of x in this interval the difference $f(x + \triangle x) - f(x)$ becomes and remains less than any given quantity for $\triangle x$ sufficiently small, whether positive or negative.[3] This definition is not essentially different from that given a little later by Cauchy and is fundamental in the calculus at the present time.

In presenting the elements of the calculus, Bolzano realized clearly that the subject was to be explained in terms of limits of ratios of

[1] See Stolz, "B. Bolzano's Bedeutung in der Geschichte der Infinitesimalrechnung." Although Bolzano was born and died at Prague, Quido Vetter ("The Development of Mathematics in Bohemia," p. 54) has called him a Milanese, inasmuch as his father was a native of northern Italy.

[2] *Rein analytischer Beweis*, pp. 9–10.

[3] See Bolzano's *Schriften*, I, 14; cf. also *Rein analytischer Beweis*, pp. 11–12.

finite differences. He defined the derivative of $F(x)$ for any value of x as the quantity $F'(x)$ which the ratio $\dfrac{F(x + \triangle x) - F(x)}{\triangle x}$ approaches indefinitely closely, or as closely as we please, as $\triangle x$ approaches zero, whether $\triangle x$ is positive or negative.[4] This definition is in essence the same as that of L'Huilier, but Bolzano went further in explaining the nature of the limit concept. Lagrange and other mathematicians had felt that the limit notion was bound up with a quotient of evanescent quantities or of zeros. Euler had explained $\dfrac{dy}{dx}$ as a quotient of zeros, and in this respect Lacroix tended to follow him. Bolzano, however, emphasized the fact that this was not to be interpreted as a ratio of dy to dx or as the quotient of zero divided by zero, but was rather one symbol for a single function.[5] He held that a function has no determined value at a point if it reduces to $\dfrac{0}{0}$. However, it may have a limiting value as this point is approached, and he correctly indicated that, by adopting the limiting value as the meaning of $\dfrac{0}{0}$, the function may be made continuous at this point.[6]

Ever since the invention of the calculus it had been felt that, inasmuch as the subject was bound up with motion and the growth of magnitudes, the continuity of a function was sufficient to assure the existence of a derivative. In 1834, however, Bolzano gave an example of a nondifferentiable continuous function. This is based upon a fundamental operation which may be described as follows: Let PQ be a line segment inclined to the horizontal. Let it be halved by the point M, and subdivide the segments PM and MQ into four equal parts, the points of division being P_1, P_2, P_3, and Q_1, Q_2, Q_3. Let P_3' be the reflection of P_3 in the horizontal through M, and let Q_3' be the reflection of Q_3 in the horizontal through Q. Form the broken line $PP_3'MQ_3'Q$. Now apply to each of the four segments of this broken line the fundamental operation as described above, obtaining in this manner a broken line of 4^2 segments. Continuing

[4] *Schriften,* I, 80–81; cf. *Paradoxien des Unendlichen,* p. 66.
[5] *Paradoxien des Unendlichen,* p. 68.   [6] *Schriften,* pp. 25–26.

this process indefinitely, the broken-lined figure will converge toward a curve representing a continuous function nowhere differentiable.[7]

This illustration given by Bolzano might have served in mathematics as the analogue of the *experimentum crucis* in science, showing that continuous functions need not, in spite of the suggestions of geometrical and physical intuition, possess derivatives. However, because the work of Bolzano did not become known at the time, such a rôle was reserved for the famous example of such a function given by Weierstrass about a third of a century later.[8]

Lagrange had held that his method of series avoided the necessity of considering the infinitesimals or limits, but Bolzano pointed out that in the case of infinite series it is necessary to consider questions of convergence. These are analogous to limit considerations, as is obvious from the statement of Bolzano that if the sequence $F_1(x)$, $F_2(x)$, $F_3(x)$, ..., $F_n(x)$, ..., $F_{n+r}(x)$, ... is such that the difference between $F_n(x)$ and $F_{n+r}(x)$ becomes and remains less than any given quantity as n increases indefinitely, then there is one and only one value to which the sequence approaches as closely as one pleases.[9] This fundamental proposition of Bolzano will be seen later to have also a significance with respect to the general definition of real number and the arithmetical continuum.

Bolzano felt, in spite of the paradoxes presented by the notions of space and time, that any continuum was to be thought of as ultimately composed of points.[10] His view in this respect resembles that of Galileo, to whom he referred in this connection.[11] Although he denied the existence of infinitely large and infinitely small magnitudes, he maintained, with Galileo, the possibility of an actual infinity with respect to aggregation. He remarked, with respect to such assemblages, the paradox which Galileo had pointed out: that the part could in this case be put into one-to-one correspondence with the whole. The numbers between 0 and 5, for example, could be paired with those between 0 and 12.[12] Bolzano's views on the infinite are substantially those which mathematicians have adopted since the

---

[7] See Kowalewski, "Über Bolzanos nichtdifferenzierbare stetige Function."

[8] Waismann (*Einführung in das mathematische Denken*, p. 122) mistakenly represents Weierstrass as the first to give such an example.

[9] *Rein analytischer Beweis*, p. 35. [10] *Paradoxien des Unendlichen*, pp. 75–76.

[11] *Ibid.*, pp. 89–92. [12] *Ibid.*, pp. 28–29.

time of Cantor, except that what Bolzano had considered as different powers of infinity have been discovered to be of the same power. Bolzano, however, sought to prove the existence of the infinite upon theological grounds, whereas later in the century the property which he and Galileo had regarded simply as a paradox was made by Dedekind, in his clarification of the calculus, the basis of a definition of infinite assemblages.

Although Bolzano's ideas indicate the direction in which the final formulation of the calculus lay and which much of the thought of the nineteenth century was to follow, they did not constitute the decisive influence in determining this. His work remained largely unnoticed until rediscovered by Hermann Hankel more than a half century later.[13] Fortunately, however, the mathematician A. L. Cauchy pursued similar ideas at about the same time and was successful in establishing these as basic in the calculus.

Cauchy rivaled Euler in mathematical productivity, contributing some 800 books and articles on almost all branches of the subject.[14] Among his greatest contributions are the rigorous methods which he introduced into the calculus in his three great treatises: the *Cours d'analyse de l'École Polytechnique* (1821), *Résumé des leçons sur le calcul infinitésimal* (1823), and *Leçons sur le calcul différentiel* (1829).[15] Through these works Cauchy did more than anyone else to impress upon the subject the character which it bears at the present time.

We have seen the notion of a limit develop gradually out of the Greek method of exhaustion, until it was expressed by Newton in the *Principia*. It was more definitely invoked by Robins, D'Alembert, and L'Huilier as the basic concept of the calculus, and as such was included by Lacroix in his textbooks. Throughout this long period, however, the limit concept lacked precision of formulation. This resulted from the fact that it was based on geometrical intuition. It could hardly have been otherwise, inasmuch as during this time the ideas of arithmetic and algebra were largely established upon those of geometrical magnitude. The calculus was interpreted by its inventors as an instrument for dealing with relationships between

[13] See Dickstein, "Zur Geschichte der Principien der Infinitesimalrechnung," p. 77; also Hankel's article "Grenze," in Ersch und Gruber's *Allgemeine Encyklopädie*.
[14] See Valson, *La Vie et les travaux du Baron Cauchy*, reviewed by Boncompagni, p. 57.
[15] The first is found in *Œuvres* (2), III; the last two in *Œuvres* (2), IV.

quantities involved in geometrical problems, and as such was largely accepted by their successors. Euler and Lagrange, in a sense represented exceptions to this rule, for they wished to establish the calculus on the formalism of their analytic function concept. However, even they rejected the limit idea. Furthermore they unwittingly deferred to the preconceptions of geometrical intuition when they uncritically inferred that their methods were applicable to all continuous curves. Although D'Alembert, L'Huilier, and Lacroix had prepared the ground for Cauchy by popularizing the limit idea in their works, this conception remained largely geometrical.

In the work of Cauchy, however, the limit concept became, as it had in the thought of Bolzano, clearly and definitely arithmetical rather than geometrical. Formerly, when illustrations of the notion were desired, the one most likely to be called to mind was that of a circle defined as the limit of a polygon. Such an illustration immediately served to bring up questions as to the manner in which this is to be interpreted. Is it the approach to coincidence of the sides of the polygon with the points representing the circle? Does the polygon ever become the circle? Are the properties of the polygon and the circle the same? It was questions such as these that retarded the acceptance of the limit idea, for they were similar to those of Zeno in demanding some sort of visualization of the passage from the one to the other by which the properties of the first figure merge into those of the second.

Quite recently the limit concept has been loosely interpreted in the assertion that whether one calls the circle the limit of a polygon as the sides are indefinitely decreased, or whether one looks upon it as a polygon with an infinite number of infinitesimal sides, is immaterial, inasmuch as in either case in the end "the specific difference" between the polygon and the circle is destroyed.[16] Such an appeal to geometrical intuition is quite irrelevant in the case of the limit concept. In giving his definition, in his *Cours d'analyse*, Cauchy divorced the idea from all reference to geometrical figures or magnitudes, saying: "When the successive values attributed to a variable approach indefinitely a fixed value so as to end by differing from it by as little as one wishes, this last is called the limit of all the others."[17]

---

[16] Vivanti, *Il concetto d'infinitesimo*, p. 39.   [17] See *Œuvres* (2), III, 19; cf. also IV, 13.

This is the most clear-cut definition of the concept which had been given up to that time, although later mathematicians were to voice objections to this also and to seek to make it still more formal and precise. Cauchy's definition appealed to the notions of number, variable, and function, rather than to intuitions of geometry and dynamics. Consequently, in illustrating the limit concept he said that an irrational number is the limit of the various rational fractions which furnish more and more approximate values of it.[18]

Upon the basis of this arithmetical definition of limit, Cauchy then proceeded to define that elusive term, infinitesimal. Ever since the time that Greek mathematical speculation had hit upon the infinitely small, this notion had been bound up with geometrical intuition of spatial properties and had been regarded as a more or less fixed minimum of extension. The concept had not thrived in arithmetic, largely because unity was considered the numerical minimum, the rational fractions having been treated as ratios of two numbers. However, the seventeenth century saw the rapid rise of algebraic methods in geometry, so that from the time of Fermat the infinitesimal had been the concern of both algebra and geometry. Newton had insisted that his method did not involve the consideration of minima sensibilia, but his interpretation of the procedure in terms of an ultimate *ratio*, rather than of a limiting *number*, made his evanescent magnitudes appear such. Liebniz had been less definite and consistent in denying the existence of actual infinitesimals, for he sometimes considered them as assignable, sometimes as inassignable, and occasionally as qualitatively zero. However, the development during the eighteenth century of the function concept, with its emphasis on the relations between variables, led Cauchy to make the infinitesimal nothing more than a variable. "One says that a variable quantity becomes infinitely small when its numerical value decreases indefinitely in such a way as to converge toward the limit zero."[19] An infinitesimal was consequently not different from other variables, except in the understanding that it is to take on values converging toward zero as a limit.

In order to make the concept of the infinitesimal more useful and to take advantage of the operational facility afforded by the Leibnizian

[18] *Œuvres* (2), III, 19, 341. [19] *Œuvres* (2), III, 19; IV, 16.

views, Cauchy added definitions of infinitesimals of higher order. Newton had restricted himself to infinitesimals, or evanescent quantities, of the first order, but Leibniz had attempted to define those of higher orders also. One of the second order, for example, he defined as being to one of first order as the latter is to a given finite constant, such as unity. Such a vague definition could not be consistently applied, but D'Alembert had sought to correct it by an interpretation in terms of limits. He had in a sense recognized that the infinitesimals were to be regarded as variables and that orders of infinitely small magnitudes were to be defined in terms of ratios of these; but his work, as we have seen, lacked the precision of statement necessary for its general acceptance. It thus remained for Cauchy to express the idea of D'Alembert in the precise symbolism of the limit concept.

Cauchy defined an infinitesimal $y = f(x)$ to be of order n with respect to an infinitesimal x if $\lim_{\substack{x \to 0 \\ y \to 0}} \left(\frac{y}{x^{n-\epsilon}} \right) = 0$ and $\lim_{\substack{x \to 0 \\ y \to 0}} \left(\frac{y}{x^{n+\epsilon}} \right) = \pm \infty$,

ϵ having its classical significance—a positive constant, however small.[20] Here again one recognizes in Cauchy's work the dominance of the ideas of variable, function, and limit. Newton, Leibniz, and D'Alembert had not distinguished clearly between independent and dependent infinitesimals, but Cauchy incorporated this in his definition when he spoke of the order of the infinitesimal, y, with respect to another, x. Thus the latter is the independent variable which may be given any sequence of values tending toward 0 as a limit, the corresponding sequence of values of y being determined from the functional relationship between y and x. Then the limits of the sequences of values of the ratios $\frac{y}{x^{n-\epsilon}}$ and $\frac{y}{x^{n+\epsilon}}$ are found, and the order of the infinitesimal y thus determined. This is substantially the same as the definition commonly given in present-day textbooks— that y is said to be an infinitesimal of order n with respect to another infinitesimal x if $\lim_{x \to 0} \frac{y}{x^{n}}$ is a constant different from zero.

In a similar manner Cauchy made rigorously clear the views on orders of infinity which D'Alembert had expressed. Whereas Bolzano,

[20] See *Œuvres* (2), IV, 281.

thinking in terms of aggregation, had asserted the possibility of an actual infinity, Cauchy, emphasizing variability, denied the possibility of this because of the paradoxes to which such an assumption appeared to lead.[21] He admitted only the potential infinite of Aristotle, and with D'Alembert interpreted the infinite as meaning simply indefinitely large—a variable, the successive values of which increase beyond any given number.[22] Orders of infinity he then defined in a manner exactly analogous to that given for infinitesimals.

Having established the notions of limit, infinitesimal, and infinity, Cauchy was able to define the central concept of the calculus—that of the derivative. His formulation was precisely that given by Bolzano: Let the function be $y = f(x)$; to the variable x give an increment $\triangle x = i$; and form the ratio $\dfrac{\triangle y}{\triangle x} = \dfrac{f(x + i) - f(x)}{i}$. The limit of this ratio ("when it exists") as i approaches zero he represented by $f'(x)$, and he called this the derivative of y with respect to x.[23] This is, of course, the differential quotient of L'Huilier, clarified by the application of the function concept of Euler and Lagrange. It is made the central concept of the differential calculus, and the expression "differential" is then defined in terms of the derivative. The differential thus represents simply a convenient auxiliary notion permitting the application of the suggestive notation of Leibniz without the confusion between increments and differentials which this symbolism had engendered. Leibniz had considered differentials as the fundamental concepts, the differential quotient being defined in terms of these; but Cauchy reversed this relationship. Having defined the derivative in terms of limits, he then expressed the differential in terms of the derivative. If dx is a finite constant quantity h, then the differential dy of $y = f(x)$ is defined as $f'(x)dx$. In other words, the differentials dy and dx are quantities so chosen that their ratio $\dfrac{(dy)}{(dx)}$ coincides with that of the "dernière raison" or the limit $y' = f'(x)$ of the ratio $\dfrac{\triangle y}{\triangle x}$.[24] This is practically the view of D'Alembert and L'Huilier, and

[21] See Enriques, *Historic Development of Logic*, p. 135.
[22] *Œuvres* (2), III, 19 ff.; cf. also IV, 16.
[23] *Œuvres* (2), IV, 22; cf. also pp. 287–89.   [24] *Œuvres* (2) IV, 27–28; 287–89.

even Leibniz had in a sense anticipated it when he said in 1684 that dy was to dx as the ratio of the ordinate to the subtangent. To make his work logical, however, Leibniz would have had to define the term "subtangent" in terms of limits, thus making the limiting ratio the derivative.

Cauchy, however, gave to the derivative and the differential a formal precision which had been lacking in the definitions of his predecessors. He was therefore able to give satisfactory definitions of differentials of higher order also. The differential $dy = f'(x)dx$ is, of course, a function of x and dx. Regarding dx as fixed, the function $f'(x)dx$ will in turn have a derivative $f''(x)dx$ and a differential $d^2y = f''(x)dx^2$.[2] In general $d^ny = f^n(x)dx^n$. Cauchy added that because the nth derivative is the coefficient by which dx^n is to be multiplied to give d^ny, this derivative is called the differential coefficient.[25]

This statement is not to be understood as indicating that derivatives of higher order are to be defined in terms of differentials of higher order. The reverse is of course the case. Differentials have no logical significance independent of that of derivatives. They were retained by Cauchy simply as an auxiliary notion offering greater operational facility than that afforded by derivatives. This fact has led the mathematician Hadamard, in connection with a discussion on the subject given in the *Mathematical Gazette* for the years 1934–36, to disparage as meaningless the use of differentials of higher order in expositions of the calculus.[26]

Cauchy's definitions of the derivative and the differential are not in any real sense new. They indicate rather a clarification, by the application of the concepts of function, variable, and limit of a variable, of others previously given. During the eighteenth century the word function had generally designated an expression which could be written down simply, in terms of variables and symbols of operations commonly employed at the time. The nondifferentiable function of Bolzano would not, of course, have been included under such an understanding. However, early in the next century the work of J. B. J. Fourier showed that quite arbitrary discontinuous curves could be represented analytically by means of infinite series of trignometric

[25] *Œuvres* (2), IV, 301 ff.
[26] Hadamard, "La Notion de différentiel dans l'enseignement."

functions. As a consequence, the attitude toward the function concept became broader.[27]

The formalistic view that a function was a simple analytic expression gave way to the understanding that it was any relationship between variables. With this attitude came also the recognition that continuity in a curve did not depend upon its being expressible by means of a single equation in continuous functions. This, in turn, led to the realization that a new definition of continuity was necessary. Cauchy's answer to the need was similar to the unnoticed one of Bolzano: the function $f(x)$ is continuous within given limits if between these limits an infinitely small increment i in the variable x produces always an infinitely small increment, $f(x + i) - f(x)$, in the function itself.[28] The expressions infinitely small are here to be understood, as elsewhere in Cauchy's work, in terms of the indefinitely small and limits: i. e., $f(x)$ is continuous within an interval if the limit of the variable $f(x)$ as x approaches a is $f(a)$, for any value of a within this interval. In this definition the view of the preceding centuries is reversed. Newton (implicitly) and Leibniz (explicitly) based the validity of the calculus on the assumption, which Greek thought had avoided, that, by a vague sense of continuity, limiting states would obey the same laws as the variables approaching them. Cauchy made the notion of continuity precisely mathematical and showed that this depends upon the limiting idea and not vice versa. Furthermore, its essence does not lie in a vague blending or unity or contiguity of parts, as intuition seems to imply and as Aristotle had stated, but in certain formal particulate arithmetical relationships, elaborated later in the theory of sets of points, which in turn led to the definition of the continuum.

With the new notion of continuity came a group of new problems. It may have been remarked that throughout the period of indecision— from Newton and Leibniz to Lagrange and Lacroix—the discussion centered about the concepts of the differential and the derivative, to the exclusion of that of the integral. The explanation of this is easily found. From the time of the Greeks down to that of Pascal, areas

<hr />

[27] See Jourdain's two articles, "Note on Fourier's Influence on the Conceptions of Mathematics," and "The Origin of Cauchy's Conceptions of a Definite Integral and of the Continuity of a Function."

[28] Œuvres (2), III, 43; IV, 19–20, 278.

had been found by various devices equivalent to summations of elements. When properly interpreted in terms of the limit concept, these methods represented the counterpart of what is now called the definite integral. With Barrow, Newton, and Leibniz, however, the remarkable discovery was made that the problem of finding areas was simply the inverse of that of determining tangents to curves. Inasmuch as convenient algorithms—those of fluxions and of differentials—were developed in connection with the latter class of problems, by the mere process of inversion the determinations of quadratures could be systematized.

The inverse of the fluxion Newton called the fluent. Leibniz had himself defined the integral as a sum of differentials, although he recognized it also as the inverse of the differential and had determined it in accordance with this fact. These inverses of the fluxion and the differential are the equivalents of what is now called the antiderivative, or the primitive, or the indefinite integral, or sometimes simply *the integral.* During the period we have been discussing, this aspect of the integral as an inverse prevailed over that of the integral as a sum. John Bernoulli, in the formal development of the *calculus summatorius* of Leibniz, gave up the definition of the integral as a sum and called it definitely the inverse of the differential. He conceived the object of the integral calculus as that of finding, from a given relation among the differentials, the relation of the quantities themselves.[29]

Euler used the sum conception to find the approximate values of definite integrals, but because he interpreted the differential as zero, he rejected the Leibnizian view of integration as a process of summation and followed John Bernoulli in defining the integral as the inverse of the differential.[30] L'Huilier went so far in emphasizing the integral as the inverse of the derivative that he suggested substituting the expression "rapport intégral" for "somme intégrale."[31] Lagrange likewise considered the problem of the integral calculus as that of determining from the "fonctions dérivées" the original function;[32] and Lacroix said that the object of the integral calculus was to determine from the differential coefficients the functions from which they were

[29] *Die erste Integralrechnung,* p. 3; cf. also p. 8; also *Opera omnia,* III, 387.
[30] *Opera omnia,* XI, 7.
[31] *Exposition élémentaire,* p. 32. Cf. p. 144.   [32] *Œuvres,* III, 443.

derived.[33] Bolzano similarly defined the integral as the inverse of the derivative.[34]

The result of this tendency in the calculus was that logically the definition of the integral during this time rested immediately upon that of the differential, and the latter became the basis of discussions on the validity of the operations and conceptions of the calculus. Developments during the early nineteenth century, however, introduced new points of view. These led to the reinstatement by Cauchy of the notion of the (definite) integral as a limit of a sum, and made necessary two independent definitions of the two fundamental concepts of the calculus, those of the derivative and the integral.

Inasmuch as the derivative has been defined as $\lim\limits_{h \to 0} \dfrac{f(x + h) - f(x)}{h}$,

we see from the definition of continuity given by Bolzano and Cauchy that the existence of this implied the continuity of the function at the value in question, although the converse is not true. The existence of an integral in the eighteenth-century sense, that is, as an anti-derivative, is therefore bound up with the question of continuity. However, even discontinuous curves apparently have an area, and so discontinuous functions may allow of an integral in the Leibnizian sense. Cauchy therefore restored the character of the definite integral as a sum. For a function $y = f(x)$, continuous in the interval from x_0 to X, he formed the characteristic sum of the products $S_n = (x_1 - x_0)f(x_0) + (x_2 - x_1)f(x_1) + \ldots + (X - x_{n-1})f(x_{n-1})$. If the absolute values of the differences $x_{i+1} - x_i$ decrease indefinitely, the value of S_n will "finally attain a certain limit" S which will depend uniquely on the form of the function $f(x)$ and on the limiting values x_0 and $X \ldots$ "This limit is called a definite integral."[35]

Cauchy cautioned that the symbol of integration $\int$ employed to designate this limit was not to be interpreted as a sum, but rather as a limit of a sum of this type.[36] Cauchy then brought out the fact that although the two operations are defined independently of each other, integration in this sense is the inverse of the process of differ-

[33] *Traité du calcul*, II, 1–2.   [34] Bolzano's *Schriften*, pp. 83–84.

[35] *Œuvres* (2), IV, 125; cf. also Jourdain, "The Origin of Cauchy's Conceptions," pp. 664 ff.; "The Theory of Functions with Cauchy and Gauss," p. 193.

[36] *Œuvres* (2), IV, 126.

entiation. He showed that if $f(x)$ is a continuous function, the function defined as the definite integral $F(x) = \int_{x_0}^{x} f(x)dx$ has as its derivative the function $f(x)$.[37] This was perhaps the first rigorous demonstration of the proposition known as the fundamental theorem of the calculus.[38]

Cauchy's definition of the definite integral allows of extension, with slight modifications, to functions which have discontinuities within the interval of definition. If, for example, the function $f(x)$ is discontinuous at the point X_0 within the interval x_0 to X, the definite integral from x_0 to X is defined as the limit, if it exists, of the sum $\int_{x_0}^{X_0 - \epsilon} f(x)dx + \int_{X_0 + \epsilon}^{X} f(x)dx$, as ϵ becomes indefinitely small.[39]

Discontinuous functions have come to play a significant rôle in mathematics and science, and the view of the integral as a sum has been that upon which the theory of integration has largely developed since the time of Cauchy. From this view, for example, the integral of Lebesgue has developed.[40] The manipulation of infinite series, such as those entering in the definition of the definite integral, had gone on for well over a century before the time of Cauchy, but the need for considering the convergence of these had not been strongly felt until about the opening of the nineteenth century. The term "convergent series" seems first to have been used, although in a somewhat restricted sense, a century and a half earlier by James Gregory;[41] but the general lack of rigor in the work of the eighteenth century was uncongenial to the precision of thought necessary to develop this idea. Varignon, at the beginning of the century, and Lagrange, at its close, had gone a step in this direction by saying that no series could safely be used unless one investigated the remainder.[42]

Nevertheless, no general definition, or theory, of convergent infinite series had been given, and Euler and Lacroix continued to employ divergent series in their work. With the turn of the next century, however, Abel, Bolzano, Cauchy, and Gauss all pointed out the need for definitions and tests of convergence of infinite series before the latter could legitimately be employed in mathematics. In this

[37] *Ibid.*, pp. 151–52.
[38] Saks, *Théorie de l'intégrale*, pp. 122–23.
[39] See Cauchy, *Œuvres* (2), I, 335 and *Œuvres* (1), I, 335.
[40] Saks, *Théorie de l'intégrale*, p. 125.
[41] See *Vera circuli et hyperbolae quadratura*, p. 10.
[42] Reiff, *Geschichte der unendlichen Reihen*, pp. 69–70, 155.

respect the work of Cauchy, in particular, laid the foundation of the theory of convergence and divergence through the wide influence which his work exerted upon his contemporaries. Cauchy defined a series as convergent if, for increasing values of n, the sum S_n approaches indefinitely a certain limit S, the limit S in this case being called the sum of the series.[43] Cauchy here showed clearly that the limit notion is involved, as it was also in differentiation and integration and in defining continuity. Furthermore, he pointed out that it is only in this sense that an infinite series may be regarded as having a sum. In other words, Zeno's paradox of the *Achilles* is to be answered in precisely such ideas, based upon the limit concept.

Cauchy went on in this work to try to prove what has become known as Cauchy's theorem—that a necessary and sufficient condition that the sequence converge to a limit is that the difference between S_p and S_q for any values of p and q greater than n can be made less in absolute value than any assignable quantity by taking n sufficiently large. A sequence satisfying this condition is now said to converge within itself. The necessity of the condition follows immediately from the definition of convergence, but the proof of the sufficiency of the condition requires a previous definition of the system of real numbers, of which the supposed limit S is one. Without a definition of irrational numbers, this part of the proof is logically impossible.

Cauchy had stated in his *Cours d'analyse* that irrational numbers are to be regarded as the limits of sequences of rational numbers. Since a limit is defined as a number to which the terms of the sequence approach in such a way that ultimately the difference between this number and the terms of the sequence can be made less than any given number, the existence of the irrational number depends, in the definition of limit, upon the known existence, and hence the prior definition, of the very quantity whose definition is being attempted. That is, one cannot define the number $\sqrt{2}$ as the limit of the sequence 1, 1.4, 1.41, 1.414, . . . because to prove that this sequence has a limit one must assume, in view of the definitions of limit and convergence, the existence of this number as previously demonstrated or defined.

Cauchy appears not to have noticed the circularity of the reasoning

[43] *Œuvres* (2), III, 114.

in this connection,[44] but tacitly assumed that every sequence converging within itself has a limit. That is, he felt that the existence of a number possessing the external relationship expressed in the definition of convergence and the sum of the series, would follow from the inner relations expressed in the Cauchy theorem. This idea may have been based upon the very thing that he and Bolzano had sought to avoid—that is, upon preconceptions taken over from geometry. The attempt to base the idea of number upon that of the geometrical line had given rise to the Pythagorean difficulty of the incommensurable and the ensuing development of the calculus. It had likewise suggested to Gregory of St. Vincent, two centuries before Cauchy, that the sum of an infinite geometrical progression could be represented by the length of a line segment and that the series could therefore be thought of as having a limit. However, in order to make the limit concept of analysis independent of geometry, mathematicians of the second half of the nineteenth century attempted to frame definitions of irrational number which did not make use of the definition of a limit.

The geometrical intuitions which intruded themselves into Cauchy's view of irrational number likewise led him erroneously to believe that the continuity of a function was sufficient for its geometrical representation and for the existence of a derivative.[45] A. M. Ampère also had been led by geometric preconceptions similar to those of Cauchy to try to demonstrate the false proposition that every continuous function has a derivative, except for certain isolated values in the interval.[46] Bolzano had in his manuscripts of about this time given an example showing the falsity of such an opinion, but it remained for Weierstrass to make this fact known.

With Cauchy, it may safely be said, the fundamental concepts of the calculus received a rigorous formulation. Cauchy has for this reason commonly been regarded as the founder of the exact differential calculus in the modern sense.[47] Upon a precise definition of the notion of limit, he built the theory of continuity and infinite series, of the derivative, the differential, and the integral. Through the popularity of his lectures and textbooks, his exposition of the calculus became

[44] Cf. Pringsheim, "Nombres irrationnels et notion de limite," p. 180.
[45] Cf. Jourdain, "The Theory of Functions with Cauchy and Gauss."
[46] See Pringsheim, "Principes fondamentaux de la théorie des fonctions."
[47] Klein, *Elementary Mathematics from an Advanced Standpoint*, p. 213.

that generally adopted and the one which has been accepted down to the present time. Nevertheless, the use of the infinitely small persisted for some time. S. D. Poisson, in his *Traité de mécanique* which appeared in several editions in the first half of the nineteenth century and which was long a standard work, used exclusively the method of infinitesimals. These magnitudes, "less than any given magnitude of the same nature," he held to have a real existence. They were not simply "a means of investigation imagined by geometers."[48] The object of the differential calculus he consequently regarded as the determination of the ratio of infinitely small quantities, in which infinitesimals of higher order were neglected;[49] and the integral was the inverse of the differential quotient.

A. A. Cournot likewise opposed the work of Cauchy, although upon somewhat different grounds. In his *Traité élémentaire de la théorie des fonctions et du calcul infinitésimal* of 1841, he asserted that his taste for the philosophy of science prepared him to treat the metaphysics of the calculus.[50] In presenting this, his attitude resembled somewhat that of Carnot. He held that the theories of Newton and Leibniz complemented each other, and that the method of Lagrange represented simply a return to the views of Newton.[51] The infinitely small *"existed in nature,"* as a mode of generation of magnitudes according to the law of continuity, although it could be defined only indirectly in terms of limits.[52]

However, Cournot protested that concepts exist in the understanding, independently of the definition which one gives to them. Simple ideas sometimes have complicated definitions, or even none. For this reason he felt that one should not subordinate the precision of such ideas as those of speed or the infinitely small to logical definition.[53] This point of view is diametrically opposed to that which has dominated the mathematics of the last century. The tendency in analysis since the time of Cournot has been toward ever-greater care in the formal logical elaboration of the subject. This trend, initiated in the first half of the nineteenth century and fostered largely by Cauchy, was in the second half of that century continued with notable success by Weierstrass.

[48] *Traité de mécanique*, I, 13–14.   [49] *Ibid.*, pp. 14–16.
[50] *Traité élémentaire*, Preface.   [51] *Ibid.*   [52] *Ibid.*, pp. 85–88.   [53] *Ibid.*, p. 72.

In spite of the care with which Cauchy worked, there were a number of phrases in his exposition which required further explanation. The expressions "approach indefinitely," "as little as one wishes," "last ratios of infinitely small increments," were to be understood in terms of the method of limits, but they suggested difficulties which had been raised in the preceding century. The very idea of a variable approaching a limit called forth vague intuitions of motion and the generation of quantities. Furthermore, there were, in Cauchy's presentation, certain subtle logical gaps. One of these was the failure to make clear the notion of an infinite aggregate, which is basic in his work in infinite sequences, upon which the derivative and the integral are built. Another lacuna is evident in his omission of a clear definition of that most fundamental of all notions—number—which is absolutely essential to the definition of limits, and therefore to that of the concepts of the calculus. The first of these points had been touched upon by Bolzano, but the theory was not further developed until much later, largely through the efforts of Georg Cantor. In the second matter the difficulty is essentially that of a vicious circle in the definition of irrational numbers, and this Weierstrass sought to resolve.

Although it was Cauchy who gave to the concepts of the calculus their present general form, based upon the limit concept, the last word on rigor had not been said, for it was Karl Weierstrass[54] who constructed a purely formal arithmetic basis for analysis, quite independent of all geometric intuition. Weierstrass in 1872 read a paper in which he showed what had been known to Bolzano sometime before—that a function which is continuous throughout an interval need not have a derivative at any point in this interval.[55] Previously it had been generally held, upon the basis of physical experience, that a continuous curve necessarily possessed a tangent, except perhaps at certain isolated points. From this it would follow that the corresponding function should in general possess a derivative. Weierstrass, however, demonstrated conclusively the incorrectness of such suggestions

[54] See Poincaré, "L'Œuvre mathématique de Weierstrass"; cf. also Pierpont, "Mathematical Rigor," pp. 34–36.
[55] This seems to have been presented by Weierstrass in his lectures as early as 1861. See Pringsheim, "Principes fondamentaux," p. 45, n.; Voss, "Calcul différentiel," pp. 260–61.

of experience. This he did by forming the nondifferentiable contin-

uous function $f(x) = \sum\limits_{n=0}^{\infty} b^n cos(a^n \pi x)$, where x is a real variable, a an

odd integer, and b a positive constant less than unity such that

$ab > 1 + \dfrac{3\pi}{2}$.[56]

Since that time many other such functions have become known, and we may even say that, in spite of geometric intuition, of all continuous functions those with tangents at some points are the exceptions.[57] Intuition has been even more discredited as a guide by the fact that we can have continuous curves *defined by motion*, which yet have no tangents.[58]

Inasmuch as it was apparent to Weierstrass that intuition could not be trusted, he sought to make the bases of his analysis as rigorously and precisely formal as possible. He did not present his work on the elements of the calculus in a number of treatises, as had Cauchy, nor even in a series of papers. His views became known, rather, through the work of students who attended his lectures.[59]

In order to secure logical exactitude, Weierstrass wished to establish the calculus (and the theory of functions) upon the concept of number alone,[60] thus separating it completely from geometry. To do this it was necessary to give a definition of irrational number which should be independent of the limit idea, since the latter presupposes the former. Weierstrass was thus led to make profound investigations into the principles of arithmetic, particularly with respect to the theory of irrationals. In this work Weierstrass did not go into the nature of the whole number itself, but began with the concept of whole number as an aggregate of units enjoying one characteristic property in common, whereas a complex number was to be thought of as an aggregate of units of various species enjoying more than one char-

[56] See Weierstrass, *Mathematische Werke*, II, 71–74; Mansion, "Fonction continue sans derivée de Weierstrass."

[57] Cf. Voss, "Calcul différentiel," pp. 261–62.

[58] Neikirk, "A Class of Continuous Curves Defined by Motion Which Have No Tangents Lines."

[59] See, for example, Pincherle, "Saggio di una introduzione alla teoria delle funzioni analitiche secondo i principii del Prof. C. Weierstrass." Cf. Merz, *A History of European Thought in the Nineteenth Century*, II, 703.

[60] Jourdain, "The Development of the Theory of Transfinite Numbers," 1908–9, n., p. 298; cf. also p. 303.

acteristic property. All rational numbers can then be defined by introducing convenient classes of complex numbers. Thus the number $3\frac{2}{3}$ is made up of 3α and 2β, where α is the principle unit and β is an aliquot part, $\frac{1}{3}$, taken as another element. A number is then said to be determined when we know of what elements (of which there is an infinite number) it is composed and the number of times each occurs. In this theory the number $\sqrt{2}$ is not defined as the limit of the sequence 1, 1.4, 1.41, . . . , nor is the idea of sequence brought in; it is simply the aggregate itself in any order 1α, 4β, 1γ, . . . where α is the principle unit and β, γ, . . . are certain of its aliquot parts, and where the aggregate is, of course, subject to the condition that the sum of any finite number of elements is always less than a certain rational number. We can now *prove*, if we wish, that this number is the limit of the variable sequence 1α; 1α, 4β; 1α, 4β, 1γ; . . . , thus correcting the logical error arising in Cauchy's theory of number and limits.[61] In a sense, Weierstrass settles the question of the *existence* of a limit of a convergent sequence by making the sequence (really he considers an unordered aggregate) itself the number or limit.

In making the basis of the calculus more rigorously formal, Weierstrass also attacked the appeal to intuition of continuous motion which is implied in Cauchy's expression—that a variable *approaches* a limit. Previous writers generally had defined a variable as a quantity or magnitude which is not constant; but since the time of Weierstrass it has been recognized that the ideas of variable and limit are not essentially phoronomic, but involve purely static considerations. Weierstrass interpreted a variable x as simply a letter designating any one of a collection of numerical values.[62] A continuous variable was likewise defined in terms of static considerations: If for any value x_0 of the set and for any sequence of positive numbers δ_1, δ_2, . . . , δ_n, however small, there are in the intervals $x_0 - \delta_i$, $x_0 + \delta_i$ others of the set, this is called continuous.[63]

Similarly, for a continuous function Weierstrass gave a definition equivalent to those of Bolzano and Cauchy, but having greater clarity and precision. To say that $f(x + \triangle x) - f(x)$ becomes infinitesimal,

---

[61] *Ibid.*, 1908–9, pp. 303 ff. Cf. Russell, *Principles of Mathematics*, pp. 281 ff.; Pringsheim, "Nombres irrationnels et notion de limite," pp. 149 ff.; Pincherle, *op. cit.*, pp. 179 ff. [62] Pincherle, *op. cit.*, p. 234. [63] *Ibid.*, p. 236.

or becomes and remains less than any given quantity, as Δx approaches zero, calls to mind either the infinitely small or else vague notions of mobility. Weierstrass defined $f(x)$ as continuous, within certain limits of x, if for any value x_0 in this interval and for an arbitrarily small positive number ϵ, it is possible to find an interval about x_0 such that for all values in this interval the difference $f(x) - f(x_0)$ is in absolute value less than ϵ;[64] or, as Heine was led by the lectures of Weierstrass to express it, if, given any ϵ, an η_0 can be found such that for $\eta < \eta_0$, the difference $f(x \pm \eta) - f(x)$ is less in absolute value than ϵ.[65]

The limit of a variable or function is similarly defined. The number L is the limit of the function $f(x)$ for $x = x_0$ if, given any arbitrarily small number ϵ, another number δ can be found such that for all values of x differing from x_0 by less than δ, the value of $f(x)$ will differ from that of L by less than ϵ.[66] This expression of the limit idea, in conjunction with Cauchy's definitions of the derivative and the integral, supplied the fundamental conceptions of the calculus with a precision which may be regarded as constituting their rigorous formulation. There is in this definition no reference to infinitesimals, so that the designation "the infinitesimal calculus," which is used even today, is shown to be inappropriate. Although a number of mathematicians, from the time of Newton and Leibniz to that of Bolzano and Cauchy, had sought to avoid the use of infinitely small quantities, the unequivocal symbolism of Weierstrass may be regarded as effectively banishing from the calculus the persistent notion of the fixed infinitesimal.

During the eighteenth century there had been a lively argument, both in connection with the prime and ultimate ratio of Newton and with the differential quotient of Leibniz, as to whether a variable which approaches its limit can ever attain it. This is essentially the crux of Zeno's argument in the *Achilles*. In the light of the precision of the Weierstrassian theory of limits, however, the question is seen to be entirely inapposite. The limit concept does not involve the idea of *approaching*, but only a static state of affairs. The single question

[64] *Ibid.*, p. 246.
[65] Heine, "Die Elemente der Funktionenlehre," p. 182; cf. also p. 186.
[66] See Stolz, *Vorlesungen über allgemeine Arithmetik*, I, 156–57; cf. also Whitehead, *An Introduction to Mathematics*, pp. 226–29.

amounts really to two: first, does the variable $f(x)$ have a limit L for the value a of x. Secondly, is this limit L the value of the function for the value a of x. If $f(a) = L$, then one can say that the limit of the variable for the value of x in question *is* the value of the variable for this value of x, but not that $f(x)$ *reaches* $f(a)$ or L, for this latter statement has no meaning.

In retrospect, it is pertinent to remark that whereas the idea of variability had been banned from Greek mathematics because it led to Zeno's paradoxes, it was precisely this concept which, revived in the later Middle Ages and represented geometrically, led in the seventeenth century to the calculus. Nevertheless, as the culmination of almost two centuries of discussion as to the basis of the new analysis, the very aspect which had led to its rise was in a sense again excluded from mathematics with the so-called "static" theory of the variable which Weierstrass had developed. The variable does not represent a progressive passage through all the values of an interval, but the disjunctive assumption of any one of the values in the interval. Our vague intuition of motion, although remarkably fruitful in having suggested the investigations which produced the calculus, was found, in the light of further elaboration in thought, to be quite inadequate and misleading. What, then, about that obscure and elusive feeling for continuity which colors so much of our thought? Is that baseless also? What about the idea of infinity upon which Bolzano had speculated and which Weierstrass had used, somewhat covertly, in his definition of irrational number? Can this be given a consistent definition? These questions were investigated largely by Dedekind and Cantor, two mathematicians who were thinking along lines similar to those which Weierstrass had followed in seeking a satisfactory definition of irrational numbers.

The year 1872 was, for a number of reasons, a significant one in the history of the foundations of the calculus. It saw, besides the presentation by Weierstrass of his continuous nondifferentiable function and the publication by one of his students of Weierstrass' lectures on the elements of arithmetic,[67] the appearance of the following: *Nouveau précis d'analyse infinitésimale* of Charles Méray; a paper in Crelle's *Journal* by Eduard Heine on "Die Elemente der Funktionenlehre";

[67] Kossak, *Die Elemente der Arithmetik.*

the first paper by Georg Cantor on the principles of arithmetic, which appeared in the *Mathematische Annalen* as "Über die Ausdehnung eines Satzes aus der Theorie der trigonometrischen Reihen"; and the *Stetigkeit und die Irrationalzahlen* of Richard Dedekind. Incidentally the work of each of these men touched upon one and the same problem—that of formulating a definition of irrational number which should be independent of that of the limit concept.[68] The work of Weierstrass in this connection has already been described. With respect to publication, this had been anticipated by Méray, who in 1869, in an article entitled "Remarques sur la nature des quantités définies par la condition de servir de limites à des variables données," sought to resolve the vicious circle in the definitions of limit and irrational number given by Cauchy. Three years later Méray further elaborated his views, in his *Nouveau précis*.

It will be recalled that Bolzano and Cauchy had attempted to prove that a sequence which converges within itself—that is, one for which, given any ϵ, however small, an integer N can be found such that for $n > N$ and for any integral value of p greater than the integer n, the inequality $|S_{n+p} - S_n| < \epsilon$ will hold—converges in the sense of external relations, that is, that it has a limit S. Méray, in this connection, cut the Gordian knot by rejecting Cauchy's definition of convergence in terms of the limit S. He called an infinite series convergent if it converged within itself, according to Cauchy's theorem. In this case one need not demonstrate the existence of an undefined number S, which may be regarded as the limit. The word number in the strict sense Méray reserved for the integers and rational fractions; the converging sequence of rational numbers, which Méray called a convergent "variante," he regarded as *determining* a number in the broad sense, rational or irrational. He was somewhat vague as to whether or not the sequence *is* the number.[69] If so, as is implied in the case of irrationals, his theory is equivalent to that of Weierstrass, although somewhat less explicitly expressed.

Attempts, similar to those of Weierstrass and Méray, to avoid the

---

[68] For a full account of this work, together with bibliographical references, see Jourdain, "The Development of the Theory of Transfinite Numbers"; see also Pringsheim, "Nombres irrationnels et notion de limite," pp. 144 ff.

[69] See "Remarques sur la nature des quantités"; also *Nouveau précis*, pp. xv, 1–7; cf. also Jourdain, "The Development of the Theory of Transfinite Numbers," 1910, pp. 28 ff.

petitio principii in Cauchy's reasoning on limits and irrational numbers were developed and published, also in 1872, by Cantor and by Heine. Méray had avoided the logical difficulty by taking $|S_{n+p} - S_n| < \epsilon$ as the definition of convergence, instead of the condition $|S - S_n| < \epsilon$, which presupposes the demonstrated existence of S. Likewise in Weierstrass' definition the irrational numbers are expressed in an analogous manner, not as limits but as entire infinite groups of rational numbers. The work of these men led Heine and Cantor to express similar views. Rather than postulate the existence of a number S, which is the limit of an infinite series which converges within itself, they considered S, not exactly as *determined* by the series, as Méray had somewhat indecisively held, but as *defined* by the series—as simply a symbol for the series itself.[70] Their definitions resemble the view of Weierstrass, with the addition of the condition of Méray that $\lim_{n \to \infty} (S_{n+p} - S_n) = 0$ for p arbitrary. This condition is equivalent to that of Weierstrass—whose aggregates were such that however summed in finite number, the sum was to remain below a certain limit—but is expressed in a rather more convenient form.

Still another attempt along these lines was made by Dedekind. Weierstrass had been lecturing on the theory of functions in 1859 and had in this way been led to investigations concerning the foundations of arithmetic. In like fashion Dedekind admitted that his attention was directed to these matters when in 1858 he found himself obliged to lecture for the first time on the elements of the differential calculus.[71] In discussing the notion of the approach of a variable magnitude to a fixed limiting value, he had recourse, as had Cauchy before him, to the evidence of the geometry of continuous magnitude. He felt, however, that the theory of irrational numbers, which lay at the root of the difficulty in the limit concept, should be developed out of arithmetic alone, if it were to be rigorous.[72]

Dedekind's approach to the problem was somewhat different from that of Weierstrass, Méray, Heine, and Cantor in that, instead of

[70] Heine, "Die Elemente der Funktionenlehre," pp. 174 ff.; Cantor, *Gesammelte Abhandlungen*, pp. 92–102, 185–86. See also the articles by Jourdain, "The Development of the Theory of Transfinite Numbers," 1910, pp. 21–43, and "The Introduction of Irrational Numbers." Cf. Russell, *The Principles of Mathematics*, pp. 283–86.

[71] Dedekind, *Essays on the Theory of Numbers*, p. 1.

[72] *Ibid.*, pp. 1–3; cf. also p. 10.

considering in what manner the irrationals are to be defined so as to avoid the vicious circle of Cauchy, he asked himself what there is in continuous geometrical magnitude which resolved the difficulty when arithmetic apparently had failed: i. e., what is the nature of continuity? Plato had sought to find this in a vague flowing of magnitudes; Aristotle had felt that it lay in the fact that the extremities of two successive parts were coincident. Galileo had suggested that it was the result of an actually infinite subdivision—that the continuity of a fluid was in this respect to be contrasted with the finite, discontinuous subdivision illustrated by a fine powder. The philosophy and mathematics of Leibniz had led him to agree with Galileo that continuity was a property concerning disjunctive aggregation, rather than a unity or coincidence of parts. Leibniz had regarded a set as forming a continuum if between any two elements there was always another element of the set.[73]

The scientist Ernst Mach likewise regarded this property of the denseness of an assemblage as constituting its continuity,[74] but the study of the real number system brought out the inadequacy of this condition. The rational numbers, for example, possess the property of denseness and yet do not constitute a continuum. Dedekind, thinking along these lines, found the essence of the continuity of a line to be brought out, not by a vague hang-togetherness, but in the nature of the division of the line by a point. He saw that in any division of the points of a line into two classes such that every point of the one is to the left of every point of the other, there is one and only one point which produces this division. This is not true of the ordered system of rational numbers. This, then, was why the points of a line formed a continuum, but the rational numbers did not. As Dedekind expressed it, "By this commonplace remark the secret of continuity is to be revealed."[75]

It is obvious, then, in what way the domain of rational numbers is to be rendered complete to form a continuous domain. It is only necessary to assume the Cantor-Dedekind axiom that the points of a line can be put into one-to-one correspondence with the real numbers. Arithmetically expressed, this means that for every division of the

[73] *Philosophische Schriften*, II, 515.
[74] *Die Principien der Wärmelehre*, p. 71.   [75] *Essays on the Theory of Numbers*, p. 11.

rational numbers into two classes such that every number of the first, *A*, is less than every number of the second, *B*, there is one and only one real number producing this *Schnitt*, or "Dedekind Cut." Thus if we divide the rational numbers into two classes *A* and *B*, such that *A* contains all those whose squares are less than two and *B* all those whose squares are more than two, there is, by this axiom of continuity, a single real number—written in this case as $\sqrt{2}$—which produces this division. Furthermore, this cut constitutes the definition of the number $\sqrt{2}$. Similarly, any real number is defined by such a cut in the rational number system. This postulate makes the domain of real numbers continuous, in the sense that the straight line has this property. Moreover, the real number of Dedekind is in a sense a creation of the human mind, independent of intuitions of space and time.

The calculus had been generally recognized as dealing with continuous magnitude, but before this time no one had explained precisely the sense in which this was to be accepted. Symbols for variables had displaced the idea of geometrical magnitude, but Cauchy implied a geometrical interpretation of a continuous variable. Dedekind showed that it was not, as had frequently been held,[76] the apparent freedom from the discreteness of the rational numbers which made geometrical quantities continuous, but only the fact that the points in them formed a dense, perfect set. On completing the number system in the manner suggested by this fact—that is, by adopting Dedekind's postulate—this system was made continuous also. Now the fundamental theorems on limits could be proved rigorously[77] and without recourse to geometry, as Dedekind pointed out, on the basis of his new definition of real number.[78] Geometry having pointed the way to a suitable definition of continuity, it was in the end excluded from the formal arithmetical definition of this concept.

The Dedekind Cut is in a sense equivalent to the definitions of real number given by Weierstrass, Méray, Heine, and Cantor.[79] Bertrand

[76] See Drobisch, "Ueber den Begriff des Stetigen und seine Beziehungen zum Calcul," p. 170.

[77] It should be noted, however, that Dedekind's definition of number has recently been criticized as involving a vicious circle. See Weyl, "Der Circulus vitiosus in der heutigen Begründung der Analysis."

[78] *Essays on the Theory of Numbers*, p. 27; cf. also pp. 35–36.

[79] See J. Tannery, Review of Dantscher, *Vorlesungen über die Weierstrassche Theorie der irrationalen Zahlen.*

Russell followed the line of thought suggested by these men, in attempting another formal definition of real number. He felt that the definitions previously given either disregarded the question of the existence of the irrational numbers or artificially postulated new numbers, leaving some doubt as to just what they are. He suggested that a real number be defined as a whole "segment" of the rational numbers. The number $\sqrt{2}$, for example, is defined as the ordered aggregate of all rational numbers whose squares are less than two. That is, instead of postulating an element dividing the rational numbers into two classes, as Dedekind did, he would merely take one of Dedekind's classes and make it, rather than the cutting element, the number.[80] This obviates the necessity of introducing any conception other than that of rational number and segment of rational numbers. According to this view, there is no need to create the irrational numbers; they are at hand in the system of rational numbers, as they had been also in the somewhat more involved doctrine of Weierstrass.

The object of all the above efforts in the establishment of the real number was to give a formal logical definition which should be independent of the implications of geometry and which should avoid the logical error of defining irrational numbers in terms of limits, and vice versa. From these definitions, then, the basic theorems on limits in the calculus can be derived without circularity in reasoning. The derivative and the integral are thus established directly on these definitions, and are consequently divested of any character connected with sensory perception, such as rate of change or surface area. Geometrical conceptions cannot be made sufficiently explicit and precise, as we have seen during our consideration of the long period of development of the calculus. Thus the required rigor was found in the application of the concept of number, made formal by divorcing it from the idea of geometrical quantity. From the definitions of number given above, we see that it is not magnitude which is basic, but order. This is brought out most clearly in the definitions given by Dedekind and Russell, these involving only ordered classes of elements. The same is true, however, of the other systems, which have the disadvantage of requiring new definitions of equality before this can be made clear. The essential characteristic of the number two is not its

[80] Russell, *Introduction to Mathematical Philosophy*, p. 72.

magnitude, but its place in the ordered aggregate of real numbers. The derivative and the integral, although still defined as limits of characteristic quotients and sums respectively, have, as a result, ultimately become, through the definition of number and limit, not quantitative but ordinal concepts. The calculus is not a branch of the science of quantity, but of the logic of relations.

Dedekind's work not only met the need for a definition of number independent of that of limit, but in addition gave an explanation of the nature of continuous magnitude. Bolzano, Cauchy, and others had given definitions of a continuous function of an independent variable. A continuous, independent variable was tacitly understood as one which could take on all values in an interval corresponding to the points of a line segment. The arithmetization of 1872, however, went beyond the geometrical picture and expressed formally, in terms of ordered aggregates, what was meant by a continuous variable or ensemble. The conditions were: first, that the values or elements should form an ordered set; second, that this should be a dense set—that is, between any two values or elements, there should always be others; and third, that the set should be perfect—that is, if the elements are divided, as in a Dedekind Cut, there should always be one which produces this cut.

This definition is far removed from any appeal to empiricism and from the picture of a smooth, unbroken "oneness" or cohesiveness, which instinctive feeling associates with the notion of continuity. It specifies only an infinite, discrete multiplicity of elements, satisfying certain conditions—that the set be ordered, dense, and perfect. This is the sense in which one is to interpret the remark that the calculus deals with continuous variables; the sense in which one is to interpret Newton's phrase "prime and ultimate ratios," or the ultimate relationship between the differentials which Leibniz thought subsisted by virtue of the law of continuity. The introduction of uniform motion into Newton's method of fluxions was an irrelevant evasion of the question of continuity, disguised by an appeal to intuition. There is nothing dynamic in the idea of continuity, nor, so far as we know, is the converse necessarily true. By sense perception we are apparently unable to conclude whether or not we are dealing in motion with a continuum. The experiments of Helmholtz, Mach, and others have shown that the physiological spaces of touch and sight are themselves discontinuous.[81]

The continuity of time which Barrow and Newton regarded as assured by its relentless even flow is now seen to be simply a hypothesis. Mathematics is unable to specify whether motion is continuous, for it deals merely with hypothetical relations and can make its variable continuous or discontinuous at will. The paradoxes of Zeno are consequences of the failure to appreciate this fact and of the resulting lack of a precise specification of the problem. The dynamic intuition of motion is confused with the static concept of continuity. The former is a matter of scientific description a posteriori, whereas the latter is a matter solely of mathematical definition a priori. The former may consequently suggest that motion be defined mathematically in terms of continuous variables, but cannot, because of the limitations of sensory perception, prove that it must be so defined. If the paradoxes of Zeno are thus stated in the precise mathematical terminology of continuous variables and of the derived concepts of limit, derivative, and integral, the seeming contradictions resolve themselves. The *dichotomy* and the *Achilles* depend upon the question as to whether or not the sets involved are perfect.[82] The *stade* is answered upon the basis of dense sets, and the *arrow* by the definition of instantaneous velocity, or the derivative.

The mathematical theory of continuity is based, not on intuition, but on the logically developed theories of number and sets of points. The latter, however, depend, in turn, upon the idea of an infinite aggregate, an idea which Zeno had invoked to fortify his arguments. Zeno's appeal to the infinite was based upon the supposed inconceivability of the notion of completing in a finite time an infinite number of steps. It is again the scientific description a posteriori which he questioned, but, so far as we know, there is no way of proving or disproving the possibility, not only of the existence of infinite aggregates in the physical sense, but also of the execution in thought of an infinite number of steps in connection with aggregates, whether finite or infinite. Since science cannot answer this point, the question may become a hypothetical mathematical one.

[81] Enriques, *Problems of Science*, pp. 211–12.
[82] Cf. Helmholtz, *Counting and Measuring*, p. xviii; Cajori, "History of Zeno's Arguments on Motion," p. 218.

Mathematics, moreover, requires a theory of the infinite in its definitions of number and continuity. The question of the mathematical existence, i. e., of the consistent logical definition, of infinite aggregates therefore remained to be answered. Galileo had suggested vaguely, and Bolzano had seen more clearly, that infinite ensembles must have the paradoxical property that a part can be put into one-to-one correspondence with the whole. This fact led Cauchy to deny their existence; and indeed upon the basis of the work of Cauchy and Weierstrass, one could have said that the infinite indicated nothing more than the potentiality of Aristotle—an incompleteness of the process in question.[83] Their infinitesimals were variables having zero as their limit, and the limit concept involved only the definition of number. However, this very definition of number implicitly presupposes the prior existence of infinite aggregates, so that this question could not indefinitely be avoided. Under the influence of Weierstrass' work in the foundations of arithmetic, Dedekind and Cantor sought a basis for the theory of infinite aggregates,[84] in order to complete this work. This they found in Bolzano's paradox. Instead of looking upon it as merely a strange property of infinite aggregates, they made it the *definition* of an infinite set. Dedekind said "A system *S* is said to be *infinite* when it is similar to a proper part of itself; in the contrary case *S* is said to be a finite system."[85] Under this definition, infinite aggregates exist as logically self-consistent entities, and the definitions of real number are completed.

Cantor, with whom Dedekind corresponded in this connection,[86] was not satisfied with merely defining infinite sets. He wished to develop the subject further. In a series of papers he reviewed the history of the infinite from the time of Democritus to that of Dedekind, and elaborated his theory of infinite ensembles, or *Mengenlehre*. Cantor's doctrine of the mathematical infinite, which has been hyperbolized as "the only genuine mathematics since the Greeks,"[87] did not refer to the potential infinity of Aristotle nor the syncategorematic infinity of the Scholastics. These were bound up always with variability.[88] Cantor referred instead to the categorematic infinity of medieval

---

[83] See Baumann, "Dedekind und Bolzano."   [84] See Hilbert, "Über das Unendliche."
[85] Dedekind, *Essays on the Theory of Numbers*, p. 63.
[86] See Georg Cantor, *Gesammelte Abhandlungen*.
[87] See Bell, *The Queen of the Sciences*, p. 104.   [88] *Ibid.*, p. 180.

philosophy—the *eigentlich-Unendlichen*. He felt, with some justice, that Scholastic thought had handled this subject more as a religious dogma than as a mathematical concept.[89] Moreover, the thought of Leibniz, whose calculus represented the most genial attempt to establish a mathematics of the infinitely large and the infinitely small, lacked resolution. Sometimes he declared against the absolute infinite, and then again he remarked that nature, instead of abhorring the actual infinite, everywhere made use of it to mark better the perfections of its author.[90]

The symbol ∞ had been used by mathematicians since the time of Wallis to represent infinity, but no definition had been given, nor had the Scholastic distinction been observed. The symbol was used by Weierstrass, for example, in the sense both of a potentiality and of an actuality: he wrote $f(a) = \infty$ to mean that $\dfrac{1}{f(a)} = 0$, and also used the expression $f(\infty) = b$ in the sense that the limit of $f(x)$ for x indefinitely large was b.[91] To avoid this confusion, Cantor chose a new symbol, ω, to represent the actual infinite aggregate of positive integers. It is to be remarked, furthermore, that whereas ∞ referred in general to magnitude, ω is to be interpreted in terms of aggregation. Wallis, Fontenelle, and others regarded ∞ variously as the largest positive integer or as the sum of all the positive integers; but the symbol ω refers to all the positive integers only in the sense that these form an aggregate of elements. This view of the infinite, as concerned with groups of elements, had previously been clearly expressed by Bolzano, but he failed to recognize what Cantor called the power of an infinite set of elements. The rational numbers can be put into one-to-one correspondence with the positive integers, and for this reason these two classes are regarded as having the same power. One of the most striking results of Cantor's *Mengenlehre*, however, is that there are transfinite numbers higher than ω. The theory of arithmetic had shown that numbers other than the rationals were needed for the continuum, and Cantor now showed that the desired set of real numbers was such that these could not be put into one-to-one correspondence with the positive integers; i. e., the set was not

[89] *Ibid.*, p. 191.   [90] *Ibid.*, p. 179; see also Leibniz, *Philosophische Schriften*, I, 416.
[91] Cajori, *A History of Mathematical Notations*, II, 45.

denumerable. It therefore represented a transfinite number of a higher power, which is often written now simply as C. Other numbers have been found above C, but the question as to whether there is one between ω and C is unanswered. However, the definition of the continuous variable, and hence of the concepts of the calculus, requires only the infinite set C.

Although Cantor's work does not clear up any objection raised on the ground of the conceptual difficulties inherent in the concept of infinity, it definitely refutes any argument raised upon the score of logical contradiction. Likewise, any criticism of the use of the infinite in defining irrational number or in the limit concept is answered by Cantor's work, which clarifies the situation.[92] As the terminus *a quo* of the investigations leading to the calculus is to be found in the Pythagorean discovery of the incommensurable and in the recognized need for satisfactory definitions of number and the infinite, so the terminus *ad quem* may be regarded as the establishment of these by the great triumvirate: Weierstrass, Dedekind, and Cantor. The fundamental notions of the calculus—the limit of a continuous variable: the derivative and the integral—have through the work of these men been given a logical rigor as impressive as that of Euclidean geometry, and a formal precision of which the Greeks had never dreamed. It has been shown that in analysis there is no need for anything but whole numbers, or finite or infinite systems of these.[93] How startlingly apropos, with respect to the development of the calculus, is the Pythagorean dictum: *All is number!*

[92] It should, perhaps, be observed at this point that the theory of infinite aggregates has resulted also in a number of puzzling and as yet unresolved antimonies. See Poincaré, *Foundations of Science*, pp. 477 ff.; Pierpont, "Mathematical Rigor," pp. 42-44. It has been suggested, in this connection, that these paradoxes may be related to the difficulties encountered in theoretical physics. See the review by Northrop, in *Bulletin, American Mathematical Society*, XLII (1936), 791-92, of Schrödinger, *Science and the Human Temperament*.

[93] Poincaré, *The Foundations of Science*, pp. 380, 441 ff.

VIII. Conclusion

THERE is a strong temptation on the part of professional mathematicians and scientists to seek always to ascribe great discoveries and inventions to single individuals. Such ascription serves a didactic end in centering attention upon certain fundamental aspects of the subjects, much as the history of events is conveniently divided into epochs for purposes of exposition. There is in such attributions and divisions, however, the serious danger that too great a significance will be attached to them. Rarely—perhaps never—is a single mathematician or scientist entitled to receive the full credit for an "innovation," nor does any one age deserve to be called the "renaissance" of an aspect of culture. Back of any discovery or invention there is invariably to be found an evolutionary development of ideas making its geniture possible. The history of the calculus furnishes a remarkably apt illustration of this fact.[1]

The method of fluxions of Newton was no more unanticipated than were his laws of motion and gravitation; and the differential calculus of Leibniz had been as fully adumbrated as had his law of continuity. These two men are to be thought of as the inventors of the calculus in the sense that they gave to the infinitesimal procedures of their predecessors the algorithmic unity and precision necessary for further development. Their work differed from the corresponding methods of their predecessors, Barrow and Fermat, more in attitude and generality than in substance and detail. The procedures of Barrow and Fermat were themselves but elaborations of the views of such men as Torricelli, Cavalieri, and Galileo, or Kepler, Valerio, and Stevin. The achievements of these early inventors of infinitesimal devices were in turn the direct results of the contributions of Oresme, Calculator, Archimedes, and Eudoxus. Finally, the work of the last-named men was inspired by the mathematical and philosophical problems suggested by Aristotle, Plato, Zeno, and Pythagoras. Without the filiation of ideas which was built up by these men and many others, the calculus of Newton and Leibniz would be unthinkable.

[1] Cf. Karpinski, "Is there Progress in Mathematical Discovery?" pp. 47–48.

If, on the one hand, mathematicians have been prone to forget the periods of suggestion and anticipation in the rise of the calculus, historians of the subject, on the other hand, have frequently failed to appreciate the significance of the later rigorous formulations. Historical accounts of the subject all too often have terminated with the work of Newton and Leibniz, even though neither of these men was able to furnish the precision of thought which was to follow two centuries later. The neglect of this later period of investigation indicates an inadequate attention to, or appreciation of, the fundamental concepts of the subject as presented by Cauchy and Weierstrass. References, which might easily have been multiplied, have frequently been indicated above, in which the views expressed in the final elaboration of the calculus have most unwarrantedly been imputed to earlier investigators in the field. Weierstrass' definition of real number has been identified with the theory of proportion of Eudoxus, and the Dedekind Cut with the speculations of Bryson. The continuum of Cantor has been viewed as expressed in the speculations of William of Occam or of Zeno. The limit concept of Weierstrass has been interpreted as identical with the prime and ultimate ratio of Newton, or even with the ancient Greek method of exhaustion. The derivative and the differential of Cauchy have been described as exactly corresponding to the related conceptions of Leibniz. The definite integral of Cauchy has been ascribed in all completeness to Fermat, or even to Cavalieri and Archimedes.

Such citations make clear how general is the tendency unguardedly to read into the minds of earlier men one's own clear thoughts on the subject, forgetting that these are the culmination of centuries of speculation and investigation. The concepts of mathematics and science are eminently cumulative in their growth—the results of a continuous effort to understand the relationships between elements and, in terms of these, to describe the confused impressions afforded by physical experience. The dynamics and astronomy of the sixteenth century, for example, were not entirely new developments, but rather grew out of medieval and ancient views on these subjects. In the same sense the use of infinitesimal conceptions during the early modern period did not proceed *de novo*, but began where the Scholastic philosophers and Greek mathematicians had left off.

Nevertheless, the fact of such progressive achievement is not to be interpreted as the unfolding of a well-conceived plan. Throughout the advance of science and mathematics, elements have constantly been discarded as well as added. For this reason no investigator is able to foresee the direction which the elaboration of his views will take. Only in retrospect can one trace the path along which such development has proceeded. Although in retracing this thread of thought one can readily recognize the notions from which the ultimate concepts have sprung, the former are not in general to be identified with the latter. Each is to be considered in the light of the mathematical and scientific milieu of the period in which it appeared. To interpret the geometric views of Archimedes and Barrow, for example, in terms of the analytical symbolism of the twentieth century is tantamount to invoking implicitly the precision and economy of thought which modern notations afford but which these earlier investigators were far from possessing.

A deeper and more sympathetic understanding among professional workers in the fields of mathematics and history might easily remove much of the misdirected thought with respect to the nature and rise of mathematical concepts. A familiarity not only with the elements of the calculus, but also with the history of its development, will serve to bring out that the question is not so much who are the founders of the subject—Weierstrass and Cauchy, or Newton and Leibniz, or Barrow and Fermat, or Cavalieri and Kepler, or Archimedes and Eudoxus—but rather in what sense each of these men may be regarded as responsible for the new analysis.

It is possible not only to trace the path of development throughout the twenty-five-hundred-year interval during which the ideas of the calculus were being formulated, but also to indicate certain tendencies inimical to this growth. Perhaps the most manifest deterring force was the rigid insistence on the exclusion from mathematics of any idea not at the time allowing of strict logical interpretation. The very concepts which gave birth to the calculus—those of variation and continuity, of the infinite and the infinitesimal—were banned from Greek mathematics for this reason, the work of Euclid being a monument to this exclusion. The work of Archimedes became most fruitful when he abandoned the Greek logical ideal and applied such forbidden concepts, but this represents an almost isolated example of their use. Similarly, in the seventeenth century a number of mathematicians, including Pascal and Barrow, avoided the use of algebra and analytic geometry as not compatible with the demands of rigor inherited from the ancient Greeks. Had they made full use of these elements, they might well have been acclaimed the inventors of the calculus. Again in the eighteenth century English mathematicians disdained, largely because of the weak logical basis (as well as for reasons of national jealousy), to use the differential method of the Continental "computists," and consequently failed to make appreciable contributions to the rapid growth of analysis characterizing that century.

It is clear that the indiscriminate use of methods and ideas which are palpably without logical foundation is not to be condoned. Such logical basis is, of course, ultimately to be sought in order to avoid hopeless confusion (as witness the eighteenth-century use of infinite series); but pending the final establishment of this, the banishment of suggestive views is a serious mistake.

On the other hand, perhaps a more subtle, and therefore serious, hindrance to the development of the calculus was the failure, at various stages, to give to the concepts employed as concise and formal a definition as was possible at the time. The paradoxes of Zeno are excellent illustrations of the obscurity which results from a failure to specify clearly and unambiguously the conditions of the problem and to give formal definitions of the terms involved. Had the Greeks demanded of Zeno a precision of statement which the mathematicians exacted of themselves, they might not have banned the concepts leading to the calculus nor disregarded almost entirely the science of dynamics. The nice distinctions of the Scholastic philosophers pointed the way to the clarification of such problems, but these men were not sufficiently familiar with the formalism of Greek geometry and Arabic algebra to be able to carry their ideas to completion.

With the decline of Scholasticism, the tendency was away from precision of thought and toward the free use of imagination, as found in the literary Renaissance. The mathematical counterpart of this is seen strongly in the works of Nicholas of Cusa and Kepler, who employed in mathematics, without seeking adequately to define them, the conceptions of infinity, the infinitesimal, motion, and continuity.

In a sense this was fortunate, in that it favored the development of methods anticipating the calculus. On the other hand, however, the lack of a sound critical attitude was to leave undefined for hundreds of years the logical bases of the procedures thus employed, while the resulting confusion of thought engendered half a dozen alternative methods. Had Newton been more precise in the statement of his limit method, and had Leibniz been more explicit in professing that he was developing an instrument of invention and not a logical foundation, the period of indecision might not have ensued. As it was, it required the work of Cauchy, Weierstrass, and others to impart to the concepts of the continuous variable, the limit, the derivative, and the integral a precision of formulation which made them generally acceptable.

In all probability, however, the chief obstacle in the way of the development of the concepts of the calculus was a misunderstanding as to the nature of mathematics. Ever since the empirical mathematics of the pre-Hellenic world was developed, the attitude has, upon occasion, been maintained that mathematics is a branch either of empirical science or of transcendental philosophy. In either case mathematics is not free to develop as it will, but is bound by certain restrictions: by conceptions derived either a posteriori from natural science, or assumed to be imposed a priori by an absolutistic philosophy. At the Egyptian and Babylonian level, mathematics was largely a body of information concerning the natural world. The early Ionians rearranged this knowledge into a deductive scheme, but the basis was still largely empirical science. The oriental mysticism of Pythagoras, however, reversed this state of affairs and gave to mathematics a supra-sensuous reality, of which the world of appearances was a counterpart. The premises were thus categorically established and all the mathematician could do was to develop the logical implications of these. This view was elaborated by Plato into an idealistic philosophy which has consistently denied the purely logical and hypothetical nature of the propositions of mathematics. At the Greek stage of the ideas leading to the calculus, however, the views of Plato exerted a favorable influence, in that they counterbalanced the Peripatetic attitude.

Aristotle considered mathematics an idealized abstraction from natural science, and as such the premises and definitions were not arbitrary, but were determined by our interpretation of the world of sense perception. The concepts allowed to geometry were only such as were consistent with this picture. The infinitesimal and the actual infinite were excluded, not so much because of any demonstrated logical inconsistency, but because of a supposed incompatibility with the world of nature, from which the entities of mathematics were regarded as derived by a disassociation of irrelevant properties. Continuity depended upon coincidence of extremities and upon oneness, as sense perception appeared to indicate. Likewise number was regarded, in conformity with the judgments of empiricism and common sense, as a collection of units, with the result that irrational magnitudes were not considered as pertaining to the realm of number. Mathematics was the logic of relations, but the nature of these was completely determined by postulates which were in turn dictated by the evidence of physical experience.

Of these two views, the Platonic and the Aristotelian, the former was for a time that under which the ideas of the calculus developed. Under this view, the conceptions of the infinite and the infinitesimal were not excluded, inasmuch as reason was not subject to the world of sensation. The entities of mathematics had an ontological reality, independent of common sense, and the postulates were discovered by reason alone. Although this made mathematics independent of natural science, it did not give it the postulational freedom it enjoys today.

While the work of Archimedes displays elements of both the Platonic and the Aristotelian views, it was the latter which triumphed in the method of exhaustion and the classic geometry of Euclid. The concepts of infinity and continuity were consequently, during the medieval period, discussed from a dialectical rather than a mathematical point of view; but when they entered into the geometry of Nicholas of Cusa, of Kepler, of Galileo, it was largely under the Platonic view of rational transcendentalism, rather than of naturalistic description. Nevertheless, other mathematicians—Roberval, Torricelli, Barrow, and Newton—were led by the natural science of the day to interpret mathematics in terms of sense perception, as Aristotle had, and to introduce motion to avoid the difficulties of the infinite and the continuous. The philosophers Hobbes and Berkeley felt the empirical tendency so strongly that they denied to mathematics the idealized concept of a point without extension, because they felt that this had no counterpart in nature.

The attitude of most of the mathematicians of the seventeenth century, however, was that of doubt. They employed infinitesimals and the infinite on the assumption that they existed, and treated the continuous as though made up of indivisibles, the results being justified pragmatically by their consistency with Euclidean geometry. In any case, the attitude was not that of an unprejudiced postulation and definition, followed by logical deduction. The investigations into the foundations of the calculus consequently took the form, during the eighteenth century, of a search for an explanation which should be intuitively plausible, rather than logically self-consistent. At this time, however, there was rapidly developing a very successful algebraic formalism, vigorously fostered by Euler and Lagrange. This led, in the nineteenth century, to a view of mathematics which non-Euclidean geometry had strongly suggested—a postulational system independent alike of the world of sense experience and of any dictates resulting from introspection. The calculus became free to adopt its own premises and to frame its own definitions, subject only to the requirement of an inner consistency. The existence of a concept depended only upon a freedom from contradiction in the relations into which it entered. The bases of the calculus were then defined formally in terms only of number and infinite aggregates, with no corroboration through an appeal to the world of experience either possible or necessary.

This formalizing tendency has not, however, been everywhere accepted. Even mathematicians have not always been in sympathy with the movement. Hermite, whose favorite idea was to compare mathematics with the natural sciences, was horrified by Cantor's work, which transcended human experience.[2] Du Bois-Reymond similarly opposed the formal definitions which were made basic in the calculus and wished instead to define number in terms of geometric magnitude, much as Cauchy had done implicitly,[3] thus retaining intuition as a guide. More thoroughgoing intuitionists like Brouwer attempt

[2] See Poincaré, "L'Avenir des mathématiques," p. 939.

[3] Pringsheim, "Nombres irrationnels et notion de limite," pp. 153 ff. See also Jourdain, "The Development of the Theory of Transfinite Numbers," 1913–14, pp. 1, 9–10.

to visualize a fusion of the continuous and the discrete, somewhat as Plato had.[4] The mathematician Kronecker opposed the work of Dedekind and Cantor, not because of its formalism, but because he thought it unnatural. These investigators had "constructed" numbers which Kronecker felt could have no existence, and he proposed instead to base everything on equations involving integers alone, a view which has not been generally shared by other mathematicians.[5]

A number of scientists and philosophers have naturally been even more hesitant than these mathematicians in giving up experience and intuition in connection with the calculus. Thoroughgoing empiricists and idealistic philosophers in particular have sought, since the time of Newton and Leibniz to read into the calculus a significance beyond that of a formal postulational system. Newton had considered the calculus as a scientific description of the generation of magnitudes, and Leibniz had viewed it as a metaphysical explanation of such generation. The formalism of the nineteenth century took from the calculus any such preconceptions, leaving only the bare symbolic relationships between abstract mathematical entities. Nevertheless, traces of the old scientific and metaphysical tendencies remained. Lord Kelvin, who considered mathematics the etherealization of common sense, once exclaimed, when he had asked what $\dfrac{dx}{dt}$ represented

and had received the answer $\lim\limits_{\Delta t \to 0} \dfrac{\Delta x}{\Delta t}$: "That's what Todhunter would say. Does nobody know that it represents a velocity?"[6] His friend Helmholtz showed a similar tendency. In his famous essay, *Ueber die Erhaltung der Kraft*, he regarded a surface as the sum of lines,[7] much as had Cavalieri in his *Exercitationes* just two centuries earlier. In another connection he asserted that incommensurable relations may occur in real objects, but that in numbers they can never be repre-

---

[4] See Helmholtz, *Counting and Measuring*, pp. xxii–xxiv; also Brouwer, "Intuitionism and Formalism," and Simon, "Historische Bemerkungen über das Continuum."

[5] See Couturat, *De l'infini mathématique*, pp. 603 ff.; cf. also Pierpont, "Mathematical Rigor," pp. 38–40; Pringsheim, "Nombres irrationnels et notion de limite," pp. 158–63; Jourdain, "The Development of the Theory of Transfinite Numbers," 1913–14, pp. 2–8.

[6] Hart, *Makers of Science*, pp. 278–79; Felix Klein, *Entwicklung der Mathematik im 19. Jahrhundert*, I, 238.

[7] *Ueber die Erhaltung der Kraft*, p. 14.

sented with exactness.[8] Mach also felt strongly the empirical origin of mathematics and held with Aristotle that geometric concepts are the product of idealization of physical experiences of space.[9] In conformity with this view, he felt that some form of geometrical meaning had necessarily to be given to the number i.[10] In this respect he is in agreement with a number of present-day scientists, who feel that $\sqrt{-1}$ simply "forms a part of various ingenious devices for handling otherwise intractable situations."[11]

The attitudes of Helmholtz and Mach are representative of the influence in science of the positive philosophy of the nineteenth century. Positivistic and materialistic thought were slow to accept the changed mathematical view and insisted that the calculus be interpreted in terms of velocities and actual intervals, corresponding to the data of experience and of ordinary algebra. Comte recognized that mathematics is not "the science of magnitudes,"[12] but he did not rise to the formal view of Cauchy. In accordance with the empirical and pragmatic attitude of Carnot, he regarded the methods of Newton, Leibniz, and Lagrange as fundamentally identical. However, because the differential calculus gave no clear conception of the infinitely small and because the method of limits apparently separated the fields of ordinary and transcendental analysis, he felt that the method of Lagrange was to be favored.[13] More strongly expressed are the views of Dühring, who, in 1872, in his classic *Kritische Geschichte der allgemeinen Principien der Mechanik*, indulged in a polemic against Gauss, Cauchy, and others who would deny the absolute truth of geometry, and who would introduce into mathematics such figments of the imagination as imaginary numbers, non-Euclidean geometry, and limits![14] Marxian materialists will not grant mathematics the independence of experience necessary for its proper development.[15] Such denial makes impossible the concept of the derivative and the scientific description of motion in terms thereof. The mathematical infinity is, in accord with

[8] Helmholtz, *Counting and Measuring*, p. 26.
[9] Mach, *Space and Geometry*, p. 94; cf. also p. 67. See also Strong, *Procedures and Metaphysics*, p. 232.
[10] *Space and Geometry*, p. 104, n.   [11] Heyl, "The Skeptical Physicist," p. 228.
[12] Comte, *The Philosophy of Mathematics*, p. 18.   [13] *Ibid.*, pp. 110–17.
[14] Dühring, *Kritische Geschichte der allgemeinen Principien der Mechanik*, pp. 475 ff., 529 ff.
[15] Engels, *Herr Eugen Dühring's Revolution in Science*, p. 47.

this view, a contradiction of the "tautology" that the whole is greater than any of its parts.[16]

If a number of philosophers were led by excessive realism to reject much of the mathematics of the nineteenth century, idealistic philosophers, following Kant, were likewise unwilling to accept the bare formalism of Cauchy and Weierstrass in the realm of the calculus. The differential had been defined by Cauchy, not as a fixed quantity, but as a variable, and Weierstrass had shown that the continuous variable depends only upon the static notion of sets of elements. Idealists attempted, nevertheless, to interpret the differential as having an intensive quality resembling the potentiality of Aristotle, the impetus of the Scholastics, the *conatus* of Hobbes, or the inertia of modern science. They wished to view the continuum, not in terms of the discreteness of Cantor and Dedekind, but as an unanalyzable concept in the form of a metaphysical reality which is intuitively perceived. The differential calculus was regarded as possessing a "positive" meaning as the generator of the continuum, as opposed to the "negation" of the limit concept.[17] As Hegel expressed it, the derivative represented the "becoming" of magnitudes,[18] as opposed to the integral, or the "has become."

Materialistic and idealistic philosophies have both failed to appreciate the nature of mathematics, as accepted at the present time. Mathematics is neither a description of nature nor an explanation of its operation; it is not concerned with physical motion or with the metaphysical generation of quantities. It is merely the symbolic logic of possible relations,[19] and as such is concerned with neither approximate nor absolute truth, but only with hypothetical truth. That is, mathematics determines what conclusions will follow logically from given premises. The conjunction of mathematics and philosophy, or of mathematics and science, is frequently of great service in suggesting new problems and points of view.

[16] *Ibid.*, pp. 48–49, 62; cf. also Bois, "Le Finitisme de Dühring," p. 95.

[17] See Kant, *Sämmtliche Werke*, XI (Part I), 270–71; cf. also II, 140–49 and *passim*. See also Cohen, *Die Princip der Infinitesimalmethode und seine Geschichte, passim;* Simon, "Zur Geschichte und Philosophie der Differentialrechnung," p. 128; Vivanti, "Note sur l'histoire de l'infiniment petit," pp. 1 ff.; Freyer, *Studien zur Metaphysik der Differentialrechnung*, pp. 23 ff.; Lasswitz, *Geschichte der Atomistik*, I, 201.

[18] Klein, *Elementary Mathematics from an Advanced Standpoint*, p. 217.

[19] Cohen, M. R., *Reason and Nature*, pp. 171–205.

Nevertheless, in the final rigorous formulation and elaboration of such concepts as have been introduced, mathematics must necessarily be unprejudiced by any irrelevant elements in the experiences from which they have arisen.[20] Any attempt to restrict the freedom of choice of its postulates and definitions is predicated on the assumption that a given preconceived notion of the nature of the relationships involved is necessarily valid. The calculus is without doubt the greatest aid we have to the discovery and appreciation of physical truth; but the basis for this success is in all probability to be found in the fact that the concepts involved were gradually emancipated from the qualitative preconceptions which result from our experiences of variability and multiplicity. Greek philosophy had attempted to separate and contrast the qualitative and the quantitative, but the later medieval and early modern period associated them through geometric representation. Even a quantitative explanation is subject to sensory notions of size, length, duration, and so forth, so that greater independence was achieved in the nineteenth century by basing the calculus upon ordinal considerations only. The history of the concepts of the calculus shows that the explanation of the qualitative is to be made through the quantitative, and the latter is in turn to be explained through the ordinal, perhaps the most fundamental notion in mathematics. As the sensations of motion and discreteness led to the abstract notions of the calculus, so may sensory experience continue thus to suggest problems for the mathematician, and so may he in turn be free to reduce these to the basic formal logical relationships involved. Thus only may be fully appreciated the twofold aspect of mathematics: as the language of a descriptive interpretation of the relationships discovered in natural phenomena, and as a syllogistic elaboration of arbitrary premises.

[20] Cf. Poincaré, *Foundations of Science*, pp. 28–29, 46, 65, 428.

Bibliography

Agardh, C. A., Essai sur la métaphysique du calcul intégral. Stockholm, 1849.

[Al-Khowarizmi] Robert of Chester's Latin Translation of the Algebra of Al-Khowarizmi, with an introduction, critical notes and an English version by L. C. Karpinski. New York and London, 1915.

Allman, G. J., Greek Geometry from Thales to Euclid. Dublin and London, 1889.

Amodeo, F., "Appunti su Biagio Pelicani da Parma." *Atti. IV Congresso dei Matematici* (Roma, 1908), III, 549–53.

Arbogast, L. F. A., Du calcul des dérivations. Strasbourg, An VIII (1800).

Archibald, R. C., Outline of the History of Mathematics. 3d ed., Oberlin, 1936.

Archimedes, Opera omnia. Ed. by J. L. Heiberg. 3 vols., Lipsiae, 1880–81.

[Archimedes] The Method of Archimedes. Recently Discovered by Heiberg. A Supplement to the Works of Archimedes, Ed. by T. L. Heath. Cambridge, 1912.

—— "Eine neue Schrift des Archimedes." Ed. by J. L. Heiberg and H. G. Zeuthen. *Bibliotheca Mathematica* (3), VII (1906–7), 321–63.

—— "A Newly Discovered Treatise of Archimedes." With commentary by D. E. Smith. *Monist*, XIX (1909), 202–30.

—— The Works of Archimedes. Ed. with notes by T. L. Heath. Cambridge, 1897.

Aristotle, the Works of. Ed. by W. D. Ross and J. A. Smith. 11 vols., Oxford, 1908–31.

Aubry, A., "Essai sur l'histoire de la géométrie des courbes." *Annaes Scientificos da Academia Polytechnica do Porto*, IV (1909), 65–112.

—— "Sur l'histoire du calcul infinitésimal entre les années 1620 et 1660." *Annaes Scientificos da Academia Polytechnica do Porto*, VI (1911), 82–89.

Ball, W. W. R., A Short Account of the History of Mathematics. London, 1888.

Barrow, Isaac, Geometrical Lectures. Ed. by J. M. Child. Chicago and London, 1916.

—— the Mathematical Works of. Ed. by W. Whewell. Cambridge, 1860.

Barry, Frederick, The Scientific Habit of Thought. New York, 1927.

See also Pascal.

Baumann, Julius, "Dedekind und Bolzano." *Annalen der Naturphilosophie*, VII (1908), 444–49.

Bayle, Pierre, Dictionnaire historique et critique. New ed., 15 vols., Paris, 1820.

Becker, Oskar, "Eudoxos-Studien." *Quellen und Studien zur Geschichte der Mathematik, Astronomie und Physik*, Part B, Studien, II (1933), 311–33, 369–87; III (1936), 236–44, 370–410.

Bell, E. T., The Handmaiden of the Sciences. New York, 1937.

—— The Queen of the Sciences. Baltimore, 1931.

—— The Search for Truth. New York, 1934.

Berkeley, George, Works. Ed. by A. C. Fraser. 4 vols., Oxford, 1901.

Bernoulli, James, "Analysis problematis antehac propositi." *Acta eruditorum*, 1690, pp. 217–19.

—— Opera. 2 vols., Genevae, 1744.

Bernoulli, John, Die Differentialrechnung aus dem Jahre 1691/92. Ostawld's Klassiker, No. 211. Leipzig, 1924.

—— Die erste Integralrechnung. Eine Auswahl aus Johann Bernoullis mathematischen Vorlesungen über die Methode der Integrale und anderes aufgeschrieben zum Gebrauch des Herrn Marquis de l'Hospital in den Jahren 1691 und 1692. Translated from the Latin. Leipzig and Berlin, 1914.

—— Opera omnia. 4 vols., Lausannae and Genevae, 1742.

See also Leibniz.

Bernstein, Felix, "Ueber die Begründung der Differentialrechnung mit Hilfe der unendlichkleinen Grössen." *Jahresbericht, Deutsche Mathematiker-Vereinigung*, XIII (1904), 241–46.

Bertrand, J., "De l'invention du calcul infinitésimal." *Journal des Savants*, 1863, pp. 465–83.

Birch, T. B., "The Theory of Continuity of William of Ockham." *Philosophy of Science*, III (1936), 494–505.

Björnbo, A. A., "Über ein bibliographisches Repertorium der handschriftlichen mathematischen Literatur des Mittelalters." *Bibliotheca Mathematica* (3), IV (1903), 326–33.

Black, Max, The Nature of Mathematics. New York, 1934.

Bledsoe, A. T., The Philosophy of Mathematics with Special Reference to the Elements of Geometry and the Infinitesimal Method. Philadelphia, 1868.

Bôcher, Maxime, "The Fundamental Concepts and Methods of Mathematics." *Bulletin, American Mathematical Society*, XI (1904), 115–35.

Bohlmann, G., "Übersicht über die wichtigsten Lehrbücher der Infinitesimalrechnung von Euler bis auf die heutige Zeit." *Jahresbericht, Deutsche Mathematiker-Vereinigung*, VI (1899), 91–110.

Bois, Henri, "Le Finitisme de Dühring." *L'Année Philosophique*, XX (1909), 93–124.

Bolzano, Bernard, Paradoxien des Unendlichen. Wissenschaftliche Classiker in Facsmile-Drucken. Vol. II, Berlin, 1889.

—— Rein analytischer Beweis des Lehrsatzes dass zwischen je zwey Werthen, die ein entgegengesetztes Resultat gewähren, wenigstens eine reelle Wurzel der Gleichung liege. Prag, 1817.

—— Schriften. Herausgegeben von der Königlichen Böhmischen Gesellschaft der Wissenschaften in Prag. 4 vols., Prag, 1930–35.

Boncompagni, B., "La Vie et les travaux du Baron Cauchy . . . par C.-A. Valson." Review. *Bullettino di Bibliografia e di Storia delle Scienze Matematiche e Fisiche*, II (1869), 1–95.

Bopp, Karl, "Die Kegelschnitte des Gregorius a St. Vincentio in vergleichender Bearbeitung." *Abhandlungen zur Geschichte der Mathematischen Wissenschaften*, XX (1907), 87–314.

Bortolotti, Ettore, "La memoria 'De infinitis hyperbolis' di Torricelli." *Archeion*, VI (1925), 49–58, 139–152.

—— "La scoperta e le successive generalizzazioni di un teorema fondamentale di calcolo integrale." *Archeion*, V (1924), 204–27.

Bosmans, Henri, "André Tacquet et son traité d'arithmétique théorique et pratique." *Isis*, IX (1927–28), 66–83.

—— "Le Calcul infinitésimal chez Simon Stevin." *Mathesis*, XXXVII (1923), 12–18, 55–62, 105–9.

—— "Un Chapitre de l'œuvre de Cavalieri (Les Propositions XVI–XXVII de l'Exercitatio Quarta)." *Mathesis*, XXXVI (1922), 365–73, 446–56.

—— "Les Démonstrations par l'analyse infinitésimale chez Luc Valerio." *Annales de la Société Scientifique de Bruxelles*, XXXVII (1913), 211–28.

—— "Diophante d'Alexandrie." *Mathesis*, XL (1926), Supplement.

—— "Grégoire de Saint-Vincent." *Mathesis*, XXXVIII (1924), 250–56.

—— "Le Mathématicien anversois Jean-Charles della Faille de la Compagnie de Jésus." *Mathesis*, XLI (1927), 5–11.

—— "Note historique sur le triangle arithmétique de Pascal." Annales de la Société Scientifique de Bruxelles, XXXI (1906–7), 65–72.

—— "La Notion des indivisibles chez Blaise Pascal." *Archivio di Storia della Scienza*, IV (1923), 369–79.

—— "Simon Stevin." Biographie Nationale de Belgique. Brussels, 1921–24.

—— "Sur l'interprétation géométrique donnée par Pascal à l'espace à quatre dimensions." *Annales de la Société Scientifique de Bruxelles*, XLII (1922–23), 337–45.

—— "Sur l'œuvre mathématique de Blaise Pascal." *Mathesis*, XXXVIII (1924), Supplement.

—— "Sur quelques exemples de la méthode des limites chez Simon Stevin." *Annales de la Société Scientifique de Bruxelles*, XXXVII (1913), 171–99.

—— "Sur une contradiction reprochée à la théorie des 'indivisibles' chez Cavalieri." *Annales de la Société Scientifique de Bruxelles*, XLII (1922–23), 82–89.

—— "Le Traité 'De centro gravitatis' de Jean-Charles della Faille." *Annales de la Société Scientifique de Bruxelles*, XXXVIII (1914), 255–317.

Boutroux, Pierre, Les Principes de l'analyse mathématique; exposé historique et critique. 3 vols., Paris, 1914–19.

Brasseur, J. B., "Exposition nouvelle des principes de calcul différentiel." *Mémoires, Société Scientifique de Liége* (2), III (1873), 113–92.

Brassine, E., Précis des œuvres mathématiques de P. Fermat. Paris, 1853.

Braunmühl, A. von, "Beiträge zur Geschichte der Integralrechnung bei Newton und Cotes." *Bibliotheca Mathematica* (3), V (1904), 355–65.

Brill, A., and M. Noether, "Die Entwickelung der Theorie der algebraischen Funktionen in älterer und neurer Zeit." *Jahresbericht, Deutsche Mathematiker-Vereinigung*, III (1892–93), 107–566.

Broden, Torsten, "Über verschiedene Gesichtspunkte bei der Grundlegung der mathematischen Analysis." Lund [1921].

Brouwer, L. E. J., "Intuitionism and Formalism." Translated by Arnold Dresden. *Bulletin, American Mathematical Society*, XX (1913), 81–96.

Brunschvicg, Léon, Les Étapes de la philosophie mathématique. Paris, 1912.

Buquoy, Georg, Graf von, Eine neue Methode für den Infinitesimalkalkul. Prag, 1821.

Burnet, John, Greek Philosophy. Part I, Thales to Plato. London, 1914.

Burns, C. D., "William of Ockham on Continuity." *Mind*, N. S., XXV (1916), 506–12.

Burtt, E. A., The Metaphysical Foundations of Modern Physical Science. London, 1925.

Cajori, Florian, "Bibliography of Fluxions and the Calculus. Textbooks Printed in the United States." *United States Bureau of Education. Circular of Information*, 1890, no. 3, pp. 395–400.

—— "Controversies on Mathematics between Wallis, Hobbes, and Barrow." *Mathematics Teacher*, XXII (1929), 146–51.

—— "Discussion of Fluxions: From Berkeley to Woodhouse." *American Mathematical Monthly*, XXIV (1917), 145–54.

—— "Grafting of the Theory of Limits on the Calculus of Leibniz." *American Mathematical Monthly*, XXX (1923), 223–34.

—— A History of Mathematical Notations. 2 vols., Chicago, 1928–29.

—— "The History of Notations of the Calculus." *Annals of Mathematics* (2), XXV (1924), 1–46.

—— A History of the Conceptions of Limits and Fluxions in Great Britain from Newton to Woodhouse. Chicago and London, 1919.

—— A History of Mathematics. 2d ed., New York, 1931.

—— "History of Zeno's Arguments on Motion." *American Mathematical Monthly*, XXII (1915), 1–6, 39–47, 77–82, 109–15, 143–49, 179–86, 215–20, 253–58, 292–97.

—— "Indivisibles and 'Ghosts of Departed Quantities' in the History of Mathematics." *Scientia* (2d series, XIX), XXXVII (1925), 303–6.

—— "Newton's Fluxions." In *Sir Isaac Newton. 1727–1927. A Bicentenary Evaluation of His Work*. Baltimore, 1928.

—— "The Purpose of Zeno's Arguments on Motion." *Isis*, III (1920), 7–20.

—— "The Spread of Newtonian and Leibnizian Notations of the Calculus." *Bulletin, American Mathematical Society*, XXVII (1921), 453–58.

—— "Who was the First Inventor of the Calculus?" *American Mathematical Monthly*, XXVI (1919), 15–20.

See also Newton.

Calculator, see Suiseth.

Calinon, A., "The Rôle of Number in Geometry." *New York Amer. Math. Society Bulletin*, VII (1901), 178–79.

Cantor, Georg, Contributions to the Founding of the Theory of Transfinite Numbers. Trans. with Introduction by P. E. B. Jourdain. Chicago and London, 1915.

—— Gesammelte Abhandlungen mathematischen und philosophischen Inhalts. Berlin, 1932.

Cantor, Moritz, "Origines du calcul infinitésimal." *Logique et Histoire des Sciences*. Bibliothèque du Congrès International de Philosophie (Paris, 1901), III, 3–25.

—— Vorlesungen über Geschichte der Mathematik. Vol. I, 2d ed., Leipzig, 1894. Vols. II–IV, Leipzig, 1892–1908.

Cardan, Jerome, Opera. 10 vols., Lugduni, 1663.

Carnot, L. N. M., "Reflections on the Theory of the Infinitestimal Calculus." Trans. by W. Dickson. *Philosophical Magazine*, VIII (1800), 222–40, 335–52; IX (1801), 39–56.

——Reflexions on the Metaphysical Principles of the Infinitesimal Analysis. Trans. by W. R. Browell, Oxford, 1832.

—— Réflexions sur la métaphysique du calcul infinitésimal. 2d ed., Paris, 1813.

Cassirer, Ernst, "Das Problem des Unendlichen und Renouviers 'Gesetz der Zahl.'" *Philosophische Abhandlungen. Hermann Cohen zum 70sten Geburtstag Dargebracht*, Berlin, 1912, pp. 85–98.

Cauchy, Augustin, Œuvres complètes. 25 vols., Paris, 1882–1932.

Cavalieri, Bonaventura, Centuria di varii problemi. Bologna, 1639.

—— Exercitationes geometricae sex. Bononiae, 1647.

—— Geometria indivisibilibus continuorum nova quadam ratione promota. New ed., Bononiae, 1653.

Chasles, Michel, Aperçu historique sur l'origine et le développement des méthodes en géométrie, particulièrement de celles qui se rapportent à la géométrie moderne. Paris, 1875.

Chatelain, Aemilio, see Denifle.

Child, J. M., "Barrow, Newton and Leibniz, in Their Relation to the Discovery of the Calculus." *Science Progress*, XXV (1930), 295–307.

See also Barrow; Leibniz.

Christensen, S. A., "The First Determination of the Length of a Curve." *Bibliotheca Mathematica*, N. S., I (1887), 76–80.

Cohen, H., Das Princip der Infinitesimalmethode und seine Geschichte. Berlin, 1883.

Cohen, M. R., Reason and Nature. An Essay on the Meaning of Scientific Method. New York, 1931.

Cohn, Jonas, Geschichte des Unendlichkeitsproblems im abendländischen Denken bis Kant. Leipzig, 1896.

Commandino, Federigo, Liber de centro gravitatis solidorum. Bononiae, 1565.

Comte, Auguste, The Philosophy of Mathematics. Trans. by W. M. Gillespie. New York, 1858.

Coolidge, J. L., "The Origin of Analytic Geometry." *Osiris*, I (1936), 231–50.

Courant, Richard, Differential and Integral Calculus. Translated by E. J. McShane. Vol. II, 2d ed., London and Glasgow, 1937.

Cournot, A. A., Traité élémentaire de la théorie des fonctions et du calcul infinitésimal. 2 vols., Paris, 1841.

Couturat, Louis, De l'infini mathématique. Paris, 1896.

Curtze, Maximilian, "Ueber die Handschrift R. 4°. 2, Problematum Euclidis explicatio der Königlich. Gymnasialbibliothek zu Thorn." *Zeitschrift für Mathematik und Physik*, XIII (1868), Supplement, pp. 45–104.

See also Oresme.

D'Alembert, Jean le Rond, Preface and articles "Exhaustion," "Différentiel," and "Limite," in *Encyclopédie ou Dictionnaire raisonné des sciences, des arts et des métiers*. 36 vols., Lausanne and Berne, 1780–82.

—— Mélanges de littérature, d'histoire, et de philosophie. 4th ed., 5 vols. Amsterdam, 1767.

Dantscher, Victor von, Vorlesungen über die Weierstrass'sche Theorie der irrationalen Zahlen. Leipzig and Berlin, 1908.

Dantzig, Tobias, Aspects of Science. New York, 1937.

—— Number, the Language of Science. New York, 1930.

Datta, Bibhutibhusan, "Origin and History of the Hindu Names for Geometry." *Quellen und Studien zur Geschichte der Mathematik, Astronomie und Physik*, Part B, Studien, I (1931), 113–19.

—— and A. N. Singh, History of Hindu Mathematics. A Source Book. Part I. Numerical Notation and Arithmetic. Lahore, 1935.

Dauriac, L., "Le Réalisme finitiste de F. Evellin." *L'Année Philosophique*, XXI (1910), 169–200.

Dedekind, Richard, Essays on the Theory of Numbers. Trans. by W. W. Beman. Chicago, 1901.

Dehn, M., "Ueber raumgleiche Polyeder." *Nachrichten von der Königlichen Gesellschaft der Wissenschaften zu Göttingen, Mathematische-Physikalische Klasse*, 1900, pp. 345–54.

De Morgan, Augustus, Essays on the Life and Work of Newton. Chicago and London, 1914.

——— "On the Early History of Infinitesimals in England." *Philosophical Magazine* (4), IV (1852), 321–30.

Denifle, Henricus, and Aemilio Chatelain, Chartularium universitatis Parisiensis. 4 vols., Paris, 1889–97.

Descartes, René, Geometria, a Renato des Cartes anno 1637 Gallice edita. With letters and notes of Hudde, de Beaune, van Schooten, and van Heuraet. 3d ed., 2 vols., Amstelodami, 1683.

——— Œuvres. Ed. by Charles Adam and Paul Tannery. 12 vols. and Supplement, Paris, 1897–1913.

Dickstein, S., "Zur Geschichte der Prinzipien der Infinitesimalrechnung. Die Kritiker der 'Théorie des Fonctions Analytiques' von Lagrange." *Abhandlungen zur Geschichte der Mathematik*, IX (1899), 65–79.

Diophantus, Les Six Livres arithmétiques et le livre des nombres polygons. Translated with introduction and notes by Paul Ver Eecke. Bruges, 1926.

D'Ooge, M. L., see Nicomachus.

Drobisch, M. W., "Ueber den Begriff des Stetigen und seine Beziehungen zum Calcul." *Berichte über die Verhandlungen der Königlich Sächsischen Gesellschaft der Wissenschaften zu Leipzig. Mathematisch-Physische Classe*, 1853, pp. 155–76.

Duhamel, J. M. C., "Mémoire sur la méthode des maxima et minima de Fermat et sur les méthodes des tangentes de Fermat et Descartes." *Mémoires de l'Académie des Sciences de l'Institut Impérial de France*, XXXII (1864), 269–330.

Duhem, Pierre, Études sur Léonard de Vinci. Vols. I and II, Ceux qu'il a lus et ceux qui l'ont lu; vol. III, Les Précurseurs parisiens de Galilée. Paris, 1906–13.

——— "Léonard de Vinci et la composition des forces concourantes." *Bibliotheca Mathematica* (3), IV (1903), 338–43.

——— "Oresme." Article in *Catholic Encyclopedia*, XI (1911), 296–97.

——— Les Origines de la statique. 2 vols., Paris, 1905–6.

Dühring, Eugen, Kritische Geschichte der allgemeinen Principien der Mechanik. 3d. ed., Leipzig, 1887.

Eastwood, D. M., The Revival of Pascal. A Study of His Relation to Modern French Thought. Oxford, 1936.

Edel, Abraham, Aristotle's Theory of the Infinite. New York, 1934.

Eneström, Gustav, "Die erste Herleitung von Differentialen trigonometrischen Funktionen." *Bibliotheca Mathematica* (3), IX (1908–9), 200–5.

——— "Leibniz und die Newtonsche 'Analysis per aequationes numero terminorum infinitas.'" *Bibliotheca Mathematica* (3), XI (1911), 354–55.

——— "Sur le Part de Jean Bernoulli dans la publication de l'Analyse des infiniment petits.'" *Bibliotheca Mathematica*, N. S., VIII (1894), 65–72.

Eneström, Gustav, "Sur un théorème de Kepler équivalent à l'intégration d'une fonction trigonométrique." *Bibliotheca Mathematica*, N. S., III (1889), 65–66.

—————— "Über die angebliche Integration einer trigonometrischen Function bei Kepler." *Bibliotheca Mathematica* (3), XIII (1913), 229–41.

—————— "Über die erste Aufnahme der Leibnizschen Differentialrechnung." *Bibliotheca Mathematica* (3), IX (1908–9), 309–20.

—————— "Zwei mathematische Schulen im christlichen Mittelalter." *Bibliotheca Mathematica* (3), VII (1907), 252–62.

Engels, Frederick, Herr Eugen Dühring's Revolution in Science. [Anti-Dühring.] [New York, no date.]

Enriques, Fedrigo, The Historic Development of Logic. Trans. by Jerome Rosenthal. New York, 1929.

—————— Problems of Science. Trans. by Katharine Royce. Chicago and London, 1914.

Euclid,The Thirteen Books of the Elements. Trans. from text of Heiberg with introduction and commentary by T. L. Heath. 3 vols., Cambridge, 1908.

Euler, Leonhard, Letters on Different Subjects in Natural Philosophy Addressed to a German Princess. 2 vols., New York, 1840.

—————— Opera omnia. Ed. by Ferdinand Rudio. 22 vols. in 23, Lipsiae and Berolini, 1911–36.

Evans, G. W., "Cavalieri's Theorem in his own Words." *American Mathematical Monthly*, XXIV (1917), 447–51.

Evans, W. D., "Berkeley and Newton." *Mathematical Gazette*, VII (1913–14), 418–21.

Fedel, J., Der Briefwechsel Johann (i) Bernoulli-Pierre Varignon aus den Jahren 1692 bis 1702. Heidelberg, 1932.

Fehr, H., "Les Extensions de la notion de nombre dans leur développement logique et historique." *Enseignement Mathématique*, IV (1902), 16–27.

Fermat, Pierre, Œuvres. Ed. by Paul Tannery and Charles Henry. 4 vols. and Supplement, Paris, 1891–1922.

Fine, H. B., The Number-System of Algebra Treated Theoretically and Historically. Boston and New York [1890].

Fink, Karl, A Brief History of Mathematics. Trans. by Beman and Smith. Chicago, 1910.

Fontenelle, Bernard, Élémens de la géométrie de l'infini. Suite des *Mémoires de l'Académie Royale des Sciences* (1725). Paris, 1727.

Fort, Osmar, "Andeutungen zur Geschichte der Differential-Rechnung. Dresden, 1846.

Freyer, Studien zur Metaphysik der Differentialrechnung. Berlin, 1883.

Froberg, J. P., De analytica calculi differentialis et integralis theoria, inventa a cel. La Grange. Upsaliae, 1807–10.

Funkhouser, H. G., "Historical Development of the Graphical Representation of Statistical Data." *Osiris*, III (1937), 269–404.

Galilei, Galileo, Opere. Edizione nazionale. 20 vols., Firenze, 1890–1909.

Galloys, Abbé Jean, "Réponse à l'écrit de M. David Gregorie, touchant les lignes appellées Robervalliennes, qui servent à transformer les figures." *Histoire de l'Académie Royale des Sciences, Mémoires*, 1703, pp. 70–77.

Gandz, Solomon, "The Origin and Development of the Quadratic Equations in Babylonian, Greek and Early Arabic Algebra." *Osiris*, III (1938), 405–557.

—— The Sources of al-Khowarizmi's Algebra. *Osiris*, I (1936), 263–77.

Garrett, J. A., see Wren.

Genty, Abbé Louis, L'Influence de Fermat sur son siècle, relativement au progrès de la haute géométrie et du calcul. Orleans, 1784.

Gerhardt, C. I., Die Geschichte der höheren Analysis. Part I. "Die Entdeckung der höheren Analysis." Halle, 1855.

—— Die Entdeckung der Differentialrechnung durch Leibniz. Halle, 1848.

—— Geschichte der Mathematik in Deutschland. München, 1877.

—— "Zur Geschichte des Streites über den ersten Entdecker der Differentialrechnung." *Archiv der Mathematik und Physik*, XXVII (1856), 125–32.

See also Leibniz.

Gibson, G. A., "Berkeley's Analyst and Its Critics: An Episode in the Development of the Doctrine of Limits." *Bibliotheca Mathematica*, N. S., XIII, (1899), 65–70.

—— "Vorlesungen über Geschichte der Mathematik von Moritz Cantor. Dritter (Schluss) Band. A Review: with Special Reference to the Analyst Controversy." *Proceedings, Edinburgh Mathematical Society*, XVII (1898), 9–32.

Ginsburg, Benjamin, "Duhem and Jordanus Nemorarius." *Isis*, XXV (1936), 340–62.

Giovannozzi, P. Giovanni, "Pierre Fermat. Una lettera inedita." *Archivio di Storia della Scienza*, I (1919), 137–40.

Giuli, G. de, "Galileo e Descartes." *Scientia*, XLIX (1931), 207–20.

Goldbeck, Ernst, "Galileis Atomistik und ihre Quellen." *Bibliotheca Mathematica* (3), III (1902), 84–112.

Görland, Albert, Aristoteles und die Mathematik. Marburg, 1899.

Gow, James, A Short History of Greek Mathematics. Cambridge, 1884.

Grandi, Guido, De infinitis infinitorum et infinite parvorum ordinibus disquisitio geometrica. Pisis, 1710.

Graves, G. H., "Development of the Fundamental Ideas of the Differential Calculus." *Mathematics Teacher*, III (1910), 82–89.

Gregory, James, Exercitationes geometricae. Londini, 1668.

—— Geometrica pars universalis. Patavii, 1668.

—— Vera circuli et hyperbolae quadratura. Patavii [1667].

Gregory of St. Vincent, Opus geometricum quadraturae circuli et sectionum coni decem libris comprehensum. Antverpiae, 1647.

320   Bibliography

Guisnée, "Observations sur les méthodes de maximis & minimis, où l'on fait voir l'identité et la différence de celle de l'analyse des infiniment petits avec celles de Mrs. Hermat et Hude." *Histoire de l'Académie Royale des Sciences*, 1706, pp. 24–51.

Gunn, J. A., The Problem of Time. An Historical and Critical Study. London, 1929.

Gunther, R. T., Early Science in Oxford. 10 vols., London, 1920–35.

Günther, S., "Albrecht Dürer, einer der Begründer der neueren Curventheorie." *Bibliotheca Mathematica*, III (1886), 137–40.

―――― "Über eine merkwürdige Beziehung zwischen Pappus und Kepler." *Bibliotheca Mathematica*, N. S., II (1888), 81–87.

Hadamard, J., "La Notion de différentiel dans l'enseignement." *Mathematical Gazette*, XIX (1935), 341–42.

Hagen, J. G., "On the History of the Extensions of the Calculus." *Bulletin, American Mathematical Society*, N. S., VI (1899–1900), 381–90.

Hankel, Hermann, "Grenze." Article in Ersch and Gruber, *Allgemeine Encyklopädie der Wissenschaften und Künste*, Neunzigster Theil, Leipzig. 1871, pp. 185–211.

―――― Zur Geschichte der Mathematik in Alterthum und Mittelalter. Leipzig, 1874.

Harding, P. J., "The Geometry of Thales." *Proceedings, International Congress of Mathematicians* (Cambridge, 1912), II, 533–38.

Harnack, Axel, An Introduction to the Study of the Elements of the Differential and Integral Calculus. Trans. by G. L. Cathcart. London, 1891.

Hart, J. B., Makers of Science. London, 1923.

Hathway, A. C., "The discovery of Calculus." *Science*, N.S., L (1919), 41–43.

―――― "Further History of the Calculus." *Science*, N. S., LI (1920), 166–67.

Heath, L. R., The Concept of Time. Chicago, 1936.

Heath, T. L., "Greek Geometry with Special Reference to Infinitesimals." *Mathematical Gazette*, XI (1922–23), 248–59.

―――― A History of Greek Mathematics. 2 vols., Oxford, 1921.

     See also Archimedes and Euclid.

Heiberg, J. L., "Mathematisches zu Aristoteles," *Abhandlungen zur Geschichte der Mathematischen Wissenschaften*, XVIII (1904), 1–49.

―――― Quaestiones archimedae. Hauniae, 1879.

     See also Archimedes.

Heine, E., "Die Elemente der Funktionenlehre." *Journal für die Reine und Angewandte Mathematik*, LXXIV (1872), 172–88.

Heinrich, Georg, "James Gregorys 'Vera circuli et hyperbolae quadratura.' " *Bibliotheca Mathematica*. (3) II (1901), 77–85.

Helmholtz, Hermann von, Counting and Measuring. Trans. by C. L. Bryan with introduction and notes by H. T. Davis. New York, 1930.

―――― Ueber die Erhaltung der Kraft, eine physikalische Abhandlung. Berlin, 1847.

Henry, Charles, "Recherches sur les manuscrits de Pierre de Fermat." *Bullettino di Bibliografia e di Storia delle Science Matematiche e Fisiche,* XII (1879), 477–568, 619–740; XIII (1880), 437–70.

Heyl, Paul, "The Skeptical Physicist." *Scientific Monthly,* XLVI (1938), 225–29.

Hilbert, David, "Über das Unendliche." *Mathematische Annalen,* XCV (1926), 161–90.

Hill, M. J. M., "Presidential Address on the Theory of Proportion." *Mathematical Gazette,* VI (1912), 324–32, 360–68.

Hobbes, Thomas, The English Works. Ed. by Sir Wm. Molesworth. 11 vols., London, 1839–45.

—— Opera philosophica quae latine scripsit omnia. 5 vols., London, 1839–45.

Hobson, E. W., "On the Infinite and the Infinitesimal in Mathematical Analysis." *Proceedings, London Mathematical Society,* XXXV (1903), 117–40.

Hoefer, Ferdinand, "Stevin." Article in *Nouvelle Biographie Générale,* Vol. XLIV, Paris, 1868, cols. 496–98.

Hogben, Lancelot, Science for the Citizen. New York, 1938.

Hoppe, Edmund, "Zur Geschichte der Infinitesimalrechnung bis Leibniz und Newton." *Jahresbericht, Deutsche Mathematiker-Vereinigung,* XXXVII (1928), 148–87.

Hudde, Johann, see Descartes.

Huntington, E. V., The Continuum and Other Types of Serial Order. With an Introduction to Cantor's Transfinite Numbers. 2d ed., Cambridge, Mass., 1917.

—— "Modern Interpretation of Differentials." *Science,* N. S., LI (1920), 320–21, 593.

Huygens, Christiaan, Œuvres complètes. Published by Société hollandaise des sciences. 19 vols., La Haye, 1888–1937.

[Ibn al-Haitham] "Die Abhandlung über die Ausmessung des Paraboloids von el-Hasan b. el-Hasan b. el Haitham." Translated with commentary by Heinrich Suter. *Bibliotheca Mathematica* (3), XII (1911–12), 289–332.

Jacoli, Ferdinando, "Evangelista Torricelli ed il metodo delle tangenti detto *Metodo del Roberval.*" *Bullettino di Bibliografia e di Storia delle Scienze Matematiche e Fisiche,* VIII (1875), 265–304.

Jaeger, Werner, Aristoteles. Grundlegung einer Geschichte seiner Entwicklung. Berlin, 1923.

Johnson, F. R., and S. V. Larkey, "Robert Recorde's Mathematical Teaching and the Anti-Aristotelian Movement." *Huntington Library Bulletin,* No. VII (1935), 59–87.

Johnston, G. A., The Development of Berkeley's Philosophy. London, 1923.

Jourdain, P. E. B., "The Development of the Theory of Transfinite Numbers." *Archiv der Mathematik und Physik* (3), X (1906), 254–81; XIV (1908–9), 287–311; XVI (1910), 21–43; XXII (1913–14), 1–21.

Jourdain, P. E. B., "The Ideas of the 'fonctions analytiques' in Lagrange's Early Work." *Proceedings, International Congress of Mathematicians* (Cambridge, 1912), II, 540–41.

———— "The Introduction of Irrational Numbers." *Mathematicai Gazette,* IV (1908), 201–9.

———— "Note on Fourier's Influence on the Conceptions of Mathematics." *Proceedings, International Congress of Mathematicians* (Cambridge, 1912), II, 526–27.

———— "On Isoid Relations and Theories of Irrational Number." *Proceedings, International Congress of Mathematicians* (Cambridge, 1912), II, 492–96.

———— "The Origin of Cauchy's Conceptions of a Definite Integral and of the Continuity of a Function." *Isis,* I (1913), 661–703.

———— "The Theory of Functions With Cauchy and Gauss." *Bibliotheca Mathematica* (3), VI (1905), 190–207.

Jurin, James, "Considerations upon Some Passages of a Dissertation Concerning the Doctrine of Fluxions, Published by Mr. Robins. . . ." *The Present State of the Republick of Letters,* XVIII (1736), 45–82, 111–79.

———— Geometry No Friend to Infidelity: or, A Defence of Sir Isaac Newton and the British Mathematicians, in a Letter Addressed to the Author of the Analyst by Philalethes Cantabrigiensis. London, 1734.

———— The Minute Mathematician: or, The Freethinker no Just-thinker, Set Forth in a Second Letter to the Author of the Analyst. London, 1735.

———— "Observations upon Some Remarks Relating to the Method of Fluxions." *The Present State of the Republick of Letters,* 1736, Supplement, pp. 1–77.

Kaestner, A. G., Anfangsgründe der Analysis des Unendlichen. 3d ed., Göttingen, 1799.

———— Geschichte der Mathematik seit der Wiederherstellung der Wissenschaften bis an das Ende des achtzehnten Jahrhunderts. 4 vols., Göttingen, 1796–1800.

Kant, Immanuel, Sämmtliche Werke. Ed. by Karl Rosenkranz and F. W. Schubert. 12 vols in 8, Leipzig, 1838–42.

Karpinski, L. C., The History of Arithmetic. Chicago and New York [c. 1925].

———— "Is There Progress in Mathematical Discovery and Did the Greeks Have Analytic Geometry?" *Isis,* XXVII (1937), 46–52.

See also Al-Khowarizmi; Nicomachus.

Kasner, Edward, "Galileo and the Modern Concept of Infinity." *Bulletin, American Mathematical Society,* XI (1905), 499–501.

Kepler, Johann, Opera omnia. Ed. by Ch. Frisch. 8 vols., Frankofurti a. M. and Erlangae, 1858–70.

See also Struik.

Keyser, C. J., "The Rôle of the Concept of Infinity in the Work of Lucretius." *Bulletin, American Mathematical Society*, XXIV (1918), 321–27.

Kibre, Pearl, see Thorndike.

Klein, Felix, Elementary Mathematics from an Advanced Standpoint. Arithmetic. Algebra. Analysis. Trans. by E. R. Hedrick and C. A. Noble. New York, 1932.

—— The Evanston Colloquium Lectures on Mathematics. New York, 1911.

—— "Ueber Arithmetisirung der Mathematik." *Nachrichten von der Königlichen Gesellschaft der Wissenschaften zu Göttingen (Geschäftliche Mittheilungen)*, 1895, pp. 82–91.

—— Vorlesungen über die Entwicklung der Mathematik im 19. Jahrhundert. Ed. by R. Courant, O. Neugebauer, and St. Cohn-Vossen. 2 vols., Berlin, 1926–27.

Klein, Jacob, "Die griechische Logistik und die Entstehung der Algebra." *Quellen und Studien zur Geschichte der Mathematik, Astronomie und Physik*, Part B, Studien, III (1936), 19–105, 122–235.

Körner, Theodor, "Der Begriff des materiellen Punktes in der Mechanik des achtzehnten Jahrhunderts." *Bibliotheca Mathematica* (3), V (1904), 15–62.

Kossak, E., Die Elemente der Arithmetik. Berlin, 1872.

Kowalewski, Gerhard, "Über Bolzanos nichtdifferenzierbare stetige Funktion." *Acta Mathematica*, XLIV (1923), 315–19.

Lacroix, S. F., Traité du calcul différentiel et du calcul intégral. 2d ed., 3 vols., Paris, 1810–19.

Lagrange, J. L., Fonctions analytiques. Reviewed in *Monthly Review*, (London), N. S., XXVIII (1799), Appendix, pp. 481–99.

—— Œuvres. 14 vols., Paris, 1867–92.

Landen, John, A Discourse Concerning the Residual Analysis. London, 1758.

—— The Residual Analysis; A New Branch of the Algebraic Art. London, 1764.

Lange, Ludwig, Die geschichtliche Entwickelung des Bewegungsbegriffes und ihr voraussichtliches Endergebniss. Leipzig, 1886.

Larkey, S. V., see Johnson, F. R.

Lasswitz, Kurd, Geschichte der Atomistik vom Mittelalter bis Newton. 2 vols., Hamburg and Leipzig, 1890.

Laurent, H., "Différentielle." Article in *La Grande Encyclopédie*, Vol. XIV.

Leavenworth, Isabel, The Physics of Pascal. New York, 1930.

Leibniz, G. W., "Addenda ad schediasma proximo." *Acta Eruditorum*, 1695, pp. 369–72.

—— The Early Mathematical Manuscripts. Trans. from the Latin Texts of C. I. Gerhardt with Notes by J. M. Child. Chicago, 1920.

—— "Isaaci Newtoni tractatus duo." Review in *Acta Eruditorum*, 1705, pp. 30–36.

Leibniz, G. W., Mathematische Schriften. Ed. by C. I. Gerhardt. Gesammelte Werke. Ed. by G. H. Pertz. Third Series, Mathematik. 7 vols., Halle, 1849–63.

—— Opera omnia. Ed. by Louis Dutens. 6 vols., Genevae, 1768.

—— Opera philosophica que exstant latina, gallica, germanica omnia. Ed. by J. E. Erdmann. Berolini, 1840.

—— Philosophische Schriften. Ed. by C. I. Gerhardt. 7 vols., Berlin, 1875–90.

—— "Responsio ad nonnullas difficultates, a Dn. Bernardo Nieuwentijt." *Acta Eruditorum*, 1695, pp. 310–16.

—— "Testamen de motuum coelestium." *Acta Eruditorum*, 1689, pp. 82–96.

—— and John Bernoulli, Commercium philosophicum et mathematicum. 2 vols., Lausannae and Genevae, 1745.

Lennes, N. J., Differential and Integral Calculus. New York, 1931.

Le Paige, M. C., "Correspondance de René François de Sluse publiée pour la première fois et précédée d'une introduction par M. C. Le Paige." *Bullettino di Bibliografia e di Storia delle scienze matematiche e fisiche*, XVII (1884), 427–554, 603–726.

Lewes, G. H., The Biographical History of Philosophy. Vol. I, Library ed., New York, 1857.

L'Hospital, G. F. A. de, Analyse des infiniment petits pour l'intelligence des lignes courbes. Paris, 1696.

L'Huilier, Simon, Exposition élémentaire des principes des calculs supérieurs, qui a remporté le prix proposé par l'Académie Royale des Sciences et Belle-Lettres pour l'année 1786. Berlin [1787].

Libri, G., Histoire des sciences mathématiques en Italie. 4 vols. Halle, s/S., 1865.

Löb, Hermann, Die Bedeutung der Mathematik für die Erkenntnislehre des Nikolaus von Kues. Berlin, 1907.

Lorenz, Siegfried, Das Unendliche bei Nicolaus von Cues. Fulda, 1926.

Loria, Gino, "Le ricerche inedite di Evangelista Torricelli sopra la curva logaritmica." *Bibliotheca Mathematica* (3), I (1900), 75–89.

—— Storia delle matematiche. 3 vols., Torino, 1929–33.

Luckey, P., "Was ist ägyptische Geometrie?" *Isis*, XX (1933), 15–52.

Luria, S., "Die Infinitesimaltheorie der antiken Atomisten." *Quellen und Studien zur Geschichte der Mathematik, Astronomie und Physik*, Part B, Studien, II (1933), 106–85.

Mach, Ernst, Die Principien der Wärmelehre. Historisch-Kritisch Entwickelt. 2d ed., Leipzig, 1900.

—— The Science of Mechanics. Trans. by T. J. McCormack. 3d ed., Chicago, 1907.

—— Space and Geometry in the Light of Physiological, Psychological and Physical Inquiry. Trans. by T. J. McCormack. Chicago, 1906.

Maclaurin, Colin, A Treatise of Fluxions. 2 vols., Edinburgh, 1742.

Mahnke, Dietrich, "Neue Einblicke in die Entdeckungsgeschichte der höheren Analysis." *Abhandlungen der Preussische Akademie der Wissenschaften*, Physikalisch-Mathematische Klasse, I (1925), 1–64.

——— "Zur Keimesgeschichte der Leibnizschen Differentialrechnung." *Sitzungsberichte der Gesellschaft zur Beförderung der Gesamten Naturwissenschaften zu Marburg*, LXVII (1932), 31–69.

Maire, Albert, L'Œuvre scientifique de Blaise Pascal. Bibliographie. Paris, 1912.

Mansion, Paul, "Continuité au sens analytique et continuité au sens vulgaire." *Mathesis* (2), IX (1899), 129–31.

——— Editorial note. *Mathesis*, IV (1884), 177.

——— "Esquisse de l'histoire du calcul infinitésimal." Appendix to Résumé du cours d'analyse infinitésimale. Paris, 1887.

——— "Fonction continue sans dérivée de Weierstrass." *Mathesis*, VII (1887), 222–25.

——— "Méthode, dite de Fermat, pour la recherche des maxima et des minima." *Mathesis*, II (1882), 193–202.

Marie, Maximilien, Histoire des sciences mathématiques et physiques. 12 vols. in 6, Paris, 1883–88.

Marvin, W. T., The History of European Philosophy. New York, 1917.

Mayer, Joseph, "Why the Social Sciences Lag behind the Physical and Biological Sciences." *Scientific Monthly*, XLVI (1938), 564–66.

Méray, Charles, Nouveau Précis d'analyse infinitésimale. Paris, 1872.

——— "Remarques sur la nature des quantités définies par la condition de servir de limites à des variables données." *Revue des Sociétés Savantes. Sciences Mathématiques, Physiques et Naturelles*, 2d. Series, IV (1869), 280–89.

Mersenne, Marin, Tractatus mechanicus theoricus et practicus. Cogitata physico-mathematica. Parisiis, 1644.

Merton, R. K., "Science, Technology and Society in Seventeenth Century England." *Osiris*, IV (1938), 360–632.

Merz, J. T., A History of European Thought in the Nineteenth Century. 4 vols., Edinburgh and London, 1896–1914.

Milhaud, Gaston, "Descartes et l'analyse infinitésimale." *Revue Générale des Sciences*, XXVIII (1917), 464–69.

——— Descartes savant. Paris, 1921.

——— Leçons sur les origines de la science grecque. Paris, 1893.

——— "Note sur les origines du calcul infinitésimal." Logique et Histoire des Sciences. *Bibliothèque du Congrès International de Philosophie* (Paris, 1901), III, 27–47.

——— Nouvelles études sur l'histoire de la pensée scientifique. Paris, 1911.

——— Les Philosophes géomètres de la Grèce. Platon et ses prédécesseurs. 2d ed., Paris, 1934.

Milhaud, Gaston, "La Querelle de Descarte set de Fermat au sujet des tangentes." *Revue Générale des Sciences*, XXVIII (1917), 332–37.

Miller, G. A., "Characteristic Features of Mathematics and Its History." *Scientific Monthly*, XXXVII (1933), 398–404.

——— "Some Fundamental Discoveries in Mathematics." *Science*, XVII (1903), 496–99.

See also Smith.

Molk, J., see Pringsheim; Schubert; Voss.

Montucla, Étienne, Histoire des mathématiques, New ed., 4 vols., Paris, [1799]–1802.

More, L. T., Isaac Newton. A Biography. New York and London, 1934.

Müller, Felix, "Zur Literatur der analytischen Geometrie and Infinitesimalrechnung vor Euler." *Jahresbericht, Deutsche Mathematiker-Vereinigung*, XIII (1904), 247–53.

Neave, E. W. J., "Joseph Black's Lectures on the Elements of Chemistry." *Isis*, XXV (1936), 372–90.

Neikirk, L. I., "A Class of Continuous Curves Defined by Motion Which Have No Tangent Lines." *University of Washington Publications in Mathematics*, II, 1930, 59–63.

Neugebauer, Otto, "Das Pyramidenstumpf-Volumen in der vorgriechischen Mathematik." *Quellen und Studien zur Geschichte der Mathematik, Astronomie und Physik*, Part B, Studien, II (1933), 347–51.

——— Vorlesungen über Geschichte der antiken mathematischen Wissenschaften. Vol. I, Vorgriechische Mathematik. Berlin, 1934.

Newton, Sir Isaac, Mathematical Principles of Natural Philosophy and System of the World. Translation of Andrew Motte Revised and Supplied with an Historical and Explanatory Appendix by Florian Cajori. Berkeley, Cal., 1934.

——— The Method of Fluxions and Infinite Series. Trans. with notes by John Colson. London, 1736.

——— La Méthode des fluxions et des suites infinies. Trans. with introduction by Buffon. Paris, 1740.

——— Opera quae exstant omnia. Ed. by Samuel Horsley. 5 vols., Londini, 1779–85.

——— Opuscula mathematica, philosophica et philologica. 3 vols., Lausannae and Genevae, 1744.

——— Two Treatises of the Quadrature of Curves, and Analysis by Equations of an Infinite Number of Terms. Explained by John Stewart. London, 1745.

[Newton, Sir Isaac] Review of Commercium Epistolicum. *Philosophical Transactions, Royal Society of London*, XXIX (1714–16), 173–224.

Nicomachus of Gerasa, Introduction to Arithmetic. Trans. by M. L. D'Ooge, with Studies in Greek Arithmetic by F. E. Robbins and L. C. Karpinski. New York, 1926.

Nieuwentijdt, Bernard, Analysis infinitorum seu curvilineorum proprietates ex polygonorum natura deductae. Amstelaedami, 1695.

—— Considerationes circa analyseos ad quantitates infinite parvas applicatae principia, et calculi differentialis usum in resolvendis problematibus geometricis. Amstelaedami, 1694.

—— Considerationes secundae circa calculi differentialis principia; et responsio ad virum nobilissimum G. G. Leibnitium. Amstelaedami, 1696.

Noether, M., see Brill.

Note, in *Histoire de l'Académie royale des sciences*, 1701, pp. 87–89.

Nunn, T. P., "The Arithmetic of Infinities." *Mathematical Gazette*, V (1910–11), 345–56, 377–86.

Oresme, Nicole, Algorismus proportionum. Ed. by M. Curtze. [Berlin, 1868.]

[Oresme] Tractatus de latitudinibus formarum. MS, Bayerische Staatsbibliothek, München, cod. lat. 26889, fol. 201^r–206^r.

Osgood, W. F., "The Calculus in Our Colleges and Technical Schools." *Bulletin, American Mathematical Society*, XIII (1907), 449–67.

[Pappus of Alexandria] Pappus d'Alexandrie. La Collection mathématique. Translated with introduction and notes by Paul Ver Eecke. 2 vols., Paris and Bruges, 1933.

Pascal, Blaise, Œuvres. Ed. by Léon Brunschvicg and Pierre Boutroux. 11 vols., Paris, 1908–14.

—— The Physical Treatises of Pascal. Trans. by I. H. B. and A. G. H. Spiers with Introduction and Notes by Frederick Barry. New York, 1937.

Perrier, Lieutenant, "Pascal. Créateur du calcul des probabilités et précurseur du calcul intégral." *Revue Générale des Sciences*, XII (1901), 482–90.

Petronievics, B., "Über Leibnizens Methode der direkten Differentiation." *Isis*, XXII (1934), 69–76.

Picard, Émile, "La Mathématique dans ses rapports avec la physique." *Revue Générale des Sciences*, XIX (1908), 602–9.

—— "On the Development of Mathematical Analysis, and Its Relations to Some Other Sciences." *Science*, N. S., XX (1904), 857–72.

—— "Pascal mathématicien et physicien." *La Revue de France*, IV (1923), 272–83.

Pierpont, James, "The History of Mathematics in the Nineteenth Century." *Bulletin, American Mathematical Society*, XI (1904), 136–59.

—— "Mathematical Rigor, Past and Present." *Bulletin, American Mathematical Society*, XXXIV (1928), 23–53.

Pincherle, Salvatore, "Saggio di una introduzione alla teoria delle funzioni analitiche secondo i principii del Prof. C. Weierstrass." *Giornale di Matematiche*, XVIII (1880), 178–254, 317–57.

Plato, Dialogues. Trans. with analyses and introductions by Benjamin Jowett. 4 vols., Oxford, 1871.

Plutarch, The Lives of the Noble Grecians and Romans. Trans. by John Dryden and revised by A. H. Clough. New York [no date].

———— Miscellanies and Essays. Comprising All his Works under the Title of "Morals." Translated from the Greek by Several Hands. Corrected and revised by W. W. Goodwin. 5 vols., Boston, 1898.

Poincaré, Henri, "Le Continu mathématique." *Revue de Métaphysique et de Morale*, I (1893), 26–34.

———— "L'Avenir des mathématiques." *Revue Générale des Sciences*, XIX (1908), 930–39.

———— The Foundations of Science. Trans. by G. B. Halsted. New York, 1913.

———— "L'Œuvre mathématique de Weierstrass." *Acta Mathematica*, XXII (1898–99), 1–18.

Poisson, S. D., Traité de mécanique. 2d ed., 2 vols., Paris, 1833.

Prag, A., "John Wallis." *Quellen und Studien zur Geschichte der Mathematik, Astronomie und Physik*, Part B, Studien, I (1931), 381–412.

Prantl, C., Geschichte der Logik im Abendlande. 4 vols., Leipzig, 1855–70.

Pringsheim, A., and J. Molk, "Nombres irrationnels et notion de limite." *Encyclopédie des Sciences Mathématiques*, Vol. I, Part 1, Fascicule 1, 133–60; Fascicule 2.

———— "Principes fondamentaux de la théorie des fonctions." *Encyclopédie des Sciences Mathématiques*, Vol. II, Part 1, Fascicule 1.

Proclus Diadochus, In primum Euclidis elementorum librum commentariorum . . . libri IIII. Ed. by Franciscus Barocius. Patavii, 1560.

Raphson, Joseph, The History of Fluxions, Shewing in a Compendious Manner the First Rise of, and Various Improvements Made in that Incomparable Method. London, 1715.

Rashdall, Hastings, The Universities of Europe in the Middle Ages. Ed. by F. M. Powicke and A. B. Emden. 3 vols., Oxford, 1936.

Rebel, O. J., Der Briefwechsel zwischen Johann (I.) Bernoulli und dem Marquis de L'Hospital. Bottrop i. w., 1934.

Reiff, R., Geschichte der unendlichen Reihen. Tübingen, 1889.

Rigaud, S. P., Correspondence of Scientific Men of the Seventeenth Century. 2 vols., Oxford, 1841.

Robbins, F. E., see Nicomachus

Robert of Chester, see Al-Khowarizmi.

Roberval, G. P. de, "Divers ouvrages." *Mémoires de l'Académie Royale des Sciences depuis 1666 jusqu'à 1699*, VI (Paris, 1730), 1–478.

Robins, Benjamin, "A Discourse Concerning the Nature and Certainty of Sir Isaac Newton's Methods of Fluxions, and of Prime and Ultimate Ratios." *Present State of the Republick of Letters*, XVI (1735), 245–270.

———— "A Dissertation Shewing, that the Account of the Doctrines of Fluxions, and of Prime and Ultimate Ratios, Delivered in a Treatise, Entitled, A Discourse Concerning the Nature and Certainty of Sir Isaac

Newton's Method of Fluxions, and of Prime and Ultimate Ratios, Is Agreeable to the Real Sense and Meaning of Their Great Inventor." *Present State of the Republick of Letters*, XVII, (1736), 290–335.

———— Mathematical Tracts. 2 vols., London, 1761.

Rosenfeld, L., "René-François de Sluse et le problème des tangentes." *Isis*, X (1928), 416–34.

Ross, W. D., Aristotle. New York, 1924.

Russell, Bertrand, The Analysis of Matter. New York, 1927.

———— Introduction to Mathematical Philosophy. London and New York [1924].

———— Our Knowledge of the External World as a Field for Scientific Method in Philosophy. Chicago, 1914.

———— The Principles of Mathematics. Cambridge, 1903.

———— "Recent Work on the Principles of Mathematics." *International Monthly*, IV (1901), 83–101.

Saks, Stanislaw, Théorie de l'intégrale. Warszawa, 1933.

Sarton, George, "The First Explanation of Decimal Fractions and Measures (1585)." *Isis*, XXIII (1935), 153–244.

———— Introduction to the History of Science. 2 vols., Baltimore, 1927–31.

———— "Simon Stevin of Bruges (1548-1620)." *Isis*, XXI (1934), 241–303.

Schanz, Der Cardinal Nicolaus von Cusa als Mathematiker. Rottweil, 1872.

Scheffers, G., see Serret.

Scheye, Anton, "Ueber das Princip der Stetigkeit in der mathematische Behandlung der Naturerscheinungen." *Annalen der Naturphilosophie*, I (1902), 20–49.

Schmidt, Wilhelm, "Archimedes' Ephodikón." *Bibliotheca Mathematica* (3), I (1900), 13–14.

———— "Heron von Alexandria im 17. Jahrhundert." *Abhandlungen zur Geschichte der Mathematik*, VIII (1898), 195–214.

———— "Leonardo da Vinci und Heron von Alexandria." *Bibliotheca Mathematica* (3), III (1902), 180–87.

Scholtz, Lucie, Die exakte Grundlegung der Infinitesimalrechnung bei Leibniz. Marburg, 1934.

Schroeder, Leopold von, Pythagoras und die Inder; eine Untersuchung über Herkunft und Abstammung der Pythagoreischen Lehren. Leipzig, 1884.

Schrödinger, Erwin, Science and the Human Temperament. New York [c 1935].

Schubert, H., J. Tannery, and J. Molk, "Principes fondamentaux de l'arithmétique." *Encyclopédie des Sciences Mathématiques*, I (Part 1, Fascicule 1), 1–62.

Sengupta, P. C., "History of the Infinitesimal Calculus in Ancient and Mediaeval India." *Jahresbericht, Deutsche Mathematiker-Vereingung*, XLI (1931), 223–27.

330   Bibliography

Sergescu, Petru, "Les Mathématiques dans le 'Journal des Savants' 1665–1701." *Archeion*, XVIII (1936), 140–45.

Serret, J. A., and G. Scheffers, Lehrbuch der Differential- und Integralrechnung. 6th ed., 3 vols., 1915–21.

Servois, F. J., Essai sur un nouveau mode d'exposition des principes du calcul différentiel; suivi de quelques réflexions relatives aux divers points de vue sous lesquels cette branche d'analise a été envisagée jusqu'ici, et, en général, à l'application des systèmes métaphysiques aux sciences exactes. Nismes, 1814.

Simon, Max, Cusanus also Mathematiker. *Festschrift Heinrich Weber zu seinem siebzigsten Geburtstag*, pp. 298–337. Leipzig and Berlin, 1912.

—— Geschichte der Mathematik im Altertum. Berlin, 1909.

—— "Historische Bemerkungen über das Continuum, den Punkt und die gerade Linie." *Atti del IV Congresso internazionale dei Matematici* (Roma, 1908), III, 385–90.

—— "Zur Geschichte und Philosophie der Differentialrechnung." *Abhandlungen zur Geschichte der Mathematik*, VIII (1898), 113–32.

Simons, L. G., "The Adoption of the Method of Fluxions in American Schools." *Scripta Mathematica*, IV (1936), 207–19.

Simplicius, Commentarii in octo Aristotelis physicae auscultationis libros. Venetiis, 1551.

Simpson, Thomas, The Doctrine and Application of Fluxions: . . . Containing . . . a Number of New Improvements in the Theory. London [1805].

Singh, A. N., see Datta.

Sloman, H., The Claim of Leibnitz to the Invention of the Differential Calculus. Trans. from the German with considerable alterations and new addenda by the Author. Cambridge, 1860.

Sluse, R. F. de, "A Method of Drawing Tangents to All Geometrical Curves." *Philosophical Transactions of the Royal Society of London* (Abridged, London, 1809), II (1672–83), 38–41.

Smith, D. E., "The Geometry of the Hindus." *Isis*, I (1913), 197–204.

—— History of Mathematics. 2 vols., New York [c. 1923–25].

—— "Lazare Nicholas Marguerite Carnot." *Scientific Monthly*, XXXVII (1933), 188–89.

—— "The Place of Roger Bacon in the History of Mathematics." In A. G. Little, *Roger Bacon Essays*. Oxford, 1914.

—— and G. A. Miller, "Was Guldin a Plagiarist?" *Science*, LXIV (1926), 204–6.

Spiers, see Pascal.

Stamm, Edward, "Tractatus de continuo von Thomas Bradwardina. Eine Handschrift aus dem XIV. Jahrhundert." *Isis*, XXVI (1936), 13–32.

Staude, Otto, "Die Hauptepochen der Entwicklung der neueren Mathematik." *Jahresbericht, Deutsche Mathematiker-Vereinigung*, XI (1902), 280–92.

Stein, W., "Der Begriff des Schwerpunktes bei Archimedes." *Quellen und Studien zur Geschichte der Mathematik, Astronomie und Physik*, Part B, Studien, I (1931), 221–44.

Stevin, Simon, Hypomnemata mathematica. 5 vols., Lugduni Batavorum, 1605-8.

Stolz, Otto, "B. Bolzano's Bedeutung in der Geschichte der Infinitesimal-rechnung." *Mathematische Annalen*, XVIII (1881), 255–79.

—— Vorlesungen über allgemeine Arithmetik. 2 parts, Leipzig, 1885-86.

—— "Zur Geometrie der Alten, insbesondere über ein Axiom des Archimedes." *Mathematische Annalen*, XXII (1883), 504–19.

Strong, E. W., Procedures and Metaphysics. A Study in the Philosophy of Mathematical-Physical Science in the Sixteenth and Seventeenth Centuries. Berkeley (Cal.), 1936.

Struik, D. J., "Kepler as a Mathematician." In *Johann Kepler, 1571–1630. A Tercentenary Commemoration of His Life and Works*. Baltimore, 1931.

—— "Mathematics in the Netherlands during the First Half of the XVIth Century." *Isis*, XXV (1936), 46–56.

Suiseth, Richard, Liber calculationum. [Padua, 1477.]

Sullivan, J. W. N., The History of Mathematics in Europe from the Fall of Greek Science to the Rise of the Conception of Mathematical Rigour. London, 1925.

Surico, L. A., "L'integrazione di $y = x^n$, per n negativo, razionale, diverso da –1 di Evangelista Torricelli. Definitivo riconoscimento della priorità torricelliana in questa scoperta." *Archeion*, XI (1929), 64–83.

Suter, H., see Ibn al-Haitham.

Tacquet, André, Arithmeticae theoria et praxis. New ed., Amstelaedami, 1704.

—— Cylindricorum et annularium libri IV. Antverpiae, 1651.

—— Opera mathematica. Lovanii, 1668.

Tannery, Jules, Review of Dantscher, "Vorlesungen über die Weierstrassche Theorie der irrationalen Zahlen." *Bulletin des Sciences Mathématiques et Astronomiques* (2), XXXII (1908), 101–5.

See also Schubert.

Tannery, Paul, "Anaximandre de Milet." *Revue Philosophique*, XIII (1882), 500–29.

—— "Le Concept scientifique du continu. Zénon d'Élée et Georg Cantor." *Revue Philosophique*, XX (1885), 385–410.

—— La Géométrie grecque, comment son histoire nous est parvenu et ce que nous en savons. Essai critique. Première partie, Histoire générale de la géométrie élémentaire. Paris, 1887.

—— "Notions historiques." In Jules Tannery, Notions de mathématiques. Paris [1902].

—— Pour l'histoire de la science hellène. 2d ed., Paris, 1930.

Tannery, Paul, Review of Vivanti, "Il concetto d'infinitesimo." *Bulletin des Sciences Mathématiques et Astronomiques* (2), XVIII (1894), 230–33.

———— Review of Zeuthen, "Notes sur l'histoire des mathématiques." *Bulletin des Sciences Mathématiques et Astronomiques* (2), XX (1896), 24–28.

———— "Sur la date des principales découvertes de Fermat." *Bulletin des Sciences Mathématiques et Astronomiques* (2), VII (1883), 116–28.

———— "Sur la division du temps en instants au moyen âge." *Bibliotheca Mathematica* (3), VI (1905), 111.

———— "Sur la sommation des cubes entiers dans l'antiquité." *Bibliotheca Mathematica* (3), III (1902), 257–58.

Taylor, Brook, Methodus incrementorum directa et inversa. Londini, 1717.

Taylor, C., "The Geometry of Kepler and Newton." *Cambridge Philosophical Society*, XVIII (1900), 197–219.

Thomson, S. H., "An Unnoticed Treatise of Roger Bacon on Time and Motion." *Isis*, XXVII (1937), 219–24.

Thorndike, Lynn, A History of Magic and Experimental Science. 4 vols., New York, 1923–34.

———— Science and Thought in the Fifteenth Century. New York, 1929.

———— and Pearl Kibre, A Catalogue of Incipits of Mediaeval Scientific Writings in Latin. The Mediaeval Academy of America. Publication No. 29. Cambridge, Mass., 1937.

Timtchenko, J., "Sur un point du 'Tractatus de latitudinibus formarum' de Nicolas Oresme." *Bibliotheca Mathematica* (3), I (1900), 515–16.

Tobiesen, L. H., Principia atque historia inventionis calculi differentialis et integralis nec non methodi fluxionum. Gottingae, 1793.

Toeplitz, Otto, "Das Verhältnis von Mathematik und Ideenlehre bei Plato." *Quellen und Studien zur Geschichte der Mathematik, Astronomie und Physik*, Part B, Studien, I (1931), 3–33.

[Torricelli, Evangelista] Opere di Evangelista Torricelli. 3 vols., Faenza, 1919.

Ueberweg, Friedrich, Grundriss der Geschichte der Philosophie. Part I. Die Philosophie des Altertums. 12th ed., ed. by Karl Praechter. Berlin. 1926.

Valerio, Luca, De centro gravitatis solidorum libri tres. 2d ed., Bononiae, 1661.

Valperga-Caluso, Tommaso, "Sul paragone del calcolo delle funzioni derivate coi metodi anteriori." *Memorie di Matematica e di fisica della Società Italiana delle Scienze*, XIV (Part I, 1809), 201–24.

Valson, C.-A., La Vie et les travaux du baron Cauchy . . . . With a preface by M. Hermite. Vol. I, Paris, 1868.

Vansteenberghe, Edmond, Le Cardinal Nicolas de Cues (1401–1464). Paris, 1920.

Ver Eecke, Paul, "Le Théorème dit de Guldin considéré au point de vue historique." *Mathesis*, XLVI (1932), 395–97.

See also Diophantus; Pappus.

Vetter, Quido, "The Development of Mathematics in Bohemia." Trans. from the Italian by R. C. Archibald. *American Mathematical Monthly*, XXX (1923), 47–58.

Vivanti, Giulio, Il concetto d'infinitesimo e la sua applicazione alla matematica. Saggio storico. Mantova, 1894.

—— "Note sur l'histoire de l'infiniment petit." *Bibliotheca Mathematica*, N. S., VIII, 1894, 1–10.

Vogt, Heinrich, "Die Entdeckungsgeschichte des Irrationalen nach Plato und anderen Quellen des 4. Jahrhunderts." *Bibliotheca Mathematica* (3), X (1910), 97–155.

—— "Die Geometrie des Pythagoras." *Bibliotheca Mathematica* (3), IX (1908), 15–54.

—— Der Grenzbegriff in der Elementar-mathematik. Breslau, 1885.

—— "Haben die alten Inder den Pythagoreischen Lehrsatz und das Irrationale gekannt?" *Bibliotheca Mathematica* (3), VII (1906–7), 6–23.

—— "Die Lebenszeit Euklids." *Bibliotheca Mathematica* (3), XIII (1913), 193–202.

—— "Zur Entdeckungsgeschichte des Irrationalen." *Bibliotheca Mathematica* (3), XIV (1914), 9–29.

Voltaire, F. M. A., Letters Concerning the English Nation. London, 1733.

Voss, A., and J. Molk, "Calcul différentiel." *Encyclopédie des Sciences Mathématiques*, Vol. II, Part 1, Fascicule 2; Part 3.

Waard, C. de, "Un Écrit de Beaugrand sur la méthode des tangentes de Fermat à propos de celle de Descartes." *Bulletin des Sciences Mathématiques et Astronomiques* (2), XLII (1918), 157–77, 327–28.

Waismann, Friedrich, Einführung in das mathematische Denken, die Begriffsbildung der modernen Mathematik. Wien, 1936.

Walker, Evelyn, A Study of the Traité des Indivisibles of Roberval. New York, 1932.

Wallis, John, Opera mathematica. 2 vols., Oxonii, 1656–57.

Wallner, C. R., "Entwickelungsgeschichtliche Momente bei Entstehung der Infinitesimalrechnung." *Bibliotheca Mathematica* (3), V (1904), 113–24.

—— "Über die Entstehung des Grenzbegriffes." *Bibliotheca Mathematica* (3), IV (1903), 246–59.

—— "Die Wandlungen des Indivisibilienbegriffs von Cavalieri bis Wallis." *Bibliotheca Mathematica* (3), IV (1903), 28–47.

Walton, J., The Catechism of the Author of the Minute Philosopher Fully Answered. Dublin, 1735.

—— A Vindication of Sir Isaac Newton's Principles of Fluxions against the Objections Contained in the Analyst. Dublin, 1735.

Weaver, J. H., "Pappus. Introductory Paper." *Bulletin, American Mathematical Society*, XXIII (1916), 127–35.

Weierstrass, Karl, Mathematische Werke. Herausgageben unter Mitwerkung einer von der Königlich Preussischen Akademie der Wissenschaften Eingesetzten Commission. 7 vols., Berlin, 1894–1927.

Weinreich, Hermann, "Über die Bedeutung des Hobbes für das naturwissenschaftliche und mathematische Denken. Borna-Leipzig, 1911.

Weissenborn, Hermann, Die Principien der höheren Analysis in ihrer Entwickelung von Leibniz bis auf Lagrange, als ein historischkritischer Beitrag zur Geschichte der Mathematik. Halle, 1856.

Weyl, Hermann, "Der Circulus vitiosus in der heutigen Begründung der Analysis. *Jahresbericht, Deutsche Mathematiker-Vereinigung,* XXVIII (1919), 85–92.

Whewell, William, History of Scientific Ideas. Being the First Part of the Philosophy of the Inductive Sciences. 3d ed., 2 vols., London, 1858.

Whitehead, A. N., An Introduction to Mathematics. New York, 1911.

Wieleitner, Heinrich, "Bermerkungen zu Fermats Methode von Aufsuchung von Extremenwerten und der Bestimmung von Kurventangenten." *Jahresbericht, Deutsche Mathematiker-Vereinigung,* XXXVIII (1929), 24–25.

——— "Das Fortleben der Archimedischen Infinitesimalmethoden bis zum Beginn des 17. Jahrhundert, insbesondere über Schwerpunktbestimmungen." *Quellen und Studien zur Geschichte der Mathematik, Astronomie und Physik,* Part B, Studien, I (1931), 201–20.

——— Die Geburt der modernen Mathematik. Historisches und Grundsätzliches. 2 vols., Karlsruhe in Baden, 1924–25.

——— "Der 'Tractatus de latitudinibus formarum' des Oresme." *Bibliotheca Mathematica* (3), XIII (1913), 115–45.

——— "Ueber den Funktionsbegriff und die graphische Darstellung bei Oresme." *Bibliotheca Mathematica* (3), XIV (1914), 193–243.

——— "Zur Geschichte der unendlichen Reihen im christlichen Mittelalter." *Bibliotheca Mathematica* (3), XIV (1914), 150–68.

Wiener, P. P., "The Tradition behind Galileo's Methodology." *Osiris,* I (1936), 733–46.

Wilson, E. B., "Logic and the Continuum." *Bulletin, American Mathematical Society,* XIV (1908), 432–43.

Windred, G., "The History of Mathematical Time." *Isis,* XIX (1933), 121–53; XX (1933–34), 192–219.

Witting, A., "Zur Frage der Erfindung des Algorithmus der Neutonschen Fluxionsrechnung." *Bibliotheca Mathematica* (3), XII (1911–12), 56–60.

Wolf, A., A History of Science, Technologv, and Philosophy in the Sixteenth and Seventeenth Centuries. New York, 1935.

Wolff, Christian, Elementa matheseos universalis. New ed., 5 vols., Genevae, 1732–41.

—— Philosophia prima, sive ontologia, methodo scientifica pertractata, qua omnis cognitionis humanae principia continentur. New ed., Francofurti and Lipsiae, 1736.

Wren, F. L., and J. A. Garrett, "The Development of the Fundamental Concepts of Infinitesimal Analysis." *American Mathematical Monthly*, XL (1933), 269–81.

Wronski, Hoëné, Œuvres mathématiques. 4 vols., Paris, 1925.

Zeller, Eduard, Die Philosophie der Griechen in ihrer Geschichtlichen Entwicklung. 4th ed., 3 vols., Leipzig, 1876–89.

Zeuthen, H. G., Geschichte der Mathematik im Altertum und Mittelalter. Kopenhagen, 1896.

—— Geschichte der Mathematik im XVI. und XVII. Jahrhundert. German ed. by Raphael Meyer. *Abhandlungen zur Geschichte der Mathematischen Wissenschaften*, XVII (1903).

—— "Notes sur l'histoire des mathématiques." *Oversigt over det Kongelige Danske Videnskabernes Selskabs. Forhandlinger*, 1893, pp. 1–17, 303–41; 1895, pp. 37–80, 193–278; 1897, pp. 565–606; 1910, pp. 395–435; 1913, pp. 431–73.

—— "Sur l'origine historique de la connaissance des quantités irrationelles." *Oversigt over det Kongelige Danske Videnskabernes Selskabs. Forhandlinger*, 1915, pp. 333–62.

Zimmermann, Karl, Arbogast als Mathematiker und Historiker der Mathematik. [No place, no date.]

Zimmermann, R., "Der Cardinal Nicolaus Cusanus als Vorläufer Leibnitzens. *Sitzungsberichte der Philosophisch-Historischen Classe der Kaiserlichen Akademie der Wissenschaften, Wien*, VIII (1852), 306–28.

Index

Abel, N. H., 280

Académie des Sciences, 241, 242, 263

Acceleration, Greek astronomy lacked concept of, 72; uniform, 82 ff., 113 ff., 130, 165

Achilles, 24–25, 116, 138, 140, 295; argument in the, 24n; refuted by Aristotle, 44

Acta eruditorum, 207, 208, 214, 221, 238

A est unum calidum, 87

Agardh, C. A., *Essai sur la métaphysique du calcul intégral*, 264

Albert of Saxony, 66, 68 f., 74

Alembert, Jean le Rond d', 237; concept of differentials, 246, 275; interpretation of Newton's prime and ultimate ratio, 247; quoted, 248; limit concept, 249, 250, 253, 257, 265, 271, 272; definition of the tangent, 251; concept of infinitesimal, 274; concept of infinity, 275

Algebra, Arabic development, 2, 56, 60, 63 ff., 97, 120, 154; contribution of Hindus, 63; symbols for quantities, 98; use of, avoided in seventeenth century, 302

Algebraic formalism fostered by Euler and Lagrange, 305

Alhazen (Ibn al-Haitham), 63

Ampère, A. M., theory of functions, 282

Anaxagoras, 40

Anaximander, 28, 40

Angle of contact, 22, 173, 174, 212

Antiphon the Sophist, 32

Apeiron, 28

Apollonius, influence in development of analytic geometry, 187

Arabs, algebra adopted from, 2; extended work of Archimedes, 56, 120; algebraic development, 60, 63 ff., 97, 154

Arbogast, L. F. A., *Du calcul des dérivations*, 263

Archimedes, 25, 29, 38, 39, 48, 49–60, 64, 65, 70, 101, 105, 112, 299, *Method*, 21, 48–51, 59, 99, 125, 139, 159; axiom of, 33, 173; use of infinitely large and infinitely small, 48 ff., 90; quadrature of the parabola, 49–53, 102, 124; volume of conoid, 53–55; spiral, 55–58, 133; determination of tangents, 56 ff.;

work on centers of gravity, 60; use of series, 76 f., 120; mensurational work, 89, 94, 107; translation and publication of works, 94, 96, 98; modifications of method of, Chap. IV; use of doctrine of instantaneous velocity, 56 ff., 130; inequalities, 159

Area, concept of, 9, 31, 34, 58 f., 62, 105, 123; application of, 18, 19, 32, 36, 96

Aristotle, 1, 20, 22, 26, 30, 46, 57, 60, 65, 66, 67, 70, 72, 79, 112, 299, 303; opposed to Plato's concept of number, 27; *On the Pythagoreans*, 37; dependence upon sensory perception, 37 ff.; considered mathematics an idealized abstraction from natural science, 37–38, 303, 307; quoted, 38, 41, 42, 43, 46; concept of the infinite, 38 ff., 69, 70, 102, 117, 151, 275, 296; concept of number, 40–41, 151, 154–55; *Physica*, 41, 43, 65, 78; theory of continuity, 41–42, 66–67, 277, 291; concept of motion, 42–45, 71 ff., 176, 178, 233; denial of instantaneous velocity, 43, 73, 82; logic of, 45, 46, 47, 61; sixteenth-century opposition to Aristotelianism, 96; views triumphed in method of exhaustion and geometry of Euclid, 45 ff., 304

Arithmetic triangle, *see* Pascal's triangle

Arrow, argument in, 24n, 295; refuted by Aristotle, 44

Assemblages, *see* Infinite aggregates

Athens, 26

Atomic doctrine, 21, 23, 38, 42, 66, 67, 84, 92, 180, 181

Atomism, mathematical, 22, 28, 50, 82, 115, 125, 134, 176, 188

Average velocity, 6, 8, 75, 83, 113 ff.

Babylonians, mathematical knowledge, 1, 14 ff., 62, 303; astronomical knowledge, 15

Bacon, Roger, 64; *Opus majus*, 66; concept of motion, 72

Barrow, Isaac, 58, 133, 209, 299, 302, 304; anticipated invention of the calculus, 164; opposed arithmetization of mathematics, 175; advocated classic concept of number

Bei Fragen zur Produktsicherheit wenden Sie sich bitte an:
If you have any questions regarding product safety,
please contact:

Walter de Gruyter GmbH
Genthiner Straße 13
10785 Berlin
productsafety@degruyterbrill.com